# Eighteenth-Century Environmental Humanities

# TRANSITS

## LITERATURE, THOUGHT & CULTURE, 1650–1850



A landmark series in long-eighteenth-century studies, *Transits* publishes monographs and edited volumes that are timely, transformative in their approach, and global in their engagement with arts, literature, culture, and history. Books in the series have engaged with visual arts, environment, politics, material culture, travel, theater and performance, embodiment, connections between the natural sciences and medical humanities, writing and book history, sexuality, gender, disability, race, and colonialism from Britain and Europe to the Americas, the Far East, the Middle/Near East, Africa, and Oceania. Works that make provocative connections across time, space, geography, or intellectual history, or that develop new modes of critical imagining are particularly welcome.

**Recent titles in the series:**

*Eighteenth-Century Environmental Humanities*
Jeremy Chow, ed.

*Families of the Heart: Surrogate Relations in the Eighteenth-Century British Novel*
Ann Campbell

*Political Affairs of the Heart: Female Travel Writers, the Sentimental Travelogue, and Revolution, 1775–1800*
Linda Van Netten Blimke

*The Limits of Familiarity: Authorship and Romantic Readers*
Lindsey Eckert

*"Robinson Crusoe" after 300 Years*
Andreas K. E. Mueller and Glynis Ridley, eds.

*Transatlantic Women Travelers, 1688–1843*
Misty Krueger, ed.

*Laurence Sterne's "A Sentimental Journey": A Legacy to the World*
W. B. Gerard and M-C. Newbould, eds.

For more information about the series, please visit www.bucknelluniversitypress.org.

# Eighteenth-Century Environmental Humanities

Edited by

JEREMY CHOW

BUCKNELL UNIVERSITY PRESS
LEWISBURG, PENNSYLVANIA

Library of Congress Cataloging-in-Publication Data

Names: Chow, Jeremy, editor.
Title: Eighteenth-century environmental humanities / edited by Jeremy Chow.
Description: Lewisburg, Pennsylvania : Bucknell University Press, 2022. | Series: Transits : literature, thought & culture 1650–1850 | Includes bibliographical references and index.
Identifiers: LCCN 2022008848 | ISBN 9781684484287 (paperback) | ISBN 9781684484294 (hardback) | ISBN 9781684484300 (epub) | ISBN 9781684484324 (pdf)
Subjects: LCSH: Human ecology and the humanities—History—18th century.
Classification: LCC GF22 .E54 2022 | DDC 304.209/033—dc23/eng20220726
LC record available at https://lccn.loc.gov/2022008848

A British Cataloging-in-Publication record for this book is available from the British Library.



♾ The paper used in this publication meets the requirements of the American National Standard for Information Sciences—Permanence of Paper for Printed Library Materials, ANSI Z39.48-1992.

www.bucknelluniversitypress.org

Distributed worldwide by Rutgers University Press

Printed and bound by CPI Group (UK) Ltd, Croydon, CR0 4YY

# CONTENTS

Introduction: Eighteenth Century + Environmental Humanities 1
Jeremy Chow

PART ONE: Eighteenth Century + Climate Change

1 Toward a Genealogy of Geoengineering: Erasmus Darwin and the Little Ice Age 23
Elliot Patsoura

2 Storm Apostrophe 38
Annette Hulbert

3 "When Stormy Winds Happen": Divine Providence, Climate Change Discourse, and the Cause of Weather Disasters 53
Adam W. Sweeting

PART TWO: Eighteenth Century + New Materialisms

4 Phillis Wheatley Peters's Niobean Soundscapes 71
Shelby Johnson

5 Syphilis and Natural History: The Ethical Limits of Human Mastery 88
Mariah Crilley

PART THREE: Eighteenth Century + Blue Humanities

6 Shore/Lines: Drawing Environmental Change on Eighteenth-Century Prince Edward Island 105
Claire Campbell

7 Of Water, Winds, and Storms: The Elemental Regimes of the Buccaneer Journal 136
Jason M. Payton

PART FOUR: Eighteenth Century + Indigeneity and Decoloniality

8 "Supporting Sinking Nations": Indigenous Women and Their Disasters in John Dennis's *Liberty Asserted* 153
Matthew Duquès

9 Imagining Decolonial Futures in William Gilbert's *The Hurricane* 170
Ami Yoon

PART FIVE: Eighteenth Century + Green Utopias

10 Slavery and Plantation Stewardship: The Eighteenth-Century Caribbean Georgics of James Grainger and Philip Freneau 189
Christopher Allan Black

11 John Thelwall and L. M. Montgomery Write the Green City 205
Kate Scarth

Acknowledgments 223

Bibliography 225

Notes on Contributors 247

Index 251

# Eighteenth-Century Environmental Humanities

# INTRODUCTION

## Eighteenth Century + Environmental Humanities

Jeremy Chow

As the first collection dedicated to the nexus of eighteenth-century studies and the environmental humanities, *Eighteenth-Century Environmental Humanities* strives to accomplish three goals:

1. To identify and further develop how the environmental humanities, as a series of methodologies and an interdisciplinary genre of scholarship, is germane to eighteenth-century studies
2. To realize and showcase how eighteenth-century studies enriches the environmental humanities, and vice versa
3. To magnify the enfolded nature of scholarship and pedagogy, and thus demonstrate how eighteenth-century environmental humanities develops from the co-constitutions of research *and* teaching

The eleven essays that follow uphold all three of these goals with an intention to invigorate conversation, scholarly engagement, and interdisciplinary teaching. The eighteenth century of *Eighteenth-Century Environmental Humanities* is not a monolith; it is as an extensive period that incorporates the Restoration, the Enlightenment, and Romanticism. This expansive purview demonstrates how both environmental humanist and eighteenth-century discourses bleed through, into, and beyond the otherwise assumed-finite boundaries of periodicity. Like those who may approach this collection as a scholarly, pedagogical, or general interest text, the authors included here have come to this interdisciplinary blending from different avenues, approaches, and schools of thought. For some, the commitment to eighteenth-century studies and the recurrent visibility of environmental concerns has opened pathways to the environmental humanities. Others here seek to remedy the opacity of eighteenth-century studies in environmental humanities publications, conferences, and professional opportunities. Collectively, though, we seek to open up realms of the environmental humanities that speak to and grow from eighteenth-century studies. We approach this intersection as teacher-scholars invested in enriching undergraduate and graduate instruction, and thus paving inroads for educational change and emerging, collaborative humanities scholarship.

This collection of essays reveals investments in the fields of environmental history, literary studies, art history, cultural studies, critical race studies, Indigenous and native studies, postcolonialism and decoloniality, performance studies, and feminist and queer theories. Yet even with these multivocal explorations, the essays found here make available only eleven visualizations of eighteenth-century environmental humanities, of which there are assuredly myriad others. There are infinite possibilities by which to imagine the intersection of eighteenth-century studies and the environmental humanities, and our work here sets out to invite other collaborative and interdisciplinary work. In this way, as collections such as Vin Nardizzi and Tiffany Jo Werth's *Premodern Ecologies in the Modern Literary Imagination* and Jeffrey Jerome Cohen's *Prismatic Ecologies* have likewise advocated, we call on our colleagues, readers, and students to approach this collection as a starting point.[1] To motivate such a call, this introduction establishes the provenance of a burgeoning and brilliant eighteenth-century environmental humanities by, first, framing axioms by which to understand and thus demystify the environmental humanities and, second, unveiling this field's imbrication within eighteenth-century studies, which this collection animates.

## DEFINING THE ENVIRONMENTAL HUMANITIES IN FOUR AXIOMS

Despite its various perspectives and applications, this collection might putatively appear to identify a singular definition of what constitutes eighteenth-century environmental humanities. Let us be clear: no such discrete definition will be proffered here. Such a distillation would, in fact, work against what we strive to accomplish. By way of defining the environmental humanities then, we propose four axioms that introduce conceptual frameworks taken up by the subsequent chapters.

### *Axiom 1: The environmental humanities is not a monolith. It goes by many names and feels.*

This first axiom bears repeating. In rejecting reduction, this collection avers that the environmental humanities is not one thing, despite our awareness that job calls and calls for papers employ the term as a seemingly homogeneous buzzword. Ursula Heise, to this point, acknowledges that the environmental humanities is not so much novel in its foundations or approaches as it is in its repurposing and recollection of humanistic perspectives, methods, and objects of study.[2] The environmental humanities, then, constitutes a provocative reshuffling of disciplinary fields, methodologies, and purviews with the goal of re-visioning what results in reorganization.

Indeed, definitions of the environmental humanities reflect varying commitments, disciplines, and social and political investments. As Serpil Opperman and Serenella Iovino contend, research in the environmental humanities is motivated by "urgent environmental problems that stretch from the geological to the biological [and] are also essentially social and cultural issues deeply interwoven with economic and political agendas and thus demand solutions on many dimensions."[3] Jeffrey Jerome Cohen and Lowell Duckert identify ecocriticism and ecotheory by these same tokens. Whereas ecocriticism "is a lively confluence of ecology, philosophy, anthropology, sociology, literature, feminism, sustainability studies, environmental justice (especially within indigenous and postcolonial studies), and queer theory," ecotheory "ranges across the environmental humanities, green studies, social activism, and the new materialisms" with the goal of "better" grappling with the worlds brought about by anthropogenic climate change.[4] As Cohen and Duckert evidence, terms like the environmental humanities, ecotheory, ecocriticism, and their epistemological kin are often used interchangeably. The term environmental humanities is preferred here because we as authors understand it as a legible invitation to scholarly and pedagogical colloquy that spans and encompasses various humanities disciplines, with a particular eye to interdisciplinary inquiry best characterized by the term "humanistic"—a loaded concept I unpack below. The definitions and applications we offer then are not exhaustive, but rather, visions of possibility for the present and future of this field.

### *Axiom 2: The environmental humanities approaches crisis; it is not exclusively tasked with remedying crisis.*

These visions of possibility are perpetually under threat from the rhetoric of crisis. In many ways, *Eighteenth-Century Environmental Humanities* approaches the ominous "crisis" in the humanities, a scare tactic bedeviling institutions of higher education across the globe, especially in the United States, to which I speak directly. Fear of the "crisis in the humanities" remains a hysteria-inducing phobia (thus the scare quotes) that, under the threat of austerity and anti-intellectual measures, requires humanities scholars to prove why their work matters—to the department, the university, the field, and ways of living in the contemporary moment. Such a repeated question assumes work in other fields is always already self-explanatory, valid, and universally justifiable. This alarm-raising rhetoric—following the financial crisis of 2008, which has reared its ugly head again at the start of the 2020s (especially in the intra-pandemic setting in which we write this)—seems to default to the fallacious logic that arts and humanities majors are less prepared for assuming "real-world" careers after graduation. In this regard, a January 2021 op-ed piece woefully conjectures that the United States is home to *too many* individuals with doctorates, especially in the humanities. The author, an assistant professor of finance, argues that doctorate-granting institutions in the humanities need to reel this output in, for in the contracting job market lies a dejection that could very well feed into anarchist wet dreams of "overthrow[ing] institutions of society."[5] But where there is a surplus in the humanities, there is a dearth in STEM. The author's solution? A government bailout that further invigorates STEM research and (further) slashes the humanities at the knees. The logical fallacies modeled by this rubbish argument exclusively uphold capitalist regimes of fiscal success wherein "social good" is implicit, which is allegedly captured by the phrase, a "good job." As Benjamin Schmidt counters, this misguided logic occludes the fact that, in our present moment, "some humanities may be demonstrating more usefulness than ever to students who seek to better understand culture from outside the dominant perspective" and thus avail the ethos of a humanities-driven education, which seeks to invigorate a "philosophy for life."[6]

In fall 2019, the *Chronicle of Higher Education* tackled the "crisis" head-on with an insert of fourteen essays, titled *Endgame: Can Literary Studies Survive?* The essays spotlighted literary studies, but the conversations and reactions begat by the collection reverberate throughout the arts and humanities. One of the fourteen, "The New Humanities," by Jeffrey J. Williams, outlines a series of "new," innovative fields, which include digital humanities, medical humanities, public humanities, and environmental humanities, that have the potential to "save" the humanities. The environmental humanities, Williams notes, is "the most socially concerned effort," which "draw[s] especially on the life sciences, but also on disciplines like geology, economics, and engineering, [and] looks at the human aspects of environmental issues—particularly climate change."[7] While Williams sees these new humanities as a remodeling of the university, he warns against the co-optation of the humanities by other fields, which, in his perspective and that of his interviewee, postcolonial environmental humanist Rob Nixon, may very well reify the neoliberalism of the university. These are accurate and timely concerns; the fear of co-optation is real, especially within the mechanized, powerhouse business the university has become.

The "crisis in the humanities" instills one form of catastrophic thinking, while the visceral realities of anthropogenic climate change and environmental degradation, captured by the term "the Anthropocene," broadens catastrophic thinking so as to reckon with another form: planetary calamity. Coined by ecologist Eugene Stoermer in the 1980s, the neologism, which translates to "the age of humans," caught fire at the start of the new millennium and has since spurred a groundswell of scholarly inquiry, political activism, and environmental advocacy. Because of its juggernaut-like momentum (recognizable in conference themes, academic book series, calls for papers, and course offerings), the Anthropocene often accompanies definitions or iterations of the environmental humanities. In response, we offer a provocation: the environmental humanities is the most generative and interdisciplinary way of understanding the premises, complications, and effects of the Anthropocene. Consider Elizabeth DeLoughrey's *Allegories of the Anthropocene*, which pinpoints the "cottage industry of new journals and publications exploring the Anthropocene" that "privilege positivist methods and have little to say about the vitality of the arts and humanities."[8] The environmental humanities center the arts and humanities with an eye to the pluralized and destructive/destructed worlds oriented by the Anthropocene.

And these approaches, for example by Stephanie LeMenager, Richard Grusin, Steve Mentz, Jussi Parikka, Anna Tsing, Charles Mann, and Kathryn Yusoff, rightly take to task the Anthropocene's totalizing perspective of planet, place, and people.[9] In other words, Anthropocene thinking too often teeters toward universalizing or homogeneity. If the Anthropocene is "our" age, then it's worth vetting who and what are included by the plural first-person pronoun, especially in the realization that collective, universalizing rhetoric has historically reinforced the erasure of Black, Indigenous, and people of color (BIPOC) and extends the insidious violence of (neo)colonialism, as Tony Birch and Kyle Powys White teach us.[10] Although not all scholars and teachers who research the Anthropocene and its effects might identify as environmental humanists, both definitionally and methodologically, the Anthropocene serves as a striking model by which to understand the interdisciplinary and multimodal approaches that the environmental humanities pioneers.

What has remained mostly under the radar is the way in which the eighteenth century and the Anthropocene work in tandem. As a geological, climatological, sociological, anthropological, and environmental phenomenon, a deeper dive into frameworks for understanding the Anthropocene necessitates attention to eighteenth-century studies. Paul Crutzen, who popularized the term Anthropocene with Stoermer, locates the eighteenth century (specifically the 1784 patent for James Watts's double-acting steam engine) as its planetary commencement, a usurpation of the Holocene (which is otherwise recognized as the current epoch, dating back from 11,500 years ago to the present). Popular focuses that accompany discussions of the Anthropocene include carbon dioxide saturation in the atmosphere, the melt of glacial caps (and the emissions trapped within), sea-level rise, extreme weather phenomena and temperatures, mass extinction, and the persistent needling that we have reached points of no return. The eschatology vibes emitted by much of Anthropocene writing are especially synced with the Industrial Revolution because, as Catherine Larrère has shown, throughout the eighteenth century, the terms "catastrophe" and "revolution" operated interchangeably.[11]

A recent international working group on the Anthropocene has fingered the "golden spike" to 1950 and thus acknowledged that the rapid carbon emissions captured in the atmosphere can be backdated to the genesis of the Industrial Revolution, which scholars of the long eighteenth century admit within their purview.[12] J. R. McNeill observes that locating alternative golden spikes (of which there are now many) runs the risk of "demoting" the eighteenth century and thus refocusing historical attention elsewhere, which would reinforce other historical and cultural indices that ignore the eighteenth century's (ironic and deleterious) import in conversations regarding climate fates and realities.[13]As contributors Elliot Patsoura, Annette Hulbert, and Adam Sweeting demonstrate, an eighteenth-century purview is germane to understanding the climactic exigencies—what Sweeting identifies as anxieties—we find ourselves ineluctably embracing now. These three authors ask, how might the writings of Erasmus Darwin, Phillis Wheatley Peters and Olaudah Equiano, and Increase Mather, respectively, allow us to situate growing eighteenth-century concerns over climate change? Indeed, the summer 2020 news curricular penned by former American Society for Eighteenth-Century Studies president Jeffrey Ravel declares, in no uncertain terms, "ASECS can no longer ignore its contributions to climate change."[14]

With that said, arguments surrounding the precision of originary dates (a sort of ill-gotten, finger-pointing primal scene) matter little to this collection not because these dates lack historical value, but because, in our opinion, they point toward the wrong question. We are not interested in whether the Anthropocene comes into its own in the eighteenth century or whether the eighteenth century nurses the Anthropocene's formative years; rather, this collection responds to how eighteenth-century culture, literature, and art, examined through the lens of the environmental humanities, can respond to, understand, and approach the Anthropocene in its polyphonic effects.

### *Axiom 3: The environmental humanities requires and inspires collaboration.*

On the heels of axiom 2, we want to tarry with these fears of crisis, first and foremost, in the environmental humanities' purported indebtedness to the social and hard sciences, or the assertion that, in drawing on ecological standpoints (which are implicitly inscribed as scientific orientations), it is somehow derivative or deferential. To study the environment is not the singular responsibility of a specific school of thought, field of study, or mode of interpretation, especially if we remind ourselves that the Greek etomyn *oikos*, from which ecology emerges, refers to the household, the family, and the family's property. *Eighteenth-Century Environmental Humanities* aspires to recognize a collective household of scholarly and pedagogical inclusivity. The authors here do not approach the environmental humanities as a means of ingratiating the humanities to the ascendency of STEM. The environmental humanities are not separable from the life sciences, and we wager that the life sciences are not isolable from the work of the humanities. Rather, our environmental humanities is one that highlights parity, cooperation, and reciprocity; we seek, in other words, a diverse, equitable, and inclusive environmental humanities, which is both a method for doing our work and also an acknowledgement of the topics we research that broach iniquities, power, and privilege—important concepts that can upheave the presumed innocence of "objectivity."

Second, this collection questions the centrality of the "human" within the humanities, which, as feminist, queer, trans, disability, critical race, and Indigenous scholars remind us, can too often ossify and erase particular identities. We cannot, furthermore, forget our commitment to other environmental inter-actors and nonhuman *kin*, a concept that Agustín Fuentes and Natalie Porter recalibrate as central to evolving our relationship with the nonhuman in the environmental humanities, generally, and animal studies, specifically. "If *kin* are those closest to us in space, time, and flesh," they write, "then *kinship*, by definition is a multispecies endeavor."[15] Or as Anna Lowenhaupt Tsing puts it, "We become who we are through multispecies aggregations."[16] In looking to coconspirators and collaborators, we are also mindful of our multispecies and environmental interlocutors that invigorate our apprehensions of embodiment, inquiry, and knowledge.

With this regard, our collection nears what Greg Garrard, for example, has offered in the process of "ecologizing humanity." Ecologizing stems from a recurrent endeavor to decenter the human "from their previously unchallenged position as the focus of exclusive interest and analysis in the humanities."[17] As Shelby Johnson and Mariah Crilley show in their contributions here, the premise of the human and its unstable boundaries are consistently under siege throughout eighteenth-century natural histories of virology, as Crilley shows, and in poetic representations of racialized maternity, as Johnson demonstrates. This process of ecologization, which disrupts definitions of the human, reveals an environmental humanities that de-hierarchizes what early moderns represented as the Great Chain of Being, or differences between and within species that have been politically, socially, and philosophically weaponized as artillery for oppression and supremacy. In its most democratizing form, to ecologize the environmental humanities is to witness forms of difference between human and nonhuman that do not necessitate subjugation, and instead visualize fecund planes of respected difference that energize critical inquiry. Lori Gruen has referred to a similar goal as "entangled empathy," which proposes a new ontological ethic that centrally addresses human and nonhuman animal engagement. In Gruen's powerful, utopian imaginary, empathic attunement can maintain discrete, nonconflated forms of subjectivity.[18]Astrida Neimanis, Cecilia Åsberg, and Johan Hedrén further invite us to consider what such an ethos might look like in the environmental humanities: "Our task is thus the tricky one of acknowledging the differences and diffractions in worldviews, histories, subjectivities, relations and practices that various communities (both human and non-human) engage in, with respect to their environment, while also cultivating an environmental humanities that is well-placed to research and analyze these differences, and remain vigilant against the risk of epistemological monoculture."[19]

To avoid an epistemological monoculture, the spirit of the environmental humanities calls on collaboration that works to dismantle finite disciplinary silos. Following the environmental humanist synchronicity of poet Harriet Tarlo and artist Judith Tucker, "our aspiration is that out of difference come connections bringing new imaginings to our audiences, imaginings that mirror our own discoveries made through collaborative cross-disciplinary practice."[20] We seek to uphold the environmental humanities, in this same mode, as a space that dwells in plurality, multiplicity, and heterogeneity. To do environmental humanities work is to do many things at once: it is to seek interdisciplinarity in its most utopian form; it is to acknowledge the humanities as multiplex, interrelational forms of being, learning, and reading; and it is to recognize environmental concerns, writ large, as those that require all the hands of the humanities on deck.

### *Axiom 4: The environmental humanities is anamorphic.*

Recognizing that the environmental humanities is not a single thing, it is likewise important to remind ourselves that work in the environmental humanities does not presume a single vision. The environmental humanities play with anamorphic possibilities. Anamorphic illustrations (such as Hans Holbein's *The Ambassadors* (1533) and Andrea Pozzo's anamorphic dome (1690) painted on the ceiling of the Church of St. Ignazio) induce optical illusions, visual distortions, and newfound clarities based on one's positionality. Anamorphosis plays with visual reliability, the limitations of imagined objectivity, and the constantly mobile ways of looking—all of which Donna Haraway addresses in her discussion of "situated knowledges." Like Haraway, the authors here set out to visualize "an account of radical historical contingency for all knowledge claims and knowing subjects, a critical practice for recognizing our own 'semiotic technologies' for making meanings," and thus elucidate and open up new epistemological vantage points.[21] The plurality of the environmental humanities—as field, method, and mode of inquiry—demands we come to terms with and likewise inhabit the unsteady and fluid ways of epistemological de/construction.

The anamorphosis of the environmental humanities similarly enfolds various genres and modes of interpretation. As axiom 1 detailed, the environmental humanities goes by many different alibis; it has many faces; its theories, conversations, and applications do not always commingle easily. What might be considered environmental humanities then? Consider, for example, ecocriticism, nature writing, literature and environment, naturalism, animism, ecospirituality, elemental thinking, new materialisms, plant and insect and animal studies, natural history and philosophy, environmental racism, ecofeminism and ecowomanism, queer ecologies, environmental justice, posthumanism, biosemiotics, environmental history, Indigenous futurisms, Afro-futurisms, and environmental arts and media and technologies, as only a handful. The environmental humanities as we understand it then functions as an interdisciplinary carrying bag that aspires to gather these diverse scholarly tactics—in tension and also together. We stand on all sides of the anamorphic environmental humanities so as to relish our own positionalities and seek insight from those who hold others.

## EIGHTEENTH CENTURY + ENVIRONMENTAL HUMANITIES

But what does this mean for eighteenth-century studies, and why might scholars of our historical period take up the environmental humanities? Answering such a question demands we come to terms with a persistent epithet that surfaces in conversations about the eighteenth century: presentism. For many, the environmental humanities (because of its emergence in the past few decades) is the wellspring of the contemporary. Presentism, for those who do historical work, epitomizes the black spot of piratical lore. Take for example, Lynn Hunt's 2002 address to the American Historical Association, titled "Against Presentism." Presentism, Hunt synthesizes, reifies contemporary ways of understanding and interpretation and skews attention away from historical periods. Hunt's bald concern is that "it threatens to put us out of business as historians."[22]

Putting the accuracy of this claim aside, we want to recognize first that, because of indictments like these, those of us pursuing the interdisciplinary manifestations of both eighteenth-century studies and environmental humanities have too often realized a double bind. Conversations with contributors have revealed an unsettling trend within the publication and professionalizing side of these fields. If eighteenth-century gatekeepers disavow the environmental humanist work because it falls into the (bad?) trap of presentism, environmental humanities gatekeepers do not make space for eighteenth-century studies precisely because it is not presentist enough. David Armitage's "In Defense of Presentism" recuperates the term as one that might more generatively embrace the various cultural histories and praxes that constitute the formation of history as something that dismantles objectivity and impassibility. The confusion about presentism, Armitage realizes, is one that hampers historical reflection and in so doing forecloses our ethical obligations to historical research, which, in its best form, mandates a temporal mediation that looks to the history and present so as to inform trajectories for the future.[23] Even more, by Joni Adamson's assessment, the roots of the environmental humanities can be and should be traced to "the earliest cosmological narratives, stories and symbols of the world's oldest cultures."[24] But theory, as we know, is different from application. We need, then, an environmental humanities that, by definition and framework, more capaciously and deliberately imagines how the work of scholars outside our current moment and century can contribute to diversifying the field. We thus need a *longue durée* of environmental humanist scholarship and teaching. This collection aids in that endeavor.

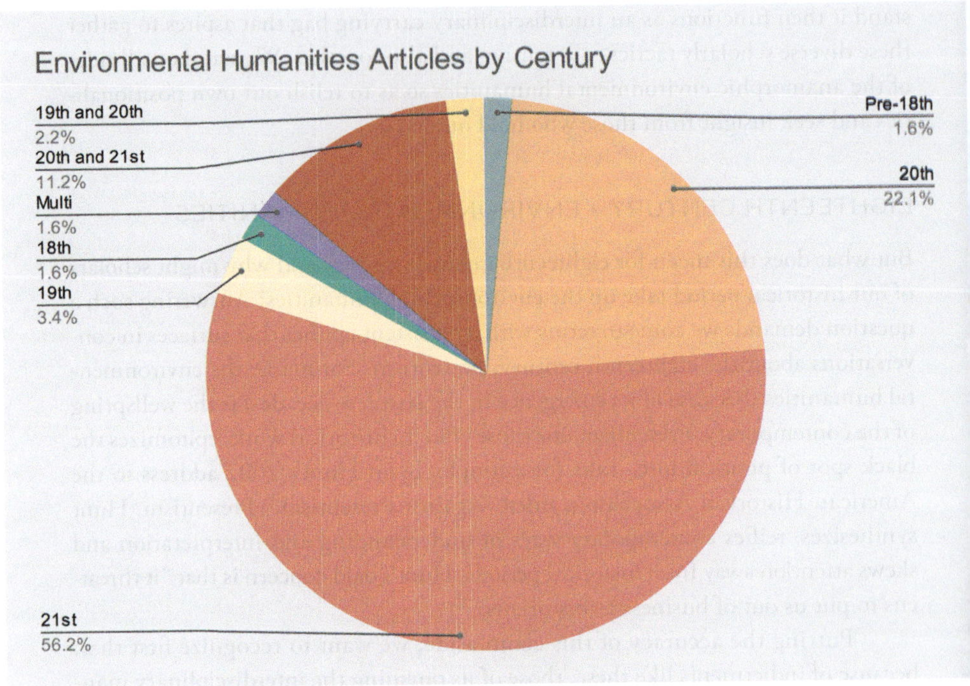


Figure I.1 Environmental humanities articles and their centuries of focus in five environmental humanities journals.

The proof is too often in the pudding. Take for example five flagship environmental humanities journals: *Environmental Humanities; Resilience: A Journal of the Environmental Humanities; Nature and Culture; Journal of Ecocriticism*; and *Green Letters*. At the time of writing this (in the throes of pandemic-situated 2020), in the collected issues, which includes roughly 550 essays, 310 (56%) explicitly address concerns of the twenty-first century, and an eye-opening 494 (89%) position environmental humanities concerns within the twentieth and twenty-first centuries. By our count, only nine essays (1.63%) take up eighteenth-century texts, topics, or histories. Figure I.1 illustrates these striking statistics.[25]

Whither the eighteenth century in these environmental humanities platforms? Eighteenth-century scholars assuredly take up environmental concerns, but perhaps out of fear of the critique of presentism, they are not couched in the framework of the environmental humanities, per se. That is, the phrase environmental humanities appears potentially illegible, or at the very least obscured. As practiced in eighteenth-century studies, it inhabits many different subgenres. We need only review myriad special issues that are framed by the prominence of animals (*Journal for Eighteenth-Century Studies*), the Anthropocene (*The Eighteenth-*

*Century Common*), eco-gothic (*Gothic Studies*), climate change (*Eighteenth-Century Studies*), plant and insect studies (*The Eighteenth Century: Theory & Interpretation*), or ecocriticism (*Eighteenth-Century Fiction*). The relationship between eighteenth-century studies and the environmental humanities, as these trends elucidate, can be tenuous, which often results in a scholarly schism that suggests a conflict between eighteenth-century environmental scholarship and the environmental humanities.

This collection strives to bridge these divides with the hope of forging intersection. We thus imagine an eighteenth-century environmental humanities that operates, as Rick Dolphijn and Iris van der Tuin describe, "transversally," so as to further develop transdisciplinary and transhistoric inquiry through generative "cuts" and grafts from which rich boundary-pushing scholarship might blossom.[26] The essays that follow do not process chronologically. This is deliberate. Such a chronology might gesture toward a historical progression or genealogy of eighteenth-century environmental humanities that would undermine our goals. Instead, the essays cut back, across, throughout, and within different historical junctures and archives found across the long eighteenth century.

What can, then, eighteenth-century studies bring to the environmental humanities? Or, put another way, how might the environmental humanities further evolve with more concerted attention to eighteenth-century studies? The authors featured here readily take up both these questions along five axes: climate change, new materialisms, the blue humanities, decoloniality and indigeneity, and green utopianism.

Chapters 1 through 3 elucidate the emergence of climate change discourses that have been brought to eighteenth-century studies and consider how eighteenth-century discourses can frame climate change knowledges. Seth Reno, Alessa Johns, Robert Markley, Dagomar Degroot, and David McCallam have repeatedly turned to climate change in both the eighteenth and twenty-first centuries to make sense of long histories of place-based knowledge and environmental disaster.[27] Degroot, for example, as a climate historian, acknowledges the issues of scale that accompany the difference between weather and climate and also cautions us, as scholars of a historical period reliant on particular archives, against "simplistic determinism," or the way in which a single force or a set of single forces predetermine and alter human history.[28] Chapters 1 through 3 reject this deterministic thinking in favor of larger cultural, literary, epistemological, and historical trends that can address a variety of climate changes rather than a singular, monolithic climate change. Disaster seems to be a particular flavor of climate change scholarship, and Patsoura, Hulbert, and Sweeting address how eighteenth-century texts and authors remain fixated on disaster and its cultural ramifications.

In chapter 1, Elliot Patsoura examines and revises genealogies of geoengineering, alongside Erasmus Darwin's *The Botanic Garden* (1791), to address the late-eighteenth-century preoccupation with and attempts to remediate the effects of the Little Ice Age. Patsoura's "Toward a Genealogy of Geoengineering: Erasmus Darwin and the Little Ice Age," sets the stage for a cluster of essays included here that speak to the intertwined scientific and supernatural logics that attempt to rationalize weather and its attendant phenomena. Patsoura identifies a fundamental irony in eighteenth-century (and contemporary) geoengineering proposals that seek to inhibit the ruptures caused by anthropogenic climate change. Namely, any attempt to imagine such hypotheses is immediately undercut by their untenability and inevitable failures. The chapter closes with a discussion of Darwin and his import in teaching British and German proto-Romantic and Romantic thought. Chapter 2, "Storm Apostrophe," by Annette Hulbert, maps a climactic literary history by comparing Phillis Wheatley Peters's storm poetry from *Poems on Various Subjects, Religious and Moral* (1773) and Olaudah Equiano's *Interesting Narrative* (1789). By locating abolitionism alongside invocations of storms wrought by the Little Ice Age, Hulbert offers environmental attunement as a corollary for moral subjecthood. In the chapter's conclusion, Hulbert acknowledges that by teaching these Black poets alongside M. NourbeSe Philip's formally experimental *Zong!* we can approach the Gordian knot of the environmental humanities, new materialisms, and literary and artistic representations of an "ontologically complex" eighteenth-century world. Adam W. Sweeting furthers the discussion of stormy weather and brings New England theology to bear on the Great Storm of 1703 in chapter 3. "'When Stormy Winds Happen': Divine Providence, Climate Change Discourse, and the Cause of Weather Disasters" reveals a striking parallelism between early-eighteenth-century theology and scientific discourses, which contribute to apprehending anthropogenic climate change. Acknowledging his own situatedness, and that of his teaching in Boston, Sweeting reflects on the historical cuts that might knit together early-eighteenth-century providential models of changing climates and the climate realities American coastal cities face now.

Chapters 4 and 5 document how new materialist frameworks, which seek to recognize how human, nonhuman, and more than human interlocutors exhibit and maintain plural forms of intra-active agency, can invigorate eighteenth-century studies. Many of the authors in this collection, not exclusively Johnson and Crilley, look to new materialist frameworks offered by Jane Bennett and Karen Barad. As political scientist and physicist, respectively, Bennett and Barad demonstrate the wide, interdisciplinary appeal of new materialisms that require our attention to alternative and additional actants, vitalities, and agencies. Catherine Keller and Mary-Jane Rubenstein, for example, situate new materialisms as a theoretical apparatus that rejects ontology as a monolith and instead welcomes "a multiplicity of experimental perspectives on mattering, and so a multiplicity of worlds."[29] The worlding opportunities of new materialisms are especially attractive to authors here because they allow us to think how worlds can be imagined through scale (zooming in and out) as well as through ecological and poetic attunement.

Shelby Johnson returns to Phillis Wheatley Peters's poetry in chapter 4, "Phillis Wheatley Peters's Niobean Soundscapes," to offer a new materialist approach that recognizes Wheatley Peters's invocation of grief shared by environmental actants and humans as an "ecological assemblage of more-than-human forms of life rather than as mere instances of pathetic fallacy." Wheatley Peters's environmental fellowship, Johnson argues, promotes ways of recuperating the eighteenth century's enmeshment/enfleshment of Black bodies and natural landscapes. Johnson powerfully ends the chapter with a reflection that situates Wheatley Peters, Black subjectivity, and Black activism in the classroom alongside the polyphonic emergence of Black Lives Matter. In chapter 5, "Syphilis and Natural History: The Ethical Limits of Human Mastery," Mariah Crilley offers an eighteenth-century new materialisms that examines the material, historical, and literary manifestations of eighteenth-century syphilis, especially as illustrated by Hans Sloane's *A Voyage to the Islands Madera, Barbados, Nieves, S. Christophers and Jamaica* (1707–1725). As Crilley suggests, we need "a method for reading for and listening to disease's agency." Crilley, in this way, imagines an important nexus of environmental humanities and medical humanities investigations that recognize the material and oftentimes invisible nonhuman, microbial worlds that participate in shaping ethical dimensions of the lived experiences of disease. Such a recognition, Crilley contends, can inject excitement, epistemological diversity, and student metacognition into an interdisciplinary studies curriculum.

Chapters 6 and 7 make headways into the blue humanities. Alternatively called the oceanic turn, critical ocean studies, and blue cultural studies, the blue humanities investigates the role and import that bodies of waters, especially global oceans, play in constructing narrative, history, and epistemology. One of the pioneers of the phrase "the oceanic turn," Hester Blum, identifies a reorientation to a "critical perspective from the vantage point of the sea, while ontologically embracing the impossibility of coming to rest on any given point within the oceanic world of relational thought."[30] Eighteenth-century studies has likewise begun to weigh what it can bring to the blue humanities, and vice versa, for example, as Siobhan Carroll, Killian Quigley, Margaret Cohen, and I have demonstrated.[31] As Cohen writes in *The Novel and Sea*, an eighteenth-century focus on bodies of water, which the blue humanities accommodates, aspires to "leave the land and embark."[32]

In chapter 6, "Shore/Lines: Drawing Environmental Change on Eighteenth-Century Prince Edward Island," Claire Campbell joins the blue humanities and Canadian environmental history to investigate map and survey records as evidence of the environmental change brought about by colonization, migration, and archive. Campbell's attention to cartography opens visual and narrative realms that can enrich an eighteenth-century environmental humanities purview, and Campbell closes by considering the successes and failures of how these materials work in the classroom and can productively unsettle the age of Google Maps—an insufficient cartography and yet its most legible form for students today. Jason M. Payton's subsequent contribution in chapter 7, "Of Water, Winds, and Storms: The Elemental Regimes of the Buccaneer Journal," envisions another eighteenth-century blue humanities purview through the writings of early-eighteenth-century Caribbean buccaneers such as Alexander Oliver Exquemelin's *The Buccaneers of America* (1678) and William Dampier's *Voyages and Descriptions* (1699). In his focus on piratical hydrography, Payton presents a "decolonial critique that recovers an understanding of water and wind as agentic"—yet a further invocation of the new materialisms proposed by Hulbert, Crilley, and Johnson, and a signal tie to Ami Yoon's decolonial reading in chapter 9. Payton ultimately shows that teaching early modern and eighteenth-century piracy can present a wider range of archives that inform global mercantilism and its environments across historical periods.

The emergent and necessary turn to Indigenous representation, decoloniality, and anticolonialism in eighteenth-century studies carries forth into chapters 8 and 9. Such representations present, as Robbie Richardson evidences, a "unique depiction of the Indian as a non-European subject of both affinity and disavowal, across a variety of genres and cultural expressions, critiqued and helped articulate evolving practices and ideas such as consumerism, colonialism, 'Britishness,' and, ultimately, the 'modern self' over the course of the century."[33] A recent issue of *Eighteenth-Century Fiction*, spearheaded by Eugenia Zuroski, takes to task Indigenous erasure in eighteenth-century literature, anthologies, and teaching and creates space for Indigenous scholars and scholars of indigeneity to imagine how to redress histories, presents, and futures of discrimination and violence. Katherine Binhammer, in that issue, provides a plan of action for decolonizing our field; Binhammer writes, "To decolonize our field, we likely need to experiment with different research methodologies than anonymous archival research on printed material from the century. It means taking seriously Indigenous knowledge practices that begin by thinking about land and relations and not about periodized time. Land-based research methodologies and community-engaged research practices do not easily fit with our field."[34] Decolonizing the eighteenth century, then, demands we grapple with the connections among our fields of research and teaching, which Duquès and Yoon pointedly address.

Further developing the focus on climate offered by the first three chapters, in chapter 8, Matt Duquès proposes a climate-history-focused reading of John Dennis's play, *Liberty Asserted* (1704), and its enmeshment with Indigenous New World sources. "'Supporting Sinking Nations': Indigenous Women and Their Disasters in John Dennis's *Liberty Asserted*" highlights Indigenous women's voices to attend to the redaction of these voices from popular eighteenth-century discourses and print materials. Dennis's Sakia (Huron) and Okima (Mohawk) illustrate an environmental discourse that makes evident disastrous upheaval to which, the play's warped logic suggests, the English are the singular remediation. Highlighting the nexus of climate change and indigeneity, Duquès closes by offering *Liberty Asserted* as a means by which to reflect on concerns in the classroom regarding Indigenous erasure, marginalization, and manifold relationships with environmental disaster in the eighteenth century. In chapter 9, "Imagining Decolonial Futures in William Gilbert's *The Hurricane*," Ami Yoon posits Gilbert's late-eighteenth-century eclogue as imagining alternative environmental futures not bound by colonial strictures. *The Hurricane* presents a decolonial future in which environmental interactors level and overwhelm imperial advances; Yoon writes, "After the hurricane's destruction, what remains will be free to grow beyond the hierarchies and oppressions of settler colonialism." As with Duquès, Yoon's pedagogical reflection demonstrates an investment in focalizing indigeneity to allow students the opportunity to cast off the blinders of an education bound by settler colonialism.

Chapters 10 and 11 address the recurrence and longevity of environmental utopias, which populate the long early modern period (from More to Bacon to Cavendish) and the long eighteenth century (from Neville to Swift to Scott to Astell). Utopianism diagnoses political, social, and cultural discontents and often implicitly and subtly invokes the environment in those diagnostics. Utopian environmental thinking, Karin Bradley and Johan Hedrén suggest, in its most ideal form, seeks to create techno-political solutions to environmental concerns, though the effects of these solutions are often deleterious and ironic, as Patsoura notices of eighteenth-century geoengineering.[35] Christopher Allan Black likewise addresses that irony, especially as eighteenth-century poets deploy the Georgic tradition and its description of Edenic landscapes to veil an apologist stance that further exploits enslaved peoples. Scarth sees a different green utopia afoot—one less ironic and more aligned with environmental justice—in the connection between eighteenth-century theory and twentieth-century fiction.

Christopher Allan Black examines the repurposing of eighteenth-century georgics as genres that highlight the co-constitution of enslaved bodies and colonial environmental commodities, such as sugar cane, in chapter 10. "Slavery and Plantation Stewardship: The Eighteenth-Century Caribbean Georgics of James Grainger and Phillip Freneau" dissects this plantation logic by comparing Grainger's "The Sugar Cane" (1765) and Freneau's "To Sir Toby" (1784) to argue that, while Grainger offers a plantation utopia in which the careful maintenance of sugar cane induces humane treatment of the enslaved, it is exclusively theory and not application. Freneau, writing in response to Grainger, acknowledges those failures. Black concludes with a reflection on how eighteenth-century Georgic traditions illustrate the exploitation of Black and captive bodies. In chapter 11, "John Thelwall and L. M. Montgomery Write the Green City," Kate Scarth revisits eighteenth-century Prince Edward Island, Canada—as Campbell does with cartography in chapter 6—with a juxtaposed reading of John Thelwall's *The Peripatetic* (1793) and L. M. Montgomery's *Anne of Green Gables* (1908). Scarth proposes Thelwall as a central interlocutor for imagining urban ecological justice and "cultural fixes," which Montgomery takes up in her popularized depiction of Anne Shirley. Chapter 11 thus bridges eighteenth-century environmental theories and their twentieth-century literary interlocutors. Scarth's pedagogy considers how the environmental humanities can make possible place-based learning that reimagines the classroom, not as an interior structure but as one that can live outside, in peripatetic fashion.

## TEACHING EIGHTEENTH-CENTURY ENVIRONMENTAL HUMANITIES

Writing about our teaching is difficult, precisely because, for most of us, we are not taught to do so. We do not write about our teaching in the same way we do our research. To be clear, this collection does not intend to equalize teaching and research. Our approaches, responsibilities, job duties, and callings to these two aspects of our work are assuredly different. But to say that teaching and researching eighteenth-century environmental humanities are two mutually exclusive experiences would be misleading.

An implicit thread throughout the chapters, especially the conclusions, is a serious question that spurs dialogue about what constitutes innovative, interdisciplinary teaching. The authors here teach in a variety of programs, departments, universities, and settings. Some of us are completing doctoral degrees and setting sail in a market characterized by extremely turbulent waters. Others of us are entering or settling into the tenure track. Some of us have long thrived in contingent faculty roles. And still others of us have reached the highest echelon of full professor. In other words, our situated knowledges about our teaching and placement in the academy vary, and this is what, we find, makes this collection a pioneering endeavor. We do not seek to define good teaching as a monolith. We do not operate under the presumption that we teach identically or that our students learn identically. We are wary of such a reduction of teaching and learning styles. Rather, readers that jump from conclusion to conclusion will realize that, just as much as the environmental humanities goes by many anamorphic appeals, faces, and feels, so various too are the opportunities for us to expand, diversify, and decolonize our syllabi; to consider new pedagogical approaches; and to actively address the challenges and spectacular successes of teaching a robust eighteenth-century environmental humanities. Teaching can be a lonely vocation. In writing about our experiences and encounters in the classroom, we seek to alleviate that loneliness through collaboration.

This is to say, then, that as much as each chapter begets a particular conversation, our pedagogical conclusions are even more of a conversation starter. The research questions we explore develop from or into teaching; our teaching generates research insights that can be found only in a classroom—whether of first-year undeclared students or seasoned graduate students. The chapters that follow here then invite us as scholars, teachers, students, and readers to reimagine what our scholarship can do in its compatibility with our teaching. To write about our teaching is not an abdication of our research. *Eighteenth-Century Environmental Humanities* is instead a collection that reminds us of the inextricable nature of pedagogy and research, which, in mutually informing each other, flourish because of their enmeshment. In colloquy lies our best, most desirable, future.

## NOTES

1. Vin Nardizzi and Tiffany Jo Werth, eds., *Premodern Ecologies in the Modern Literary Imagination* (Toronto: University of Toronto Press, 2019); Jeffrey Jerome Cohen, ed., *Prismatic Ecologies: Ecotheory beyond Green* (Minneapolis: University of Minnesota Press, 2013).
2. Ursula Heise, "Introduction: Planet, Species, Justice—and the Stories We Tell about Them," in *The Routledge Companion to the Environmental Humanities*, ed. Ursula Heise, Jon Christensen, and Michelle Niemann (New York: Routledge, 2017), 1–10.
3. Serpil Opperman and Serenella Iovino, *Environmental Humanities: Voices from the Anthropocene* (New York: Rowman and Littlefield, 2017), 3.
4. Jeffrey Jerome Cohen and Lowell Duckert, "Introduction: Welcome to the Whirled," in *Veer Ecology: A Companion for Environmental Thinking*, ed. Jeffrey Jerome Cohen and Lowell Duckert (Minneapolis: University of Minnesota Press, 2017), 2.
5. Noah Smith, "America Is Pumping Out Too Many Ph.Ds," *Bloomberg Opinion*, January 4, 2021, https://www.bloomberg.com/opinion/articles/2021-01-04/america-is-pumping-out-too-many-ph-d-s?srnd=opinion.
6. Benjamin Schmidt, "The Humanities Are in Crisis," *The Atlantic*, August 23, 2018, https://www.theatlantic.com/ideas/archive/2018/08/the-humanities-face-a-crisisof-confidence/567565/.
7. Jeffrey J. Williams, "The New Humanities," *Chronicle Review*, November 14, 2019, 26, http://connect.chronicle.com/rs/931-EKA-218/images/ChronicleReview_Endgame.pdf.
8. Elizabeth DeLoughrey, *Allegories of the Anthropocene* (Durham, NC: Duke University Press, 2019), 2.
9. Stephanie LeMenager, "The Humanities after the Anthropocene," in *The Routledge Companion to the Environmental Humanities*, ed. Ursula Heise, Jon Christensen, and Michelle Niemann (New York: Routledge, 2017), 473–481; Richard Grusin, ed., *Anthropocene Feminism* (Minneapolis: University of Minnesota Press, 2017; Steve Mentz, *Break Up the Anthropocene* (Minneapolis: University of Minnesota Press, 2019); Jussi Parikka, *The Anthrobscene* (Minneapolis: University of Minnesota Press, 2014); Anna Tsing, *The Mushroom at the End of the World* (Princeton, NJ: Princeton University Press, 2015); Charles Mann, *1493: Uncovering the New World That Columbus Created* (New York: Vintage, 2011); Kathryn Yusoff, *A Billion Black Anthropocenes or None* (Minneapolis: University of Minnesota Press, 2018).
10. Tony Birch, "'The Lifting of the Sky': Life Outside the Anthropocene," in *Humanities for the Environment*, ed. Joni Adamson and Michael Davis (New York: Routledge, 2016), 195–209; Kyle Powys White, Chris Caldwell, and Marie Schaefer, "Indigenous Lessons about Sustainability Are Not Just for 'All Humanity,'" in *Sustainability: Approaches to Environmental*

*Justice and Social Power*, ed. Julie Sze (New York: New York University Press, 2018), 149–179.

11. Catherine Larrère, "Un catastrophisme absolu?," in *L'invention de la catastrophe au XIIIe Siècle*, ed. Anne-Marie Mercier-Faivre and Chantal Thomas (Geneva: Droz, 2008), 143–152.
12. Jan Zalasiewicz et al., "The Working Group on the Anthropocene: Summary of Evidence and Interim Recommendations," *Anthropocene* 19 (2017): 55–60.
13. J. R. McNeill, "Introductory Remarks: The Anthropocene and the Eighteenth Century," *Eighteenth-Century Studies* 49, no. 2 (2016): 124.
14. Jeffrey Ravel, "President's Column," *American Society for Eighteenth-Century Studies News Circular* 175 (June 2020): 2. https://5a6f0c4a-6896-4292-a494-72f07a565f1b.filesusr.com/ugd/acf0d2_9b0a142af9a04121927554afbe0b4099.pdf.
15. Agustín Fuentes and Natalie Porter, "Kinship," in *Critical Terms for Animal Studies*, ed. Lori Gruen (Chicago: University of Chicago Press, 2019), 183 (emphasis original).
16. Anna Tsing et al., "Strathern beyond the Human: Testimony of a Spore," *Theory, Culture & Society* 31, no. 2–3 (2014): 230.
17. Greg Garrard, "Environmental Humanities: Notes Towards a Summary for Policymakers," in *The Routledge Companion to the Environmental Humanities*, ed. Ursula Heise, Jon Christensen, and Michelle Niemann (New York: Routledge, 2017), 463.
18. Lori Gruen, *Entangled Empathy: An Alternative Ethic for Our Relationships with Animals* (New York: Lantern Books, 2016).
19. Astrida Neimanis, Cecilia Åsberg, and Johan Hedrén, "Four Problems, Four Directions for Environmental Humanities: Toward Critical Posthumanities for the Anthropocene," *Ethics and the Environment* 20, no.1 (2015): 69.
20. Harriet Tarlo and Judith Tucker, "'Drawing Closer': An Ecocritical Consideration of Collaborative, Cross-Disciplinary Practices of Walking, Writing, Drawing and Exhibiting," in *Extending Ecocriticism: New Readings in Literature, Visual Arts, Performance and On-Site Interpretation*, ed. Peter Barry and William Welstead (Manchester, UK: Manchester University Press, 2017), 65.
21. Donna Haraway, "Situated Knowledges: The Science Question in Feminism and the Privilege of Partial Perspective," *Feminist Studies* 14, no. 3 (1988): 579.
22. Lynn Hunt, "Against Presentism," *Perspectives on History*, May 1, 2002, https://www.historians.org/publications-and-directories/perspectives-on-history/may-2002/against-presentism.
23. David Armitage, "In Defense of Presentism," in *History and Human Flourishing*, ed. Darrin McMahon (Oxford: Oxford University Press, forthcoming), 20.
24. Joni Adamson, "Introduction," in *Humanities for the Environment*, ed. Joni Adamson and Michael Davis (New York: Routledge, 2016), 5.
25. I am indebted to Madison Weaver for this visual as well as her careful and attentive examination of these journals.
26. Rick Dolphijn and Iris van der Tuin, *New Materialisms: Interviews & Cartographies* (Ann Arbor, MI: Open Humanities Press, 2012). http://dx.doi.org/10.3998/ohp.11515701.0001.001.
27. Seth Reno, "The New Volcanoes of Industry," *18th-Century Common*, February 22, 2019, https://www.18thcenturycommon.org/volcano/; Alessa Johns, ed., *Dreadful Visitations: Confronting Natural Catastrophe in the Age of Enlightenment* (New York: Routledge, 1999); Robert Markley, "'Casualties and Disasters': Defoe and the Interpretation of Climactic Instability," *Journal for Early Modern Cultural Studies* 8, no. 2 (2008): 102–124; Dagomar Degroot, *The Frigid Golden Age: Climate Change, the Little Ice Age, and the Dutch Republic, 1560–1720* (Cambridge, UK: Cambridge University Press, 2018); David McCallam, *Volcanoes in Eighteenth-Century Europe: An Essay in Environmental Humanities* (Oxford: Oxford University Press, 2019).
28. Degroot, *Frigid Golden Age*, 16.
29. Catherine Keller and Mary-Jane Rubenstein, eds., *Entangled Worlds: Religion, Science, and New Materialisms* (New York: Fordham University Press, 2017), 3.

30. Hester Blum, "Introduction: Oceanic Studies," *Atlantic Studies* 10, no. 2 (2013): 155.
31. Siobhan Carroll, *An Empire of Air and Sea: Uncolonizable Space in the British Imagination, 1750–1850* (Philadelphia: University of Pennsylvania Press, 2015); Killian Quigley, "The Pastoral Sublime: William Diaper and Eclogue's Marine Frontier," *Eighteenth-Century Studies* 53, no. 1 (2019): 109–128; Margaret Cohen, *The Novel and the Sea* (Princeton, NJ: Princeton University Press, 2010); Jeremy Chow, "Taken by Storm: *Robinson Crusoe* and Aqueous Violence," *Robinson Crusoe after 300 Years*, ed. Glynis Ridley and Andreas Mueller (Lewisburg, PA: Bucknell University Press, 2021): 108–134.
32. Cohen, *Novel and the Sea*, 3.
33. Robbie Richardson, *The Savage and Modern Self: North American Indians in Eighteenth-Century British Literature and Culture* (Toronto: University of Toronto Press, 2018).
34. Katherine Binhammer, "Is the Eighteenth Century a Colonizing Temporality?" *Eighteenth-Century Fiction* 33, no. 2 (2021): 202.
35. Karin Bradley and Johan Hedrén, *Green Utopianism: Perspectives, Politics and Micro-Practices* (New York: Routledge, 2014).

# Part One

## EIGHTEENTH CENTURY + CLIMATE CHANGE

1

# TOWARD A GENEALOGY OF GEOENGINEERING

## Erasmus Darwin and the Little Ice Age

ELLIOT PATSOURA

THROUGHOUT THE EIGHTEENTH CENTURY, Britain witnessed an unprecedented rise of public interest in the documentation of the weather. First initiated by the widespread damage to land- and sea-based assets in the Great Storm of 1703, this national fixation sought to register not simply the extremities of weather brought on by multiple phases of the Little Ice Age—a period extending from the fourteenth century through to the mid-nineteenth that saw fundamental changes to the behavior and size of glaciers and a lowering of average temperatures in the Northern Hemisphere[1]—but also the consistency and sheer banality of the ongoing, daily weather. "Regular recording drew out the routine features of the British atmosphere," Jan Golinski observes, "rather than focusing exclusively on extreme peculiarities like the 1703 storm. It tended to 'normalize' the weather, making it a quotidian process that went on all the time and not just when dramatic events drew special attention. Viewed in this way, the national climate appeared generally benevolent—both moderate overall and gently variable in temperature and precipitation, . . . assets to the agriculture, commerce, health, and character of the nation."[2] A reaction to (or manifestation of) eighteenth-century climate anxiety born of such outlying climactic events, the subjection of British weather to this "civilizing" process molded it into an emblem of a "civilized" society—"an appropriate attribute of a nation that prided itself on its reason, refinement, and sensibility" undertaken by "enlightened investigators [working] to subdue it by scientific reason," ameliorating the icy extremities of climate in the public imagination.[3]

The attempted civilization of the weather over the course of the eighteenth century was led by a combination of institutional innovation and public enthusiasm. Golinski documents an attempt on the part of the Royal Society, for instance, "to coordinate weather recording" by "circulat[ing] a printed invitation to prospective observers" of the weather in the 1720s. Overwhelmed by the data that consequently flowed in over the coming decades from "dozens of weather recorders . . . inspired by a sense of public-spiritedness to contribute to this project," for some time the Society was unable to appropriately channel said data into a steady publication stream. It was only in 1774 that the Society was able "to compile and publish its own weather record, using instruments kept at its London premises"; by the 1780s this trend had reached a rather bland (if reassuring) zenith in the broader publication of "books consist[ing] solely of weather records."[4]

The national fervor for the civilization of British weather also leached into the Lunar Society of Birmingham, a scientific society that boasted as its members figures crucial to the rise of industrialized techno-scientific modernity: Samuel Galton, Joseph Priestley, Josiah Wedgwood, and Benjamin Franklin were all proud "Lunatics," as was Erasmus Darwin, one of the society's cofounders. While the documentation of the weather and the growing set of instruments possible for its measure occupied the minds of most members, Darwin's interest in the weather itself was especially, and notably, strong.[5] Beginning in 1763 with "speculat[ions] about the effect of weather on health," and taking a broader, "philosophical" turn in 1783, with the entry of a meteor into the atmosphere, Darwin would come to be "the only Lunar member to publish anything significant in the way of meteorological studies."[6] Darwin's self-termed "anemology"—what has been described as "a complete, ingenious, and alas over-simplified and incorrect theory of wind"[7]—circulated not only in scientific publications and correspondence with his fellow Lunatics, but at length in "The Economy of Vegetation," published as the first part of Darwin's immensely popular didactic book-length poem, *The Botanic Garden,* in 1791.[8]

Herein, Darwin's anemology forms the explanatory backdrop to a wide-ranging explication of global weather patterns, and a specific and quite infamous proposal to actively *reshape* these patterns. Disgruntled by the increasingly frosty climate of his native Britain during what would turn out to be the final stage of the Little Ice Age,[9] and unnerved by the interruption of colonial shipping by rogue icebergs,[10] Darwin proposed, with a curious blend of heroic couplet, exploratory footnote, and extensive, thematically organized "additional notes"—credited, notably, with "initiat[ing] the dialogue between modern science and literature in Britain"[11]—a coordinated effort to transport entire "islands of ice" from the North Pole into the tropical regions of the Earth where, as Thomas De Quincey rather sardonically put it in one of a number of subsequent dismissals of Darwin's scheme, they "would disappear as rapidly as sugar-candy in children's mouths."[12] "If the nations who inhabit this hemisphere of the globe, instead of destroying their seamen, and exhausting their wealth in unnecessary wars, could be induced to unite their labour to navigate these immense masses of ice into the more southern oceans," opines Darwin, "two great advantages would result to mankind; the tropic countries would be much cooled by their solution, and our winters, in this latitude, would be rendered much milder, for perhaps a century or two, till the masses of ice become again enormous."[13]

Today, the heretofore unrealized project of so "civilizing" global weather patterns, rendering them amenable to the whims of industrial and colonial interests by way of techno-scientific intervention, is widely constructed as more "desirable" and "feasible" than it has ever been. As a globally coordinated response to the climate crisis appears increasingly unlikely, "geoengineering" schemes designed to lower global carbon dioxide levels by actively extracting it from the atmosphere, or to directly lower the heat of the globe by reflecting solar radiation back into the cosmos, are no longer limited to the literary domain (of science fiction or, indeed, of romantic didactic poetry), nor easily dismissed as remnants of speculative Cold War programs for the control of large-scale weather patterns.[14] Rather, geoengineering is steadily gaining traction in the public imagination as a necessary response to the climate crisis,[15] a "plan B"[16] that not only "troubles our 'common sense' about environmentalism" but "raises profound, even existential, questions about both the human and the more-than human, and the relationship between them."[17]

Following efforts in the environmental humanities to interrogate prior instances of climate anxiety as a means of critically apprehending our own,[18] this chapter examines how fundamental shifts in our technological capacities and the increasing enthusiasm for the manipulation of global weather systems are preceded by an eighteenth-century emergence. In locating the literary inscription of a deep skepticism toward the feasibility of geoengineering alongside Darwin's unabashed advocacy for the undertaking in his poetic work, I want to outline how the capacity to intervene in said systems may in fact be accompanied by an immanent and necessarily repressed sense of the inevitable shortcomings of such interventions—both now and in the eighteenth century. I explore how Darwin gives expression to the incapacity of nature to be fully domesticated, precisely in and among a proposal propounding the very necessity of its domestication. I first examine the anemology that informs Darwin's understanding of the climactic problem his proposal is designed to resolve,[19] then turn to the particulars of Darwin's presentation of the proposal in the first canto of "The Economy of Vegetation," outlining how, with a comparatively more subtle set of allusions to the misguided nature of such proposals, Darwin's geoengineered "wrath against the icebergs"[20] is informed as much by a desire to craft the climate as it is to facilitate an opportunity for nature to reassert itself against its attempted domestication. Given Darwin's status as an eighteenth-century "prophet for the scientific worldview that came to dominate the nineteenth and twentieth centuries," this is an especially (and admittedly) striking claim.[21] However, it is all the more instructive for it, as it directs our critical attention not only to the means by which the environmental or commercial feasibility of a given geoengineering proposal is constructed, but also to how the ostensible support for geoengineering in a time of climate anxiety—whether in Darwin's time or our own—may be little more than a carrier for a latent and very human desire to stage its own failure: specifically, to facilitate an expression of nature's sovereignty in the last instance from techno-scientific capture. As such, I take a first step toward a larger genealogy of geoengineering that would chart the differing means by which this latent desire has been textually inscribed since the eighteenth century without explicitly undercutting the surface consistency of a given geoengineering proposal.

Darwin's anemology is explicated across numerous modes in "The Economy of Vegetation," but none are perhaps so in tune with the eighteenth-century British enthusiasm for the tedious documentation of weather as the thirty-third entry in Darwin's "additional notes." Darwin's entry on "Winds" is by far the largest among a series of notes further expounding topics already versified and explicated in footnote in the preceding pages. It consists of a cataloguing of the various types of wind according to this theory, an overview of meteorological instrumentation, and finally an "application of the preceding Theory to some Extracts from a Journal of the Weather": a non-linear series of entries dating from 1790, 1785, 1799, and 1732.[22] Herein, Darwin stipulates that the winds of the Earth both originate and dissipate in what he terms "an officina aeris, a shop where air is both manufactured and destroyed in the greatest abundance."[23] Three such "workshops" exist, for Darwin: the largest two are located in the North and South Poles, respectively, and a third (comparatively under-remarked-on) "within the tropics, or at the line, though in a much less quantity than at the poles."[24] The only "two original winds" in existence, from which all other winds are derived, originate in these first two locales: "one consisting of air brought from the north, and the other from air brought from the south." Crucially, among Darwin's extensive list (and explication) of such "deflections or retrogressions" from the original winds in this note (southwest winds, northeast winds, southeast winds, northwest winds, trade winds, monsoons and tornadoes, and land and sea breezes) are "the north-east winds" of his own country "consist[ing] of regions of air from the north, travelling sometimes at the rate of about a mile in two minutes, during the vernal months, for several weeks together, from the polar regions toward the south."[25] These winds of northern origination underwrite Darwin's distaste for the effects of the Little Ice Age, for they "consist of air greatly cooled by the evaporation of the ice and snow over which it passes, and, as they become warmer by their contact with the earth of this climate, are capable of dissolving more moisture as they pass along, and are thence attended with frosts in winter, and with dry hot weather in summer."[26] With this crucial component of his anemology, Darwin is able to account for the severity of the cold he personally experiences during the Little Ice Age as an index of the quantity of ice to be found at the point of the northern wind's origination in the North Pole.

In so doing, however, Darwin cannot evade a fundamental lacuna in his account. This lacuna is alluded to by Darwin himself at the outset of the anemological note (where he introduces his "theory of the winds [a]s yet very imperfect") and is able to be identified and quantified in the first instance on account of the newly available public distribution of climactic data. According to Darwin, the imperfection of his theory is due less to "the want of observations sufficiently numerous of the exact times and places where they begin and cease to blow" than to "our yet imperfect knowledge of the means by which great regions of air are either suddenly produced or suddenly destroyed."[27] Darwin appears little bothered by a small sample size when pontificating on the origin and ends of wind. Near the close of the note on winds, he compares meteorological data compiled in Edinburgh to that compiled by the Royal Society in the more southerly locale of London and fixates on a difference of temperature variations recorded therein. Darwin then uses this difference to suggest that the closer the documented changes are to the North Pole, the greater the variation in such changes:

> *August* 28, 1732. Barometer was at 31, and *Dec.* 30, in the same year, it was at 28 2-tenths. Medical Essays, Edinburgh, vol. ii. P.7. It appears from these journals that the mercury at Edinburgh varies sometimes nearly three inches, or one-tenth of the whole atmosphere. From the journals kept by the Royal Society at London, it appears seldom to vary more than two inches, or one-fifteenth of the whole atmosphere. The quantity of the variation is said still to decrease nearer the line, and to increase in the more northern latitudes; which much confirms the idea that there exists, at certain times, *a great destruction or production of air within the polar circle.*[28]

From such empirical observations, Darwin derives the problematic keystone of his anemology; namely, the claim that

> one fifteenth part of the atmosphere is occasionally destroyed, and occasionally reproduced, by unknown causes. These causes are brought into immediate activity over a great part of the surface of the earth, at nearly the same time, but always more powerful to the northward than to the southward of any given place, and would hence seem to have their principle effect in the polar circles; existing, nevertheless, though with less power, toward the tropics or at the line.[29]

In both of the above passages, Darwin locates this destruction of wind geographically, and quantifies such destruction with the help of publicly available data, such that he can conclude with full confidence that "there must be, at this time, a great and sudden absorption of air in the polar circle, by some unknown operation of nature."[30]

Darwin's comparative and highly civilized quantification of the outstanding problem of his anemology stands in stark contrast to its explanatory amelioration, and this in a manner highly suggestive of a compulsion to register the ineluctable capacity of nature to avoid techno-scientific capture. Indeed, the vertiginous nature of this amelioration of a spatially delimited and clearly quantified problem unashamedly signals to the reader that Darwin is well aware that neither he nor, perhaps, anyone else can fundamentally resolve the lacuna lying at the very basis of not only his anemology but also, and by extension, his rendition of the climactic challenges motivating his geoengineering proposal manifest in other parts of his text. "Though the immediate cause of the destruction or re-production of great masses of air at certain times . . . cannot yet be ascertained," Darwin writes in the final paragraph of the note's "Conclusion" section, "we may still suspect that there exists in the arctic and antarctic circles, a BEAR or DRAGON, yet unknown to philosophers, which, at times, suddenly drinks up, and as suddenly, at other times, vomits out one-fifteenth part of the atmosphere; and hope that this or some future age will learn how to govern and domesticate a monster which might be rendered of such important service to mankind."[31] Owing perhaps as much to its location in the backwaters of Darwin's text as to its patent ridiculousness, Darwin's eighteenth-century appropriation of the medieval institution of using the monstrous to mark the cartographically unknown has, to date, gone almost entirely unremarked on in the scholarly literature. With the proviso that "one should not underestimate the degree to which the eighteenth-century mind could juggle supernatural and scientific credulity in the marvelous," Max F. Schulz's lone and considered reading holds that "the contradictory assertiveness and tentativeness of Dr. Darwin's words" means that "one cannot be sure whether he is devising an explanation he means his readers to take literally or symbolically, or both at once."[32] Darwin's approach—which is, in essence, a "practice of blurring, and confusing, fact and symbol"—is thought symptomatic of an effort "to make divine sense out of a world rapidly undergoing quantification."[33]

I want to suggest that what Schulz here terms a "duality of attitude in coping with the earthly experience,"[34] might be better glossed as a duality of attitude in coping with climate anxiety. Darwin's effort might be reconceptualized as an effort to make sense of a world rapidly undergoing quantification, and one that is simultaneously well aware of the inevitable insufficiency of such an undertaking. For Darwin appears to be humorously provoking his reader to acknowledge that this origin of the wind will, in fact, not *ever* be fully and sufficiently accounted for; the wind will continue to outrun any attempted scientific capture.

Darwin's staging of this quality of nature at the height of its attempted quantification ensures a jarring experience for a reader otherwise acclimatized to Darwin's faith in the capacity of techno-scientific modernity. To seriously, faithfully render the origin of wind symbolically in a gesture of humble civility would be to entertain the prospect of its future scientific capture, to allude to a nature that is only temporarily *hidden* from thought, rather than radically foreclosed to it.[35] In stark contrast to this, the blatant extremity of Darwin's proposition is plain for all to see: it derives an entirely surprising quality from its location among an otherwise dry and extensive documentation of wind varietals, and it subverts the clear demarcation of symbol from fact, verse from footnote, with which the reader of "The Economy of Vegetation" would, at this point of the text, be well familiar.[36] Holding out the prospect only of "govern[ing] and domesticat[ing] a monster" severely and compulsively intoxicated by the atmosphere, Darwin makes a gentle yet definitive mockery of the capacity of human understanding to ever tame the originating forces of nature. Nature, Darwin senses, will inevitably continue to undercut the British fantasy of domestication and remain ultimately foreclosed to a scientific thought anxiously fixated on its full explication.

Darwin's jarring gesture in the section of "additional notes" necessitates a reexamination of similarly monstrous symbols in the poem itself, particularly those that pertain to the geoengineering proposal found at the close of the poem's first canto. In this way, we find that Darwin's blurring of fact and symbol is less an isolated indulgence than a symptomatic breaking of the excessively blatant demarcations of his didactic poem that his reader is driven toward from the very outset of the text. Darwin's opening "Apology" to the reader openly foregrounds the divide between fact and symbol that organizes his rendition of the economy of nature. Following the lead of Alexander Pope's *The Rape of the Lock*, the classical elements around which each canto of "The Economy of Vegetation" is organized are symbolized in accord with the Rosicrucian doctrine of elements.[37] The primary agent of the poem's first canto, for instance, is the element of heat, manifest as a coterie of "*Nymphs of primeval Fire.*"[38] Treated by Darwin as metonyms as much for natural forces as for the newly founded forces of industry, these nymphs are called upon in the versification of Darwin's geoengineering proposal at the close of the first canto to "expand a thousand sails" on "ice-built isles and "steer" this "floating frost," these "melting treasures," to "cool, with Arctic snows, the tropic year."[39] Such machinery is thought "proper" by Darwin for a text containing "conjectures on some articles of natural philosophy" that are "not . . . supported by accurate investigation or conclusive experiments."[40] "Extravagant theories," Darwin advises his reader, "in those parts of philosophy where our knowledge is yet imperfect, are not without their use."[41]

In two successive, and related, instances that precede the narration of the elimination of frost, the lacuna manifest at the center of Darwin's anemology begins to be subtly inscribed into the geoengineering proposal. The first such instance is found alongside Darwin's direction of his reader toward the additional note on winds. The fiery nymphs are here called on to attack "the *Fiend of Frost*" on account of the detrimental effects of frost on vegetation expounded in an accompanying footnote, which consists of a series of ways in which "frost destroys those plants of our island." The means of this attack is an awakening of the *south*western winds responsible for thawing the frost brought to the British Isles by the *north*eastern wind:

> Unite, *illustrious Nymphs!* your radiant powers,
> Call from their long repose the *Vernal Hours.*
> Wake with soft touch, with rosy hands unbind
> The struggling pinions of the *Western Wind;*
> . . . . . . . . . . . . . . . . . . . . . . . . . . . . . . . . .
> Call your bright myriads, trooping from afar,
> With beamy helms, and glittering shafts of war;
> In phalanx firm, the *Fiend of Frost* assail,
> Break his white towers, and pierce his crystal mail;
> To Zembla's moon-bright coasts the Tyrant bear,
> And chain him, howling, to the Northern Bear.[42]

On the surface of a mere description of the broader "economy of nature" with which the entire poem is concerned is simultaneously the narration of a conscription of nature to the human cause. For in the proposition articulated here of heat's seasonal returning of a howling, tyrannical "Fiend of Frost" to the northern reaches of the Russian Arctic (the area now referred to as the Novaya Zemlya archipelago), Darwin describes and prescribes the redirection of the northern wind toward its point of origination without a human in sight—what amounts in Darwin's poem to a self-inhibition of nature in line with human whims. As with his treatment of the lacuna at the heart of his anemology, the duality of representation remains crucial here: in the midst of a spectacle of a nature aligning itself with human whim, Darwin *overstates* the extent to which a perverse nature will ever operate in line with human desire in order to subtly inscribe its very impossibility.

The unbecoming nature of this self-inhibition, and the degree of misplaced human investment in such a notion, is made more blatant in the poem with the explicit introduction of human agency to a structurally analogous scenario that similarly narrates the redirection of frost. In the immediately subsequent stanza, the fiend of frost is no longer anthropomorphized by Darwin as a figure in battle against symbolic instantiation of the element of heat, but rather rendered in the form of an "enormous GRAMPUS" (killer whale), that is similarly redirected back toward "the frozen pole":

> So when enormous GRAMPUS, issuing forth
> From the pale regions of the icy North,
> Waves his broad tail, and opes his ribbed mouth,
> And seeks on winnowing fin the breezy South;
> From towns deserted rush the breathless hosts,
> Swarm round the hills, and darken all the coasts;
> Boats follow boats along the shouting tides,
> And spears and javelins pierce his blubbery sides,

> Now the bold Sailor, raised on pointed toe,
> Whirls the wing'd harpoon on the slimy foe;
> Quick sinks the monster in his oozy bed,
> The blood stain'd surges circling o'er his head,
> Steers to the frozen pole his wonted track,
> And bears the iron tempest on his back.[43]

Consistent with the incapacity of human intervention against frost that is subtly signaled in the previous stanza, Darwin here explicitly introduces the capacity of a human retort to this mobility, followed by that of a singular heroic effort. This retort is not simply that of an absolute negation that would presumably take the form of the elimination of the grampus. The human cannot be so crude, after all, as to eliminate a seasonal dynamic of nature outright. Rather, we are again confronted with a ridiculous spectacle, perhaps misidentified as such on account of the now-explicit role humans play. Despite appearances, this passage is a failed reenactment of the attempted chaining of frost to its northern origination, a staging of the inevitable failure of an intervention designed to reorient anemological processes, and similarly looks forward to the geoengineering proposal that follows in the later stages of the canto.

This much is evident when considering what might be understood as the unacknowledged debt owed by Darwin's passage to Ludovico Ariosto's *Orlando Furioso*.[44] Darwin can be read here as expanding on Ariosto's playful association of the whale with the island:

> Offshore, as we walked toward her, we saw a whale,
> the largest ever beheld in all creation.
> Anybody seeing that thing would fail
> to guess what it could be. In our estimation,
> it was an island to which one could sail
> for a holiday picnic. To our great consternation,
> this misprision of ours was to have no small effect.
> It was a pity that we were incorrect.[45]

This lends an excessive quality otherwise missing from the reenactment's initial presentation as a spectacle of human command over the economy of nature. By reading Darwin's grampus as a symbol of an ice island that forms the object of the geoengineering proposal to come, it appears that the humans partaking in the spectacle are subject to a double misidentification in a scene ostensibly attesting to their superiority. From the preceding scene, this redirection of frost to its point of origination (bearing, this time, a marker of purported human conquest) is as inadmissible in the context of Darwinian anemology as it is undesirable—the Little Ice Age being, after all, a phenomenon of increasing concentration of glaciers at the poles of the Earth. From an awareness of Ariosto's comedy, we might further suggest that such actions on the part of humans are fundamentally, and unknowingly, misdirected from the outset. "Incorrect" humans are here engaging a symbolic figment of their desire, and indeed leaving their own mark on it—a sailless mast on a vessel misperceived as sinking. The iron tempest for Darwin serves not only as a playful allusion to the boatlike mobility of the grampus vessel, foreshadowing the spectacle of ice-islands to come in the poem, but also, and perhaps more tellingly, serves as a marker of the incapacity of human intervention against a nature impervious to full capture.

Encoded within the canto's late geoengineering proposal "to navigate these immense masses of ice into the more southern oceans," is therefore a blurring of feasibility and impossibility.[46] Darwin's wrath against the icebergs expresses not simply a desire to shape the world according to human whim, but a desire to register the very impossibility of ever fully domesticating the weather. In situating the explicit call to shape the world according to human whim by such deceptively intuitive means as towing islands of ice alongside a reminder of the basis of such an endeavor in human hubris and misrecognition, Darwin registers the very impossibility of ever fully domesticating the weather in and as a call to complete such an undertaking. Geoengineering is in this respect proposed by Darwin as a litmus test of humanity's techno-scientific powers that will fail.

Yet if Darwin's wrath against the icebergs manifests not simply a desire to shape the world but a desire to register the very impossibility of ever fully domesticating the weather, the question remains as to what might motivate this duality of representation. I hold that Darwin actively seeks to facilitate the inevitable failure of a geoengineering proposal, a failure equivalent to the reassertion of nature's sovereignty over techno-scientific capture. Read in this way, it is not simply inconsistency that prevents the explicit incorporation of such motivations into the geoengineering proposal itself, but rather the risk of inhibiting the future facilitation of such an expression of nature's sovereignty. Geoengineering proposals such as Darwin's might in this respect be more fruitfully approached as necessarily subtle expressions of a human desire comparable to a Strangelovian desire for nuclear catastrophe; here we find less a desire for outright destruction and more a desire for facilitating an expression of nature's sovereignty in the last instance over techno-scientific capture.

Darwin's proposal—located as it is at the openings of a techno-scientific modernity increasingly predisposed to the possibility of geoengineering—points to the necessity of reading geoengineering proposals of all stripes *symptomatically*. Indeed, any analogies to nuclearist enthusiasm might be more blatant in comparable proposals to come in the "tragicomedy of overreaching, hubris, and self-delusion"[47] that is the modern history of geoengineering proposals. Some fifty years after Darwin's *Botanic Garden*, the American meteorologist James Bollard Espy (also known as "the Storm King") made a similarly intuitive appeal to replicate the effect of volcanic eruptions on precipitation, proposing to light a series of enormous bonfires (the equivalent of artificial volcanoes) on the east coast of the United States to bring about rain.[48] While never fully realized, the idea was later revivified in 1961 in the form of literal flying nuclear reactors that would direct "huge amounts of heat" toward "a rain cloud, or to a moist air mass with rain potential . . . to alter the natural precipitation in a given geographical region by increasing the buoyancy of the cloud or air parcel."[49] With the duality of Darwin's approach in mind, such instances might be fruitfully reassessed in a manner sensitive to the potentially unconscious motivations expressed in and through geoengineering proposals of striking boldness and consequence, specifically by looking for comparable traces of duality in such proposals. At the very least, Darwin's approach compels us to revise our own approach when critiquing geoengineering proposals: specifically, to reformulate a critical enterprise heretofore based on assessing geoengineering proposals on the basis of environmental or commercial feasibility, on misguided attempts to fully quantify the risks involved with a given proposal, or to reject proposals by way of the invocation of the unknown (as much an appeal for many as a ground for refutation). The critique of geoengineering might thereby be rendered further determinate, if not more consequential, with a focus on the symptoms of the latent desires motivating such proposals and perhaps expressed through literary devices, if not a sensitivity to the lengths the human mind might go to when driven by climate anxiety.

## PEDAGOGICAL SUPPLEMENT

I introduce students to *The Botanic Garden* as a major influence on the course of British and German Romanticism. Both the content and the style of Darwin's work had a remarkable (if not always long-lasting) impact on figures such as William Wordsworth and Samuel Taylor Coleridge, as well as directly inspiring proto-Romantic works such as Johann Wolfgang von Goethe's "Die Metamorphose der Pflanzen."[50] Introducing *The Botanic Garden* in this way can prove fruitful, as it allows for deep reflection on how Darwin's curious blend of enthusiasm for and skepticism of the techno-scientific capture of nature evident in the elements of the text under consideration in this chapter compares to the attitudes of such well-known figures. Moreover, this approach allows for ample consideration of how the dominant theme in German Romantic poetry and Idealist *Naturphilosophie* of an ontological continuity between nature and self—the understanding of nature as "a mirror of the spirit"[51]—might be thought alongside Darwin's staging of the projective qualities of human renditions of nature found in the aforementioned rendition of the grampus.

Examining Darwin's geoengineering proposal elaborated in the first canto of "The Economy of Vegetation" in this context, however, can still pose its own problems. For it is all too easy to misread Darwin's rendition of nature's sovereignty from techno-scientific capture, his gentle yet definitive mockery of human hubris, as reproducing long-running attitudes concerning the "veiling of nature." As Pierre Hadot's seminal history of "unveiling nature" has outlined, many a philosopher and poet has sought to elevate their discourse as capable of finally unveiling a nature otherwise hiding from the gaze of human understanding.[52] Such an approach, however, runs counter to the meaning of Darwin's presentation of geoengineering, for it would keep open the possibility of a sufficiently guileful technoscience one day "lifting the veil" and dissipating nature's monstrosity by way of fully quantifying it.

The distinction between the "hidden" and the "radically foreclosed," when considering the continuity between human and nature, is therefore a crucial one if Darwin's skepticism is not to be fully assimilated into more conventional philosophical or poetic motifs. I have found a useful resource reflective of this distinction to be Anthony Paul Smith's introduction of the non-philosophy of François Laruelle into the environmental humanities. Smith speaks of "the perversity of nature" as that manifest in the realization "that nature is stranger than any one regional knowledge, be it philosophy, theology, or scientific ecology, can capture."[53] Citing the current climate's determination by the quantitative approach of neoliberal capital, Smith observes nature's perversity in its "purposely deviating from what is accepted as good, proper, or reasonable in capitalist society," instead "inhering to every part of culture and within every human person" and "resist[ing] bowing before capitalism's demand, to be measured as something relative rather than the radical condition for any relative measurement."[54] From my own experience, it is this non-philosophical modeling of the relations between nature and humans—a model of "continuity without capture," rather than a mirroring of the spirit—that affords a full consideration of the implications of Darwin's proposal for the history of geoengineering proposals, insofar as it allows for nature's perversity to be enacted in and through human actions themselves. Smith's claim that "we know the perversity of nature" as "it is present in our bones, the aches some get when a storm is coming and the way that weather is no longer a matter of mere conversation but of life and death concern," that "we are witnesses to the perversity of nature as we are an instance of its perversity," has made for especially fruitful discussions.[55]

## NOTES

1. Jean M. Grove notes that "the term refers to the behaviour of glaciers, not directly to the climatic circumstances that caused them to expand," adding that "glaciers enlarged and their fronts oscillated about forward positions without retracting as far as the positions they occupied before the initial advance." See "The Onset of the Little Ice Age" in *History and Climate: Memories of the Future?*, ed. P. D. Jones, A.E.J. Ogilvie, T. D. Davies, and K. R. Briffa (New York: Springer Science+Business Media, 2001), 153. For a detailed examination of the Little Ice Age's vast impact on Europe through the rise of multiple famines, see

*Famines during the "Little Ice Age" (1300–1800): Socionatural Entanglements in Premodern Societies*, ed. Dominik Collet and Maximilian Schuh (Cham: Springer, 2018).

2. Jan Golinski, *British Weather and the Climate of Enlightenment* (Chicago: University of Chicago Press, 2007), 56–57.
3. Golinski, *British Weather*, 42–44.
4. Golinski, *British Weather*, 55–56.
5. "Almost any Lunar member can be found to have had an interest in meteorological observation or instrumentation," observes Robert E. Schofield. "Records have been found showing extensive weather observations by Watt and by Withering, while correspondence of Boutlon, Watt, Priestley, Edgeworth, and Whitehurst shows continuing interest in weather instruments: barometers, hygrometers, wind and rain gauges." See *The Lunar Society of Birmingham: A Social History of Provincial Science and Industry in Eighteenth-Century England* (Oxford: Clarendon Press, 1963), 284.
6. Schofield, *Lunar Society*, 284.
7. Schofield, *Lunar Society*, 287.
8. Erasmus Darwin, *The Botanic Garden. A Poem, in Two Parts, with Philosophical Notes*, 2nd ed. (New York: T. & J. Swords, 1807).
9. Dagomar Degroot notes how, prior to the "final cold phase of the Little Ice Age . . . called the Dalton Minimum (1760–1850)"—after which "the world escaped . . . into the human-made crisis of global warming"—the prior phase of lower than average temperatures persisted in the Northern Hemisphere until "around 1720," and "only frigid winters in the early 1740s interrupted a warmer and in many places drier and more tranquil climatic regime" until the 1760s. See *The Frigid Golden Age: Climate Change, the Little Ice Age, and the Dutch Republic, 1560–1720* (Cambridge, UK: Cambridge University Press, 2018), 41.
10. Darwin (*Botanic Garden*, 35–36) cites a 1718 "voyage from Jamaica to England, in the beginning of June, [that] met with ice-islands coming from the north, which were surrounded with so great a fog, that the ship was in danger of striking upon them, and that one of them measured sixty miles in length," and another "instance of ice-islands brought from the southern polar regions, on which the Guardian struck at the beginning of her passage from the Cape of Good-Hope towards Botany-Bay, on December 22, 1789. These islands were involved in mist, were about one hundred and fifty fathoms long, and about fifty fathoms above the surface of the water. A part from the top of one of them broke off, and fell into the sea, causing an extraordinary commotion in the water, and a thick smoke all round it." Included in *The Botanic Garden* is a short poem—"Visit of Hope to Sidney Cove near Botany Bay"—composed by Darwin as an accompaniment to commemorative medallions created by Darwin's close friend Josiah Wedgwood, from clay transported by Sir Joseph Banks from the Sydney shoreline; a poem "representing HOPE encouraging ART and LABOUR, under the influence of PEACE, to pursue the employments necessary for rendering an infant colony secure and happy" (268). The confronting sublimity of the ice island would soon after be registered in William Cowper's 1799 poem "On the Ice Islands Seen Floating in the German Ocean," in *The Poetical Works of William Cowper. Complete Edition, with Memoir, Explanatory Notes, Etc* (New York: John Wurtele Lovell, n.d.), 508–509, in which the prospect of human intervention is emphatically denied; these structures being "self-launched" and "unremovable by skills, / Or force of man." I would like to thank Jeremy Chow for bringing this text to my attention.
11. Michael R. Page, *The Literary Imagination from Erasmus Darwin to H. G. Wells: Science, Evolution, and Ecology* (Farnham, UK: Ashgate, 2012), 17.
12. Thomas De Quincey, "On Christianity as an Organ of Political Movement" in *Theological Essays and Other Papers*, vol. 1 (Boston: Ticknor, Reed, and Fields, 1854), 35. On the reception of Darwin's proposal by, first, anti-Jacobite critics and later representatives of British Romanticism, see Siobhan Carroll, "On Erasmus Darwin's *The Botanic Garden*, 1791–1792," *BRANCH: Britain, Representation and Nineteenth-Century History*, ed. Dino Franco Felluga (2016).

13. Darwin, *Botanic Garden*, 35.
14. See Chris Pak's *Terraforming: Ecopolitical Transformations and Environmentalism in Science Fiction* (Liverpool: Liverpool University Press, 2016), alongside an enthralling history of weather control schemes extending from ancient myth through to contemporary examples in James Rodger Fleming's *Fixing the Sky: The Checkered History of Weather and Climate Control* (New York: Columbia University Press, 2010).
15. See Katherine Ellison, "Why Climate Change Skeptics Are Backing Geoengineering," *Wired*, March 28, 2018, https://www.wired.com/story/why-climate-change-skeptics-are-backing-geoengineering/; and Sherryn Groch, "Space Mirrors, Fake Volcanoes: The Radical Plans to Fix the Climate," *Sydney Morning Herald*, February 14, 2020, https://www.smh.com.au/national/space-mirrors-fake-volcanoes-the-radical-plans-to-fix-the-climate-20200122-p53tq3.html.
16. See Ryan Gunderson, Diana Stuart, and Brian Petersen, "The Political Economy of Geoengineering as Plan B: Technological Rationality, Moral Hazard, and New Technology," *New Political Economy* 24, no. 5 (2018): 696–715.
17. Jeremy Baskin, *Geoengineering, the Anthropocene and the End of Nature* (Cham: Palgrave Macmillan, 2019), 2 and viii.
18. Notable examples of which include Lucian Boia's *The Weather in the Imagination*, trans. Roger Leverdier (London: Reaktion Books, 2005) and Gillen D'Arcy Wood's *Tambora: The Eruption That Changed the World* (Princeton, NJ: Princeton University Press, 2014).
19. Darwin, *Botanic Garden*, 245 and 227.
20. De Quincey, "On Christianity," 36.
21. Page, *Literary Imagination*, 19.
22. Darwin, *Botanic Garden*, 227–245.
23. Darwin, *Botanic Garden*, 229.
24. Darwin, *Botanic Garden*, 229.
25. Darwin, *Botanic Garden*, 238.
26. Darwin, *Botanic Garden*, 231.
27. Darwin, *Botanic Garden*, 227.
28. Darwin, *Botanic Garden*, 244; emphasis added.
29. Darwin, *Botanic Garden*, 238.
30. Darwin, *Botanic Garden*, 230.
31. Darwin, *Botanic Garden*, 240.
32. Darwin, *Botanic Garden*, 240.
33. Max F. Schulz, *Paradise Preserved: Recreations of Eden in Eighteenth- and Nineteenth-Century England* (Cambridge, UK: Cambridge University Press, 1985), 34.
34. Schulz, *Paradise Preserved*, 34.
35. "Nature is not veiled," Anthony Paul Smith reminds us; "nature does not 'love to hide,' but thinking this allows our regional knowledges to think that they can unveil nature, that they can touch and circumscribe nature with thought and thereby either exploit her for our own gain or save her." See *A Non-Philosophical Theory of Nature: Ecologies of Thought* (New York: Palgrave Macmillan, 2013), 14.
36. On Darwin's undermining of the easy generalizations about his verse that he himself invites, see the first chapter of Devin Griffiths's *The Age of Analogy: Science and Literature between the Darwins* (Baltimore, MD: Johns Hopkins University Press, 2016), 51–82.
37. Alexander Pope, *The Rape of the Lock and a Key to the Lock* (London: Hesperus Press, 2004), 3–4.
38. Darwin, *Botanic Garden*, 8. Italic font in quotes from *Botanic Garden* in this chapter is original unless otherwise noted.
39. Darwin, *Botanic Garden*, 35–36.
40. Darwin, *Botanic Garden*, 35–36.
41. Darwin, *Botanic Garden*, 3.
42. Darwin, *Botanic Garden*, 29–30.

43. Darwin, *Botanic Garden*, 30–31.
44. Ludovico Ariosto, *Orlando Furioso: A New Verse Translation*, trans. David R. Slavitt (Cambridge, MA: Belknap Press, 2009), 96.
45. Darwin (*Botanic Garden*, 4) refers to the grampus as a "whale" in the summary of the argument of the first canto: "Frost assailed. Whale attacked." This is an important detail, as the grampus is comedically distinguished in *Furioso* (95–96) from the whale mistaken for an iceberg, the former being among a series of fish ("grouper, grampus, hake / and other varieties") that "were leaping / out of the water eager to fry or bake or broil for her. Thus, all of us were keeping / our distance—because, at such a sight, you take / your time."
46. Darwin, *Botanic Garden*, x.
47. Fleming, *Fixing the Sky*, 2.
48. Fleming, *Fixing the Sky*, 7.
49. Fleming, *Fixing the Sky*, 210.
50. See Desmond King-Hele's *Erasmus Darwin and the Romantic Poets* (Basingstoke, UK: Macmillan, 1986). King-Hele describes the near-simultaneous publication of Darwin's *The Loves of the Plants* (the second part of *The Botanic Garden*, first published individually in 1789) and "Die Metamorphose der Pflanzen" as the product of a "commonality of interest" in his brief exploration of the links between the two thinkers (169–171), but Ann Shteir has since drawn attention to Goethe's acknowledgment of Darwin's *Loves* as having in fact inspired the latter work; see "Narrating Nature in Erasmus Darwin's *The Botanic Garden*" in *Fact and Fiction: Literary and Scientific Cultures in Germany and Britain*, ed. Christine Lehleiter (Toronto: University of Toronto Press, 2016), 79.
51. Pierre Hadot, *The Veil of Isis: An Essay on the History of the Idea of Nature*, trans. Michael Chase (Cambridge, MA: Belknap Press, 2006), 273.
52. Hadot, *Veil of Isis*.
53. Smith, *Non-Philosophical Theory of Nature*, 13.
54. Smith, *Non-Philosophical Theory of Nature*, 14–15.
55. Smith, *Non-Philosophical Theory of Nature*, 15.

# 2
# STORM APOSTROPHE

ANNETTE HULBERT

Figure 2.1 J.M.W. Turner's *The Slave Ship: Slavers Throwing Overboard the Dead and Dying—Typhon* [*sic*] *Coming On*, 1840. Courtesy of Museum of Fine Art, Boston.

The 1781 *Zong* massacre, in which the crew of the slave ship *Zong* threw overboard at least 130 enslaved Africans in order to collect on their insurance policies, was famously memorialized in J.M.W. Turner's 1840 oil painting *The Slave Ship* (figure 2.1), originally titled *The Slave Ship: Slavers Throwing Overboard Dead and Dying—Typhon* [*sic*] *Coming On.*[1] The painting depicts the approaching storm running into a blood-red sunset on the horizon, matching the color of the ship's masts and the bloodied water. In the bottom third of the painting, a number of bodies, some with chains still attached, drift in the ocean while fish, birds, and sea creatures feed on severed limbs. The storm noted in the title is a blur of brushstrokes merging into the rest of the scene, making it difficult to distinguish sky, ocean, or ship. Turner's painting similarly blurs the distinction between social and environmental concerns, a phenomenon that Joni Adamson and Scott Slovic term environmental justice ecocriticism.[2] Adamson and Slovic begin their field-shaping introduction to the 2009 *MELUS* issue, "Ethnicity and Ecocriticism," by showing how Hurricane Katrina gave rise to a "growing consensus that social justice and environmental issues are linked" before noting that "the roots of the environmental movement can be traced back to the abolition movement."[3] This essay examines how the storm-tossed sea becomes intertwined with the violent legacy of slavery in the eighteenth century, forming a literary climatic history that Turner unspools in his painting.

In order to understand the intersection of storms and slavery as Turner understands it in 1840, I take up a constellation of texts that first made these connections in the eighteenth century. I trace the origins of this climatological archive to Phillis Wheatley Peters and Olaudah Equiano's strikingly different firsthand descriptions of enslavement, both of which frequently draw on both literary and historical accounts of storms. Wheatley and Equiano never crossed paths, but their work was literally bound together in an 1814 edition of Equiano's *The Interesting Narrative of the Life of Olaudah Equiano, or Gustavus Vassa the African, Written by Himself* (originally published in 1789). Reading their work side by side, it becomes apparent that both authors rely on the device of apostrophe—typically the address of a dead, absent, or inanimate being by a first-person speaker—to probe whether providence in the shape of a storm can render justice. What they find, I argue, is that the effect of calling on a storm provides variable and unpredictable results, showing how justice and the environment are complexly interrelated in ways that fail to bring clarity.

Turner reportedly drew inspiration for the painting from Thomas Clarkson's account of the *Zong* massacre in *The History of the Rise, Progress, and Accomplishment of the Abolition of the African Slave-Trade by the British Parliament*, first published in 1808 and reissued in 1839. According to Clarkson, the *Zong* captain reported that he threw the captured Africans overboard after missing his port and subsequently discovering that he had only two hundred gallons of water aboard, short of what he believed was necessary for the entire group's survival. In recounting the trial that ensued after insurers refused to pay out the claim, Clarkson notes the dramatic moment in court when "Providence" counters the captain's narrative and offers "an unequivocal proof of guilt": "A shower of rain fell and continued for three days immediately after the second lot of slaves had been destroyed, by means of which they might have filled many of their vessels with water, and thus have prevented all necessity for destruction of the third."[4]

For Clarkson, the rainstorm serves both as proof of guilt and a sign from providence showing that the slavers were not making rash choices in fear of survival but instead coldly calculating that they could collect insurance on enslaved persons "lost at sea" but not on those who died aboard. Though Clarkson's intent is to show the storm as arbiter of the crime, he replicates dehumanizing language that refers to the murders in terms of property loss.

The verse-tag Turner composed to accompany the painting's entry in the Royal Academy exhibition catalog extends Clarkson's narrative and envisions a different role for the storm in the massacre:

> Aloft, all hands, strike the top-masts and belay;
> Yon angry setting sun and fierce-edged clouds
> Declare the Typhon's coming.
> Before it sweeps your deck throw overboard
> The dead and dying—ne'er heed their chains
> Hope, hope, Fallacious Hope!
> Where is thy market now?[5]

Turner's interpretation implies a strange temporality: the slavers are responding to the storm by throwing the captive Africans overboard, in contrast to Clarkson's narrative about the storm's arrival after the massacre. The shift toward apostrophic address at the end of the verse-tag sheds light on this strangeness. Jonathan Culler writes, of the weirdness of apostrophe, that "what is at issue is not a predictable relation between a signifier and a signified, a form and its meaning, but the uncalculable force of an event."[6] In the case of the *Zong*, the force of the event ruptures traditional narrative forms: Turner's verse makes the massacre a response to the storm and the storm a response to the massacre.

Turner's typhoon provides a fitting introduction to my concerns in this essay because Turner's *Slave Ship* remediates not only the *Zong*, but a variety of literary and historical characterizations from across the eighteenth-century abolitionist movement, each centrally featuring a storm. Leo Costello terms Turner's method "dialectical history painting."[7] This concept of multiple timescales compressed together on the canvas calls to mind the work of Christina Sharpe, who writes that Turner's decision to leave details unspecified is a refusal to "collapse a singularity into a ship named the Zong; that is, Turner's unnamed slave ship stands in for the entire enterprise."[8] One of the central terms Sharpe develops in *In the Wake: Blackness and Being* (2016) is "Weather," an "atmospheric condition of time and place"—the time and place of slavery—that "produces a pervasive climate of anti-blackness."[9] Sharpe draws attention, for instance, to the Fugitive Slave Act, which denied "free air" of a "free state" to enslaved people. In the hold of the slave ship, where kidnapped Africans found it difficult to breathe, slavery permeates the atmosphere. Yet Sharpe also suggests that "the shipped, the held, and those in the wake also produce out of the weather their own ecologies."[10] In reading these produced ecologies, I consider not only the work the storm does as a metaphor for forms of resistance, but also the limits of the storm as metaphor. As Sharpe implies, before the "Weather" is deployed within a rhetoric of emancipation, it enables the conditions that make slavery and colonial capitalism possible. In each of the case studies I provide in this essay, writers draw attention to the limits of metaphor by acknowledging a lack of clear causality.

The storm, the unforeseen obstacle that traditionally threatens to thwart the maritime trade, became a popular trope in the later eighteenth and early nineteenth centuries. Bridget Keegan has charted a sharp rise in poems about shipwrecks during this period, corresponding with Britain's rise to dominance as a sea power. Poetry about shipwreck, Keegan writes, is "always political," frequently providing commentary on "the work of colonization and military conquest" from the perspective of laboring-class sailors.[11] A variation of this image, the eighteenth-century abolitionist trope of the slave ship caught in (or punished by) a storm, took hold in the eighteenth century and was well-established by the time Turner was painting. John McCoubrey and Costello cite a 1727 passage from James Thomson's "Summer" as one of the earliest and most important models for Turner's painting. In the poem, a slave ship caught in an Atlantic hurricane is pursued by sharks:

> Lured by the scent
> Of steaming crowds, of rank disease, and death,
> Behold! he rushing cuts the briny flood.
> Swift as the gale can bear the ship along;
> And from the partners of that cruel trade
> Which spoils unhappy Guineas of her sons
> Demands his share of prey—demands themselves.
> The stormy fates descend: one death involves
> Tyrants and slaves; when straight, their mangled limbs
> Crashing at once, he dyes the purple seas
> With gore, and riots in the vengeful meal.[12]

Tellingly, this passage was also reprinted in Clarkson's *History*, where Turner may well have encountered it alongside the description of the *Zong*'s rainstorm. Thomson's vision of a storm that mingles with other oceanic elements to avenge perpetrators of the slave trade has clear resonance for Turner's *Slave Ship*, where sea, storm, and bloodthirsty sea creatures similarly merge together on the canvas. Costello believes that Thomson's storm offered the abolitionist movement a historical model for seeing political change as ordained and enacted by divine forces, a model that Turner recognized and incorporated into his painting.[13] But the storm is an ambiguous metaphor at best: in both Turner's painting and Thomson's poem, "tyrants and slaves" are equally prey to shark and storm. Though Thomson is writing long before the British abolitionist movement existed as such, it is clear he presages how the natural violence of the storm at sea informs the social violence of the slave trade.

Turner's *Slave Ship* portrays a world of submerged meaning activated by the storm. He hints in his verse-tag that the storm possesses a reformist politics of its own, though he hints that the justice rendered by climate is imperfect. John Durham Peters writes in *The Marvelous Clouds* that "we can never separate the weather and the gods," wryly noting that "irregular patterns of blessing and bane" have conditioned us to view the natural world as a capricious parent.[14] Jane Bennett's *Vibrant Matter* provides insight into Turner's personifications as well as other storm texts that portray the natural world revolting against slavery: "A touch of anthropomorphism, then can catalyze a sensibility that finds a world filled not with ontologically distinct categories of beings (subjects and objects) but with variously composed materialities that form confederations."[15] Moments of apostrophe and personification, as Heather Keenleyside has observed, "might help us to imagine the sort of world that scholars like Bruno Latour and Jane Bennett envision: one that is not divided into persons and things, agents and instruments, animate and inanimate beings."[16] Of course, slavery has a long history of denying Africans their humanity by relegating them to the category of objects. Acknowledging these dimensions, Monique Allewaert argues that "diasporic Africans forged a mode of personhood that emerges from the production and dissolution of assemblages that reject both the antinomianism of spirit and matter and that of subjects and objects," making possible a new "mode of personhood" that "ameliorates but does not eliminate the losses that accumulated in the Atlantic world."[17] In Allewaert's reading, an assemblage is a powerful antidote to the concept that a person is an emptiness.

I turn now from Thomson and Turner, both frequently cited as contributors to the European tradition of the sublime, to Wheatley Peters, who draws on deep classical knowledge to rewrite the sublime in her apostrophe to the storm. For Wheatley Peters, like Turner, a storm is not merely a representation of the Middle Passage, but a web of ecological relations that calls to mind the atemporal disaster of Atlantic slavery—a fine point that Shelby Johnson echoes in chapter 4. But whereas Thomson and Turner write of how the storm at sea reveals the injustice of the slave trade, Wheatley Peters seldom suggests such a direct correlation. The framework environmental justice ecocriticism provides is helpful here, showing how Wheatley Peters explores the historical dynamics of capitalism, colonialism, and slavery through references to real storms.

"To MAECENAS," the prefatory piece that opens Phillis Wheatley Peters's 1773 *POEMS ON VARIOUS SUBJECTS, RELIGIOUS AND MORAL*, establishes Wheatley Peters as a learned reader of literary storms. The poem directly addresses Maecenas, a historical figure best known as the patron of Virgil and Horace and counselor of Octavian, the first emperor of Rome. But Maecenas is only one among a number of addressees in a poem that takes an unconventional approach to apostrophe—typically an address of a single absent, dead, or inanimate being by a first-person speaker—by addressing Homer and Virgil while borrowing language from John Milton and Alexander Pope. Wheatley intends not only to cultivate appreciative readers, but also to suggest that she will continue a classical tradition of storm writing that begins with *The Iliad*.

Wheatley Peters follows up the initial address of Maecenas by abruptly cutting to Homer and suggesting that his ability to render the storm is the greatest measure of his literary mastery. The second stanza describes the speaker's rapturous reaction to stanzas from *The Iliad*:

> Great Sire of verse, before my mortal eyes,
> The lightnings blaze across the vaulted skies,
> And, as the thunder shakes the heav'nly plain,
> A deep-felt horror thrills through all my veins.[18]

Wheatley Peters, who owned a copy of Alexander Pope's translation of Homer's *The Iliad*, pays homage to both writers with this image of a solitary individual who bears witness to a sublime spectacle unfolding in the sky. In citing the storm in this context, Wheatley Peters mirrors Pope's language comparing Homer's talent with the power of Jupiter, the deity of storms. Pope writes in the preface to *The Iliad* that Homer "seems like his own Jupiter in his terrors, scattering the lightnings, and firing the heavens."[19] The storm in both Wheatley Peters's poem and Pope's preface is not only a pyrotechnic production designed to thrill the reader, but a staging of the poet's power to animate and reanimate through speech.

Wheatley Peters is particularly interested in the storm as rhetorical tool, though the lines immediately following the description of Homer first appear to imply that she fears she will fall short of the intended task:

> But here I sit, and mourn a grov'ling mind
> That fain would mount and ride upon the wind.[20]

Wheatley Peters participates in a well-established literary tradition of establishing an affinity with her predecessors while modestly protesting that she cannot do them justice (as Pope does in the preface to *The Iliad*). Her collected poetry repeatedly returns to this tactic. The mention of a "grov'ling mind" might also be read as an example of the phenomenon Allewaert notices in Wheatley Peters's poetry, the "contracting and stalling" of "classical and neoclassical forms" as a refusal to engage with the logic that governed "commodity production and exchange in the eighteenth-century colonial world."[21] I want to draw attention, however, to the second line of the stanza, in which Wheatley Peters announces her ambition to "ride upon the wind," effectively setting her intention to claim the neoclassical storm in her own poetics. Though she follows in the same tradition as Homer and Pope,

Wheatley Peters implies that her own goals differ, since her version of mastering of the storm genre looks less like hurling lightning bolts than sitting astride the storm itself. To ride on the winds or ride on the clouds are expressions that appear in Psalms to convey a display of God's power ("He makes the clouds his chariot and rides on the wings of the wind") and in Revelations to suggest retribution ("Behold, he cometh with clouds, and every eye shall see him, and they *also* which pierced him").[22] Pope alludes to the same Scriptural passages and connects the verse more directly to a storm in *Essay on Man, Epistle II*: "Nor God alone in the still calm we find, / He mounts the storm, and walks upon the wind."[23] Riffing on Pope, Wheatley Peters indicates the magnitude of her ambition by comparing herself to God "mounting" the storm. Paula Bennett points out that "Maecenas" was likely composed at the same time as Wheatley Peters's famous "oral examination" or trial, in which eighteen prominent Boston citizens ascertained for themselves that Wheatley Peters had indeed authored her own poems. "Maecenas" can thus be read, Bennett suggests, as a response to the humiliation of the examination.[24] Wheatley's desire to "mount" the storm, read in this context, can reassert absolute control over her poetics, countering condescending accusations that she was not the sole author of her works.

The question then turns to what is at stake for Wheatley Peters in returning to the storm image throughout the rest of *POEMS*, particularly in conjunction with the apostrophe form. A provisional answer might be found in what Henry Louis Gates Jr. calls the "trope of the talking book."[25] Gates's *Signifying Monkey* examines the genre of slave writing to show how Black authors announce their status as subject in response to European readings of blackness as absence. For Gates, Wheatley Peters and Equiano, among other enslaved people writing during the period, conspicuously display their knowledge of the relationship between subjects and objects that exist in commodity cultures in order to prompt readers to recognize "the process by which the master endows his commodities with the reflection of subjectivity."[26] Equiano, for instance, animates a watch, a portrait, and, finally, a book, giving them voice and agency. In writing these objects into being, Equiano narrates his own transition from "slave-object to author-subject."[27]

Returning to Wheatley Peters, we can see how her choice to animate the storm might serve a similar purpose, demonstrating her mastery of literary technique as she grants the storm a subjectivity of its own. Yet if the storm can be read in line with the "trope of the talking book," it also complicates that trope. While the watch, the portrait, and the book that Gates cites are recognizable objects of Western culture, a stormy sea is an image reminiscent of the Middle Passage and the brutal conditions aboard the slave ships that crossed the Atlantic. Ide Corley suggests, for instance, that the storm in a slave narrative "erupts as an icon of the absence of the subject, an icon of trauma."[28] An integral part of the European tradition of the sublime, storms serve as a double articulation of the spaces of slavery and remind the reader of the erasure of the enslaved subject. As Wheatley Peters makes clear in her preface, writing the storm signals both her participation in and deviation from European literary traditions.

From an early age, Wheatley appears determined to work through the semiotics of the storm: how it registers transatlantic trauma even as it produces a site of resistance. Wheatley's first published poem, written when she was only thirteen years old, is an experiment in determining the poetic form best suited to describing a storm. "On Messrs. Hussey and Coffin" which appeared December 21, 1767, in the *Newport Mercury*, is a reflection on death averted, relaying the story of a pair of merchants caught in a sudden storm off the shores of Cape Cod. The poem was based on a snippet of a story Wheatley overheard at her enslaver John Wheatley's dinner table, while serving the Hussey and Coffin of the poem's title. The first lines of the poem interrogate Messrs. Hussey and Coffin about their state of mind during their near-drowning experience: "Did Fear and Danger so perplex your Mind, / As made you fearful of the whistling wind?"[29] Though Wheatley Peters will expand on the theme of storm repentance in "TO A LADY ON HER REMARKABLE PRESERVATION IN AN HURRICANE IN NORTH-CAROLINA," here, she is less concerned about the terror Hussey and Coffin likely feel than with the collection of forces that animate the storm:

> Was it not Boreas knit his angry Brow
> Against you? or did Consideration bow?
> To lend you Aid, did not his Winds combine
> To stop your Passage with a churlish Line,
> Did haughty Eolus with Contempt look down
> With aspect windy, and a study'd Frown?
> Regard them not;—the Great Supreme, the Wise,
> Intends for something hidden from our Eyes.[30]

As so often happens in Wheatley Peters's poetry, all obstacles vanish when confronted with the divine. Boreas, the north wind, and Eolus, the king of the four winds, are revealed to be impotent in comparison to the "Great Supreme" as the reader is instructed to "regard them not," implying that any death that occurs in the midst of a storm is fated. Rather, the reader is asked to assume a godlike position themselves in order to judge whether Hussey and Coffin will ascend to heaven or descend to "Beds down in the Shades below" if the "raging Sea" claims their lives.[31] The poem is most interested in the question of which entity is allowed to decide Hussey and Coffins' fates, portraying the storm as a divine tussle over their souls.

Early in her poetic career, then, Wheatley Peters develops a reading of the storm that resists the conventional function of the trope. Once again, the watches and portraits Gates analyzes provide a useful counterpoint. Unlike these objects, the storm already possesses some degree of agency as a "natural" entity, agency that the poet is able to direct for her own purposes. Yet Wheatley Peters animates the storm in order to deliver a judgment against Hussey and Coffin, then deftly transfers control of the storm to the "Great Supreme." Wheatley Peters ends the poem decisively with a prose apostrophe to God:

> Had I the Tongue of a Seraphim, how would I exalt thy
> *Praise; thy Name as Incense to the Heavens should fly, and the*
> *Remembrance of thy Goodness to the shoreless Ocean of Beatitude!*
> *Then should the Earth glow with seraphick Ardour.*
> Blest Soul, which sees the Day while Light doth shine,
> To guide his Steps to trace the Mark divine.[32]

The break from addressing Hussey and Coffin to God is abrupt enough that Eric Wertheimer wonders whether it suggests a "case of navigation gone awry at the behest of some unseen storm," an end that would imagine merchants and cargo sunk beneath the waves.[33] Here, as in "Maecenas," Wheatley Peters uses the storm to rapidly shift her object of address—in this case, seemingly tracing Hussey and Coffin's thought process as they realize the divine author of the storm. Corley similarly shows how, in Equiano's use of apostrophe, "the dynamics of lyrical address become messy," confusing the reader as to "who animates what or where the agency of the text lies."[34] Here, Wheatley Peters's apostrophe purposely directs the reader to consider where agency lies in the text. Wheatley Peters seems acutely aware that she gains power over the storm by transferring between objects of address. Apostrophe signals these shifts.

"On Messrs. Hussey and Coffin," for all of its complexity, never made it into Wheatley Peters's first poetry collection, *Poems on Various Subjects* (1773), though a version of the poem, "On Two Friends, Who Were Cast Away," was promised in Wheatley's 1772 *Proposals*.[35] A similar poem, again based on a real storm, was probably written to commemorate the catastrophic Great Chesapeake Bay Hurricane that swept through eastern North Carolina and the Chesapeake Bay area in 1769. In this poem, however, the poet is less of a commanding presence in the first lines, not directly controlling the storm but waiting for its arrival. The resulting poem is more subtle than "On Messrs. Hussey and Coffin," gradually showing the poet to be a divine agent, in much the same way as in "Maecenas." Gates writes of how the trope of the Talking Book evolves as Equiano "narrates his own movement from slave-object to author-subject."[36] "TO A LADY ON HER REMARKABLE PRESERVATION IN AN HURRICANE IN NORTH-CAROLINA" can be read as the production of a more mature poet who is able to master the form of apostrophe while also demonstrating the process by which she gains the authority to shape the climate.

The poem begins with a storm that is out of the frame, one that exists both for the subject and the speaker only as a barely detectable sound:

> Though thou did'st hear the tempest from afar,
> And felt'st the horrors of the wat'ry war,
> To me unknown, yet on this peaceful shore,
> Methinks I hear the storm tumultuous war,
> And how stern *Boreas* with impetuous hand
> Compell'd the *Nereids* to usurp the land.[37]

The subject of the poem—named later as "Maria," a married mother separated from her spouse and daughter on a perilous journey—hears the storm approaching before the speaker of the poem, far away from the tumult, intuits its arrival. It is unusual for Wheatley Peters's speaker to initially lack sensory understanding, since her poetry typically emphasizes the speaker's control of material elements. Read in dialogue with her other storm poems, however, particularly "Maecenas," we can see the poet-speaker narrating her gradual ascension to godlike status as she first hears and then commands the storm.

As in "Hussey and Coffin," the classical powers raise the storm that will subsequently be dismissed by an omnipotent Christian god. Wheatley Peters describes how Maria is instructed during the moment of crisis by "some heav'nly oracle" to consider "things of eternal consequence," not only of a personal nature, but "of the future doom, / And what the births of the dread world to come."[38] A storm-repentance is crucial to Maria's survival. However, the catch is that Maria must consider not only her own sins, but the "births of the dread world to come." Though the reader is not privy to Maria's thoughts, the poet-speaker—who, as previously established, is watching the storm and Maria's conflict play out from a distant shore—is subsequently empowered to welcome Maria back to shore: "From tossing seas I welcome thee to land/ Resign her, *Nereid*" 'twas thy God's command."[39] Maria's storm-repentance has been deemed worthy—though it is not clear by whom—so she is spared from the "tossing seas" and welcomed to shore by Wheatley Peters. Wheatley Peters draws attention to God's command to Nereid to "resign" Maria, in so doing creating a neat hierarchy: Nereid answers to God, who then spares Maria and releases her back to Wheatley. Wheatley does not downplay her own role in ensuring Maria's safe arrival, performing the ritualistic welcome the line before God commands the storm to cease. In this sense, Wheatley reveals her own authority over the disastrous weather that threatens ships during transatlantic journeys. Wheatley occupies multiple roles in each of her poems: moving from attentive reader to masterful poet in "Maecenas," from Hussey and Coffins' conscience to the "tongue of a seraphim," from helpless spectator on the shore to divine intermediary. Each of these moments is precipitated by a storm, delivering on her promise in the preface to make a storm the controlling metatrope of her poetics.

Whereas Wheatley Peters invites spectators to envision themselves as sufferers in the midst of the storm, Equiano muses about whether the storm will bring him justice or death. He first envisions the storm mercifully striking him down as an alternative to being sold as a slave on the island of Montserrat in the West Indies: "At the sight of this land of bondage, a fresh horror ran through my frame, and chilled me to the heart. My former slavery now rose in dreadful review in my mind, and displayed nothing but misery, stripes, and chains; and, in the first paroxysm of my grief, I called upon God's thunder, and in his avenging power, to direct the stroke of death to me, rather than permit me to become a slave, and to be sold from lord to lord."[40]

As a sailor, Equiano knew the dangers of a seafaring life well. Though there are no exact statistics for how many enslaved people perished during the deadly hurricane season of 1780, existing records suggest not only that their deaths numbered in the thousands, but also that they were particularly vulnerable to the aftereffects of storms: food shortages, contaminated water, and disease.[41] Having called for providential thunder, Equiano begins to fear that the storm he has called for will *not* be literal, a sailor's honorable death in a rough sea, but a miserable end being worked to death in the West Indies. As he prepares the ship to sail back to England (believing that the crew will abandon him there), he states his mounting conviction that "Fate's blackest clouds were gathering over my head," accompanied by his expectation that "their bursting would mix me with the dead."[42] The unusual phrasing here evokes a muddy composite of rain and bodies, a storm that does not outright kill Equiano, but "mix[es]" him with other bodies in imagery that evokes a mass grave. Although the comparison is initially puzzling, Equiano's previous paragraph alludes to just this sort of mixture, his first exposure to laboring on the ship in the "scorching West-India sun . . . while the dashing surf would toss the boat and the people in it frequently above high-water mark. Sometimes our limbs were broken by this, or even attended with instant death, and I was day by day mangled and torn."[43] Equiano's assertion that a storm will "mix [him] with the dead," besides recalling Turner's painting, is an example of how the atmospheric condition of slavery produces "new ecologies," to return to one of Sharpe's evocative terms. For Equiano, producing "new ecologies" incorporates and reframes imagery from a well-known abolitionist poem written by abolitionist Thomas Day to describe his own terrifying experience of Montserrat's famously unpredictable climate, thus twisting Day's words to show the ethical insufficiency of the storm as an abolitionist metaphor.

Equiano draws these lines directly from Day's book-length abolitionist poem *The Dying Negro* (1773). In Day's poem, the section Equiano borrows is preceded by the speaker's realization that he has been kidnapped from his coastal town and trapped into slavery:

> Ye Gods of Afric! in that dreadful hour
> Where your thunders and avenging pow'r!
> Did not my pray'rs, my groans, my tears invoke
> Your slumb'ring justice to direct the stroke?
> No pow'r descended to assist the brave,
> No light'nings flash'd, and I became a slave.
> From lord to lord my wretched carcase sold,
> In Christian traffic, for their sordid gold:
> Fate's blackest clouds were gather'd o'er my head;
> And, bursting now, they mix me with the dead.[44]

The enslaved man in Day's poem first expects African gods to send a storm to save him, but soon sees the storm as a metaphor for his own bad luck in becoming ensnared by the slavers. "No light'nings flashed, and I became a slave" interrogates the absence of divine interference and draws attention instead to a Christianity that trades in "sordid gold." Both Day and Equiano emphasize that the storm can unpredictably change path. By the end of the poem, however, the narrative is revised, and the promised storm arrives in a passage that echoes language from Revelation 8:3:

> —Thanks, righteous God!—Revenge shall yet be mine;
> Yon flashing lightning gave the dreadful sign.
> I see the flames of heav'nly anger hurl'd,
> I hear your thunders shake a guilty world.
> The time has come, the fated hour is nigh,
> When guiltless blood shall penetrate the sky
> . . . . . . . . . . . . . . . . . . . . . . . . . . . . . . . . . . . . . . . . . . . . .
> For Afric triumphs!—his avenging rage
> No tears can soften, and no blood assuage.
> He smites the trembling waves, and at the shock
> Their fleets are dash'd upon the pointed rock.
> He waves his flaming dart, and o'er their plains,
> In mournful silence, desolation reigns—.[45]

The biblical storm is portrayed as a victory, a relentless and unforgiving vengeance that wrecks the slavers' ships and gives rise to a new era where "souls are free, and men oppress no more!"[46] Read in the context of Day's poem, Equiano's elusive storm becomes a deferred promise of justice that will someday be fulfilled. Just as Wheatley Peters draws on the language of the Scriptures filtered through the lens of Pope, Equiano supplements Day's biblical language to suggest that the effect of the poet calling on the storm is unclear. Read as a rhetorical appeal, we can see how Equiano draws together several threads in his revision of Day: placing his work in a storied literary tradition of storms and pinpointing the complex anxieties of a mercantile empire. Though Equiano has called on a storm to deliver him, he grows concerned that his prayer will be misinterpreted, that providence will grant him the watery death he anticipates on land rather than at sea. But by broadening the definition of the storm to include Montserrat's deadly environment, Equiano prompts readers to interrogate the climate that makes slavery possible.

Equiano's assemblage of storms returns us to the layered history that Turner intends to depict in his painting: each storm as an organism interacting as part of the ecosystem of the abolitionist movement. The storms that Wheatley and Equiano use to respond to the atmospheric pressures of slavery while inventively departing from the European tradition of the sublime become reintegrated into that same tradition in Turner's painting. Linking Turner's *Slave Ship* to storm apostrophe—and attending to the rhetorical strangeness, the weird temporalities, and the ethical insufficiency of the storm as a symbol—focuses our attention back on the eighteenth-century Black-authored texts that initiate the tradition Turner continues into the nineteenth century.

## TEACHING THE STORM

Sharpe writes that teaching about slavery is a process marked by "absences in the archives" and "accumulated erasures, projections, fabulations, and misnamings."[47] I ask students to imagine the stories erased by the transatlantic slave trade by introducing M. NourbeSe Philip's 2008 *Zong!*, a series of interconnected poems written about the massacres aboard the *Zong*.[48] The poems in "Os," *Zong!*'s first section, consist of words found in the *Zong* legal case, *Gregson vs. Gilbert*, combined with new words derived from words in the case. Reading excerpts of *Zong!* before engaging with Equiano's account primes students to consider how Equiano strategically rearranges the language from Day's poem and other sources. We discuss how Equiano is not simply borrowing from Day, but instead, as Laura Doyle argues, that Equiano "flaunts his mastery of the British code, and he unpacks it by choosing to move along its fault lines, to surface its repressions."[49] Reading Equiano in a course on Atlantic survival narratives, students are quick to notice what Doyle has written about at length: Equiano explicitly engages with survival tropes that appear in Daniel Defoe's *Robinson Crusoe*.

From Thomson's suggestion that "tyrant and slave" both perish in the storm to Equiano's fear that he has accidentally condemned himself to a stormy death, students notice while reading these texts that nature fails to bring a satisfying justice—a point that Sharpe implies in her overview of the quotidian nature of Black death. Building on these ideas, I ask students to think about what kinds of residual animism still exist today. One example is the idea that storms or other natural catastrophes are punishments for our environmental misdeeds. It is tempting to find in ecologically motivated new materialist scholarship evidence that seeing animacy in nature can clarify social and ethical relations, but the inconsistencies located in the eighteenth-century storm apostrophe provide a productive challenge to the idea that nature abides by a moral order. Instead, locating the tension between the storm as a metaphor and the storm as a weather event offers a vital conceptual and ecocritical frame for understanding an ontologically complex natural world.

## NOTES

1. J.M.W. Turner, *The Slave Ship*. 1840. Oil on canvas. Museum of Fine Arts, Boston.
2. Joni Adamson and Scott Slovic, "The Shoulders We Stand On: An Introduction to Ethnicity and Ecocriticism," *MELUS: Ethnicity and Ecocriticism* 34, no. 2 (Summer 2009): 5–24.
3. Adamson and Slovic, "Shoulders We Stand On," 5–6.
4. Thomas Clarkson, *The History of the Rise, Progress, and Accomplishment of the Abolition of the African Slave-Trade by the British Parliament* (Shoe-Lane: R. Taylor and Company, 1808; repr. London: Frank Cass, 1968). Citations refer to the reprint edition.
5. Leo Costello, *J.M.W. Turner and the Subject of History* (Surrey, UK: Ashgate, 2012), 205.
6. Jonathan Culler, "Apostrophe," *Diacritics* 7 no. 4 (Winter 1977): 68.
7. Costello, *J.M.W. Turner*, 208.
8. Christina Sharpe, *In the Wake: On Blackness and Being* (Durham, NC: Duke University Press, 2016), 38.
9. Sharpe, *In the Wake*, 106.
10. Sharpe, *In the Wake*, 104.
11. Bridget Keegan, *British Labouring-Class Nature Poetry, 1730–1837* (New York: Palgrave, 2008), 138–139.
12. James Thomson, *The Seasons* (London: printed for J. Millan, 1726–1746; repr. Oxford: Clarendon, 1972), 1015–1026. Citations refer to the reprint edition.
13. Costello, *J.M.W. Turner*, 210. In addition to Thomson's "Summer," McCoubrey identifies Thomas Day's 1773 *The Dying Slave*, William Cowper's *The Negro's Complaint* (1792), *Song on the Wreck of a Slave Ship* (1806) by one "Marius," and James Montgomery's *Abolition of the American Slave Trade* (1814), each of which depict a storm wrecking a slave ship. John McCoubrey, "Turner's *Slave Ship*: Abolition, Ruskin, and Reception," *Word and Image* 14, no. 4 (1998): 319–353.
14. John Durham Peters, *The Marvelous Clouds: Toward a Philosophy of Elemental Media* (Chicago: University of Chicago Press, 2015), 244.
15. Jane Bennett, *Vibrant Matter: A Political Ecology of Things* (Durham, NC: Duke University Press, 2010), 32.
16. Heather Keenleyside, "The Rise of the Novel and the Fall of Personification," in *Eighteenth-Century Poetry and the Rise of the Novel Reconsidered*, ed. Kate Parker and Courtney Weiss Smith (Lewisburg, PA: Bucknell University Press, 2014), 126.
17. Monique Allewaert, *Ariel's Ecology* (Minneapolis: University of Minnesota Press, 2013), 138.
18. Phillis Wheatley Peters, "To Maecenas," in *The Collected Works of Phillis Wheatley*, ed. John C. Shields (New York: Oxford University Press, 1988), 14.
19. Alexander Pope, *The Iliad*, 1715 (London: J. Bumpus, 1820), 35.
20. Wheatley Peters, "To Maecenas," 28–30.
21. Allewaert, *Ariel's Ecology*, 117.

22. Holy Bible (New York: Oxford University Press, 1952). Authorized King James Version, Psalm 104:3, Psalm 18:10, Revelation 1:7.
23. Alexander Pope, *Essay to Man, Epistle II* (London: J. Wilford, 1733), 10.
24. Paula Bennett, "Phillis Wheatley's Vocation and the Paradox of the 'Africa Muse,'" *PMLA* 113, no. 1 (January 1998): 74.
25. Henry Louis Gates Jr., *The Signifying Monkey: A Theory of African-American Literary Criticism* (Oxford: Oxford University Press, 1988), 136.
26. Gates, *Signifying Monkey*, 157.
27. Gates, *Signifying Monkey*, 170.
28. Ide Corley, "The Subject of Abolitionist Rhetoric: Freedom and Trauma in 'The Life of Olaudah Equiano,'" *Modern Language Studies* 32, no. 2 (Autumn 2002): 154.
29. Phillis Wheatley Peters, "On Messrs. Hussey and Coffin," in *Phillis Wheatley: Biography of a Genius in Bondage*, ed. Vincent Caretta (Athens: University of Georgia Press, 2014), 1.
30. Wheatley Peters, "Messrs. Hussey and Coffin," 3–10.
31. Wheatley Peters, "Messrs. Hussey and Coffin," 12–15.
32. Wheatley Peters, "Messrs. Hussey and Coffin," 26–28.
33. Eric Wertheimer, *Underwriting: The Poetics of Insurance in America, 1722–1872* (Stanford, CA: Stanford University Press, 2006), 65.
34. Corley, "Subject of Abolitionist Rhetoric," 154. Corley has suggested that the storm sequence in *An Interesting Narrative* ends with Equiano's apostrophe to his former life as a slave.
35. Caretta, *Phillis Wheatley*, 66.
36. Gates, *Signifying Monkey*, 170.
37. Phillis Wheatley Peters, "TO A LADY ON HER REMARKABLE PRESERVATION IN AN HURRICANE IN *NORTH-CAROLINA,*" in *The Collected Works of Phillis Wheatley*, ed. John C. Shields (New York: Oxford University Press, 1988), 1–6.
38. Wheatley Peters, "TO A LADY," 20–24.
39. Wheatley Peters, "TO A LADY," 26.
40. Olaudah Equiano, *The Interesting Narrative of Olaudah Equiano*, ed. Vincent Caretta (New York: Penguin Classics, 1995), 98.
41. Matthew Mulcahy, "Weathering the Storms: Hurricanes and Risk in the British Greater Caribbean," *Business History Review* 78, no. 4 (2004): 635–663.
42. Equiano, *Interesting Narrative*, 99.
43. Equiano, *Interesting Narrative*, 99.
44. Thomas Day, *The Dying Negro: A Poem*, 3rd ed. (London: W. Flexney, 1775), 12–13.
45. Day, *Dying Negro*, 22–23.
46. Day, *Dying Negro*, 24.
47. Sharpe, *In the Wake*, 12.
48. M. NourbeSe Philip, *Zong!* (Middletown, CT: Wesleyan University Press, 2008).
49. Laura Doyle, *Freedom's Empire: Race and the Rise of the Novel in Atlantic Modernity, 1640–1940* (Durham, NC: Duke University Press, 2008), 193.

3

# "WHEN STORMY WINDS HAPPEN"

## Divine Providence, Climate Change Discourse, and the Cause of Weather Disasters

ADAM W. SWEETING

IN THIS ESSAY I CONSIDER how the common colonial New England assumption that God directly intervened in the weather might shed light on the ways we describe extreme weather events in the twenty-first century. This approach brings contemporary discussions concerning the implications and causes of anthropogenic climate change into conversation with late-seventeenth- and early-eighteenth-century environmental humanities scholarship. I am particularly interested in the theological readings early New England ministers applied to dangerous weather and the extent to which naturalistic explanations of storms overlapped with their belief in God's providential control over the skies. Such mixed explanations, I argue, prefigure our current discussions about the role anthropogenic climate change plays in contributing to or causing particular weather events. In both cases, investigators first ask what constitutes normal weather and then examine how and why dangerous storms might be tied to human misdeeds, either through sinful behaviors or the excess burning of fossil fuels. Whereas a Puritan divine might have viewed torrential rain as a sign of God's displeasure, a contemporary climatologist might ask whether human-induced climate change contributed to the deluge. That is, to what extent are we culpable for meteorological havoc?

In the discussion that follows, I compare representative colonial New England writings on the causes of severe weather to recent attempts to attribute dangerous storms to climate change. In the first section I trace the intersection of faith in providential intervention and the rise of modern scientific discourse in the seventeenth and eighteenth centuries. Next, I look to two catastrophic snow events in eastern New England separated by three centuries to explore different conceptions of who or what caused these events. In the subsequent section, much of the focus is on the meteorological writings of Increase Mather (1639–1723), including his response to the disastrous English storms of November 1703. To be sure, the climatological differences between Mather's time and ours could not be more different. But, as I argue in the final section of this essay, we wrestle with similar anxieties over how much our actions have contributed to stormy weather. Other scholars have explored conceptions of climate in early America, and I make no claims to offering a comprehensive review of colonial New England meteorological writing. In an influential 1982 article, for example, Karen Ordhal Kupperman outlined the ways New England's extreme winters surprised early colonists who assumed that similar climates prevailed at similar latitudes across the globe.[1] More recently, Anya Zilberstein traced the ways elites on both sides of the Atlantic shaped the colonial enterprise by describing a welcoming climate during a period we now understand to have been among the coldest stretches of the Little Ice Age.[2] But rather than recounting descriptions of early colonial New England climate, I am intrigued by how questions about the causes of remarkable weather have been framed, then and now.

I am interested, too, in how weather events, particularly in their most dramatic manifestations, function as signs. By way of example, the first generations of Puritan colonists wondered whether dangerous or unusual meteorological events signaled God's displeasure. They assumed that an all-powerful deity orchestrated meteorological events with a specific purpose in mind. Biblical stories such as Noah's flood showed that God readily employed ferocious storms to punish his people. He revealed his anger, too, when admonishing Job through the whirlwind. Sudden and torrential downpours or spectacularly violent thunderstorms were thus never simple occurrences; rather, they occasioned intense examination of God's aims. And so it is today. With each passing storm, the signals concerning human behavior become more clear. For Puritans and us alike, the weather tells a story about human behavior. The signs need to be read.

* * *

Early New Englanders assumed that God expressed his will through divine providence, the means by which he intervened into the events of the world. They distinguished between "general" providence, which can be understood as the regular working of creation, and "special providences" such as lightning strikes of churches, freakish births, earthquakes, and disastrous storms. By their very strangeness, such special providences revealed God's presence. They were communications from an unseen realm. Whatever the intention behind such divine interventions, by interrupting the ordinary workings of the world that God himself created, they cried out for explanation.[3]

By the late seventeenth century, the rise of the new empirically based mechanistic sciences in Europe led to a revised conception of God's providential role. The new science gave rise to a conception of a universe designed by an all-powerful God who set in motion a series of regular astronomical, terrestrial, and biological patterns that could be understood through the processes of human reason. The transition did not happen overnight, particularly in meteorology. Nor did providential beliefs simply vanish with the rise of Newtonian science. Indeed, it is something of a false distinction to draw clear lines between Enlightenment-era weather science and providence, since leading intellects of the day moved back and forth across the disciplines without necessarily seeing a conflict between providence and design.[4] In so doing, they fused the empiricism of Francis Bacon with a philosophical tradition rooted in Aristotle to distinguish between types of causes. This system revealed that God stood as the first cause of all things; he, in turn, set in motion the secondary causes by which natural processes occur. Such accounts, however, left room for the occasional interruption of normal processes through providential intervention. Thus, while an ordinary spring rain could be explained by the secondary cause of seasonal cycles and normal precipitation patterns, the occasional torrential downpour that led to disastrous floods raised the specter of special providence, in which the more powerful first cause overrides the natural order of things. In such a view, God did not merely passively observe his own creation; rather, he periodically intervened in ways that were consistent with the Puritan conception of an active and engaged deity.[5] Weather explanations, it seemed, could be scientific *and* theological, bridging causes both natural *and* supernatural.

* * *

We can trace the way coastal New Englanders from different centuries examined the human fingerprint in severe weather by looking to the responses to two similar snow catastrophes separated by nearly three hundred years. The first occurred in February 1717. Over the course of eleven days, four major blizzards buried the region in snow. New Englanders of the time were of course familiar with cold and snowy winters, but nothing quite prepared them for the onslaught of what came to be called the Great Snow of 1717. Precisely how much snow fell remains unclear. Reports from the interior north suggest upward of five to six feet may have fallen; in regions where forests had been cleared, powerful winds created drifts that reached still higher. As the environmental historian Thomas Wickman notes, for many people the most immediate concerns were logistical and practical. How, for example, would they get food or tend to their animals in the wake of the snow? That is, if they even had flocks to tend, since hundreds, if not thousands, of cattle and sheep were buried alive or froze to death in the ensuing weeks of intense cold. Remarkably, there were no reports of human deaths directly related to the snows.[6]

Despite the immediate practical concerns, the storms called out for theological interpretation. The Connecticut minister Eliphalet Adams, for example, noted that "When the Calamity is General, we ought to Consider what the Lord's Intention therein may be, to hearken to his Voice and set in with his Design."[7] Like many of his contemporaries, Adams acknowledged that under normal circumstances weather events could be explained by natural causes. He describes, for example, the formation of whirlwinds as being "occasioned by the meeting of Opposite winds, which prevent each other passage and are therefore necessarily carried Sideways in a Circular Motion."[8] Such a statement suggests that regularly occurring meteorological events could be understood through observation and science. Occasionally, however, God "Executes his good, wise, and just Purposes" by directly intervening in the weather to send a message.[9] Such intervention demanded an explanation of God's aims. Typically, those calls pointed to God's anger at human failure. For Adams, several aspects of the 1717 snows underscored the Lord's displeasure with New England. For starters, they happened "so very late in the year when it was the less to be Expected" that residents were caught off guard. Moreover, the "*Snow Descended in so great a Quantity, far beyond what is Usual*" in the "*Memory of Man.*"[10] Making matters worse, the storms came so quickly and were "So Terribly Repeated" that attendance at church services became well-nigh impossible, the parishioners having been "*driven out of the House of God thereby.*"[11] Clearly, the "Hand of the Lord was Visible" in shaping the storms.[12]

Nearly three hundred years later, a similar series of storms pummeled the Boston area, shutting down the region in the process as public transit ground to a halt and schools remained shuttered for weeks. All told, a record 108 inches of snow fell at Logan Airport in the early weeks of 2015, much of it falling in rapidly repeating blizzards in late January through mid-February. As in 1717, New Englanders understood that snow comes with the territory, but with the possible exception of the Blizzard of 1978, nothing prepared residents for such a meteorological onslaught. Indeed, a *New York Times* opinion piece described the storms as a "slow-motion natural disaster of historic proportions."[13] At the same time, many wondered what role anthropogenic climate change played in shaping this "natural disaster." To be sure, it may initially seem counterintuitive that global warming could produce violent snowstorms, but that is almost certainly what happened in Boston in 2015. Record sea-surface temperatures off the New England coast that winter helped saturate the air with moisture, which was then returned to the ground as snow when that same air collided with cooler spiraling storms. The sharp temperature contrast between the sea surface and the storms likely further fueled the snow's fury.[14] The world was warmer and the weather responded accordingly. In this sense, human action (the reliance on nonrenewable energies) and the otherwise natural weather joined forces in a new type of agency capable of inflicting meteorological violence. Whereas Reverend Adams was certain that God directly caused the severe snows of 1717, a new causal agent—anthropogenic climate change—may have been at least partly responsible for the 2015 disaster.

Until quite recently, climatologists were reluctant to connect a particular meteorological event to climate change, on the safe assumption that weather patterns have always varied.[15] The 1995 report of the Intergovernmental Panel on Climate Change, for instance, found insufficient data "to determine whether consistent global changes in climate variability or weather extremes have occurred over the 20th century."[16] By this reasoning, a single record temperature year may be a naturally occurring anomaly that could be explained without pointing to human-induced climate change. Such a cautious view, however, is no longer tenable. In recent years researchers have grown increasingly confident in their ability to identify the signs of climate change within individual extreme events. With increased computing power and ever-more-sophisticated climate modeling, that confidence will continue to grow.[17] Colonial New Englanders, of course, had no such instruments at their disposal. They did, however, share an intense faith in God and the belief that he providentially intervened into earthly affairs. That he "caused" both ordinary and extraordinary weather was beyond dispute. So, too, was the assumption that such weather arrived infused with theological significance.

For both colonial Puritans and contemporary New Englanders, the distinction between regular and unusual meteorological events raises an analogous distinction between climate and weather. Contemporary climatologists typically use at least a thirty-year time frame to establish averages and means for variables such as temperature and precipitation. Such long-term patterns are what they call "climate." Thus, we might claim that the New England climate is distinguished by four distinctive seasons that over time give rise to predictable patterns of heat or cold. The term "weather," on the other hand, denotes specific conditions such as a snowstorm, a few days of rain, or a weeklong heat wave.[18] Yet, while we may claim to understand the difference, our sense of such events often straddles the intellectual and temporal divides between climate and weather. In their sheer power and effect on the landscape, snowstorms such as those described by Adams remind us that weather occurs in particular places at particular times. These events are not mere data points within the abstracted climate averages. Extreme events puncture our sense of the ordinary. Like a warm spell in December, they capture our attention precisely because they scramble our expectations for what climate *should be*. They push us to reflect on the contours of local climate and call out for theological and meteorological analysis.

* * *

The full title of a 1662 poem by the physician and minister Michael Wigglesworth speaks to Puritan efforts to understand the causes of dangerous weather, in this case a severe lack of rain: "God's Controversy with New-England, Written at the Time of the Great Drought Anno 1662, By a Lover of New-England's Prosperity." Droughts, of course, differ from powerful storms in the insidious way they cause harm over the course of long, dry growing seasons. Unlike a blizzard or wildfire, at first they may not even be visible. Eliphalet Adams knew exactly when to preach his sermon on the snows of 1717. All he had to do was look out his window. But at what point to preach a sermon on drought? They are, in Benjamin Cook's words, "slow moving disasters, with effects that accumulate incrementally over weeks, months, and even years."[19] For colonial New Englanders (indeed, for people across the globe), prolonged periods without rain could be a matter of life and death as crops withered in the fields. At such times, prayers for relief might be in order. Yet prayers do not necessarily point blame. They simply ask God to show mercy. Wigglesworth, however, directly attributed the drastic conditions of 1662 to the collective guilt of his flock and, by extension, the rest of New England.

The poem traces the history of colonial settlement, beginning with the "waste and howling wilderness" that greeted the first Europeans (25).[20] As a sign of his blessing, "The Lord had made (such was his grace) / "For us a Covenant" (101–102). At first, "Our morning stars shone all day long," but over time God grew dissatisfied with his people and altered the weather for the worse (133). "The air became tempestuous" and the wilderness began to quake, at which point God spoke through powerful thunderstorms to warn colonists that they had become morally complacent and unduly appreciative of all he had done for them (153). When the backsliding continued, God switched the mode of punishment by withholding the rain until droughty conditions prevailed, putting the entire colony at risk: "The cattell mourn, and hearts of men / Are fill'd with fear and anguish" (389–390). Unfortunately, it is a disaster the colonists brought on themselves by turning away from the Lord through "riot and excess" (408). At the same time, the poem warns of "worser things" than the drought if New Englanders continued down their wayward path (416).

As the ecocritic Scott Slovic has noted, jeremiads such as Wigglesworth's often bear a striking rhetorical resemblance to classic environmental texts such as Rachel Carson's *Silent Spring.* In both cases, human action threatens an environmental crisis.[21] To be sure, unlike the profligate spraying of DDT that Carson describes, the sins outlined by Wigglesworth were not themselves environmental in nature; rather, they exemplified the "cold-Dead heartedness" and "dissembling shows of Holiness" he saw all around him (221, 233). Unfortunately, the poem's speaker fears that still more dreadful conditions lie ahead. In fact, God can bring them "To Morrow" should he choose (418). Today's storms and droughts send a similar message: More of this to come.

Perhaps no New England divine reflected more deeply on the meaning and causes of severe weather than the Reverend Increase Mather. Though often remembered as a dour figure in American history, Mather closely followed European scientific developments from his home in Boston and briefly struggled to reconcile the new science with his faith in providence. In a story that has been well told by many others, he ultimately could not mesh the two, though certainly not for lack of trying.[22] In *The Doctrine of Divine Providences* (1684), for instance, he outlined a hierarchy of God's causal actions that allowed for distinctions between normally occurring events and those that were truly providential. Under the "Ordinary Providence of God" he listed the "Laws and Course of Nature" that God "hath established in the world." These are the laws of time, the cycle of seasons, and ordinary meteorological processes such as wind, snow, and rain. This was the world of the new science and mechanistic philosophy then emerging in Europe. It assumes an orderly cosmos of regular and measurable patterns that reveal God's wisdom and the "great and marvelous work" behind his creation. Occasionally, however, God resorted to "Extraordinary Providences" that violated Nature's laws. Such events stand "above and beyond the Constituted order of nature."[23]

By way of contrast, in a second book from 1684, *Essay for the Recording of Illustrious Providences*, Mather collected firsthand testimony from those who had experienced providential events in New England. Here, he tried to merge empirical data with supernatural accounts, a project in keeping with contemporaneous efforts to place Puritan cosmology within an increasingly mechanistic view of the universe.[24] To be sure, Mather rarely directs his concern toward the weather as a system of atmospheric forces. He, instead, typically addresses a particular weather event that occurred independently of larger natural systems. His accounts fall into what Vladimir Janković describes as the seventeenth- and eighteenth-century British tradition of "meteoric reportage," by which natural history writers, many of them ministers, recorded and described unusual and place-specific weather events rather than offering systematic descriptions of weather science.[25] Ferocious weather teased out the specific from the general, much like a providential intervention separated second causes from the first. Nor should we place Mather's descriptions within what we now call empirical science, though he was certainly aspiring to empirical mastery of the available data. As he writes in the preface to *An Essay for the Recording of Illustrious Providences*, "I have often wished, that the Natural-History of New England might be written and published to the world" according to the "rules and methods described by that learned and excellent person Robert Boyle," the Irish-born chemist, who, like Mather, sought to reconcile science with a belief in God's providential intervention in the world.[26] In the end, however, he would not have been troubled by the distinction between the two.[27]

Despite his early interest in collecting empirical data, by the time he wrote *The Voice of God in Stormy Winds* (1704), Mather had shifted his focus toward the role of divine intervention, albeit without completely abandoning the new science.[28] The work concerns the catastrophic November 1703 storm and resulting floods in southern England, a historic event that Daniel Defoe described in *The Storm*. It was a weather disaster unlike any earlier event in England. According to modern climatological reconstructions, an extratropical cyclone wrapped around a zone of extreme low pressure in the North Atlantic resulted in several days of pummeling winds and waves in Wales and southern England and sent torrential rains east toward Germany, the Netherlands, and France. More than eight thousand people lost their lives while entire seaside towns were overrun by quickly rising water levels. Queen Anne called it "a Calamity so Dreadful and Astonishing" the like of which "hath not been seen or felt, in Memory of any Person living in our Kingdom."[29] To this day, it remains the benchmark of severity for judging all major English storms.

News of the storm arrived in New England by late March 1704.[30] Although it remains unclear whether Mather ever read contemporaneous accounts in the Royal Society's *Philosophical Transactions* that placed the storm's formation and path within a largely naturalistic context, his two sermons on the subject never waver from his insistence that human misbehavior had unleashed God's providential handiwork.[31] Sins such as excessive drink and pride, he concludes, "has made the world to be a Tempestuous Stormy World," a claim at once spiritual and meteorological in nature.[32] Although some might dismiss providence in favor of purely naturalistic accounts, they forget that storms that "tear Trees up by the Roots and overturn Houses" cannot "come to pass without the Providence of God." In other words, "Second Causes are Overruled by Him who is the First Cause of all Things."[33] Severe weather, Mather assumes, plays a role in a grand meteorological narrative that operates within a teleological frame. Ultimately, he concludes, natural causes alone cannot explain the most severe events; such explanatory power can be ascribed only to the direct intervention of the supernatural.

In his earlier meteorological work Mather focused on New England. Here, however, he reached across the Atlantic to connect events in England with the situation in Boston. Although the storm did not directly hit New England, Mather noted that the next day (a Sunday) "a more than ordinary snow storm" in the region prevented many from attending worship. He noted, too, that "several of our Neighbors perished" when their ship, which had been anchored off the English shore, went down on "that doleful night."[34] His local weather, he concluded, functioned within a larger system that rippled across vast spaces: "God can soon do to Boston as He has done to London, to Bristol, to Salisbury, and to many other Towns in England." New England had been placed on warning: "This awful Providence should put us upon Earnest Prayer to God, that the wrath of Heaven may not fall upon our Nation."[35] That is, we've brought these storms on ourselves and, please Lord, may it not be too late to amend our ways.

Like Wigglesworth before him, Mather should not be read as a colonial prophet of climate change. He would have believed it impossible that human action alone could overwhelm the natural order of things. Only God had such power. Still, aspects of his account prefigure contemporary discussions concerning the fingerprints of human action in our climate system. On who or what can we blame a storm? "When stormy winds happen," Mather claims, scientific "men are apt to impute it to Second Causes only, to the Disposition of the Air, the Seasons of the Year, or the Influence of the Constellations in Heaven."[36] That is, they assume nothing unusual propels even the most destructive storms. Like Mather, but for different reasons, I have little patience for claims that strong storms are purely natural events. We know that humans have warmed the climate with the same certainty that Mather assumed his fellow New Englanders had damaged the moral climate of the region. Mather, of course, functioned within a theological framework in which God both creates *and* violates natural processes. Logic and faith thus demanded that he look beyond ordinary nature toward God whenever violent storms occurred. Mankind's sinful ways may be the proximate cause for a disaster, but God always retains his role as first cause. No amount of science could strip it away.

Our current efforts to understand extreme weather can be couched in terms similar to Mather's, albeit without the direct providential intervention. Like Mather, we watch the skies and delve into our history to tease out which storms count as "normal" and which ones are extreme. We wonder how much more we must endure. Like his fellow Puritans, Mather firmly believed a future judgment awaits us all. His faith amounted to a form of certainty. We might be less certain about the future than he was, but the picture gets clearer with each new study and the evidence of our eyes. By excessively burning fossil fuels and clearing forests we have created an overheated atmosphere that veers toward a new first cause—us—albeit one that functions without the teleological purpose of the Puritan God.

* * *

In this final section I briefly take up recent efforts to attribute extreme weather events to anthropogenic climate change. Such efforts, I argue, can lead us to reconfigure our conception of causal agents. In *The Voice of God in Stormy Winds*, Mather asserted that "Men can't cause storms" but the "Great God of Heaven can do it, and *does* do it" (emphasis added).[37] In Mather's reading, God (first cause) can interrupt normal natural processes (second cause) to punish us for our sins. By this reckoning, had the colonists remained true to their covenant, violent storms would be much less likely. When we, however, point to anthropogenic climate change as a contributing cause to a particular storm or as a portent for more extreme weather down the line, we identify a potential new first cause that interrupts the secondary causes of normal weather patterns. We have, in effect, become like Mather's God—capable of shaping extreme weather events. But whereas the Puritan God meteorologically reprimanded his flock for sinful behaviors, we now punish ourselves for our own sins. We've become the simultaneous penitent and judge. Of course, not everyone is equally responsible for the current situation, which should give us pause when assigning all of humankind into an undifferentiated causal agent born into the Anthropocene.[38] As Rob Nixon points out, the "slow violence" of pollution, toxicity, and climate change disproportionately affects the world's poorest inhabitants even though they are far less culpable in creating them in the first place.[39] In fact, since the worst effects of climate change are likely to be felt in the least culpable places on Earth, our new first cause apparently does not care whether the appropriate sinners are being punished. That represents a big change from the Puritan world of providence.

The first major study to pin a specific event to anthropogenic climate change was a 2004 report in the journal *Nature* that showed climate change doubled the odds for the occurrence of the catastrophic 2003 European heat wave, a disaster that left upward of seventy thousand people dead.[40] Since then, attribution studies have improved to the point that leading journals now regularly publish papers that seek to link specific events to our overheated atmosphere. The *Bulletin of the American Meteorological Society*, for example, publishes an annual collection of studies titled "Explaining Extreme Events from a Climate Perspective." At first, they detected signals consistent with what we would expect from climate change but were hesitant to link particular events to human activity, despite the findings of the 2004 *Nature* study. That changed, however, with the January 2018 edition, which found incontrovertible evidence that at least three such events from 2016 could *only* be explained by pointing to climate change, including the record-breaking planetary warmth of that year. As the *Bulletin's* editor noted in the accompanying press release, the report makes "clear that we're experiencing new weather, because we've made a new climate."[41] In keeping with the conditions of the Anthropocene, we've created a new climatological first cause.

In the early stages of storm attribution studies, researchers assumed a counterfactual state of affairs to model the kind of weather we would expect had there been no anthropogenic climate change. It's an approach rooted in the so-called frequentist methodology of classical statistics. Unlike their colleagues in other sciences, climatologists cannot establish a "control" Earth. Instead, they utilize computer models to simulate how climate systems will respond to a range of potential greenhouse gas emissions scenarios. As part of that process they simulate a planet whereon the industrial revolution never happened, a place where carbon dioxide, methane, and other greenhouse gases circulate entirely by natural processes. This, then, becomes the baseline for measuring the effects of anthropogenic atmospheric forcing.[42] It is the null hypothesis—a world where human activity does not significantly shape the climate. Or, as Increase Mather might have imagined, the world we should expect if God were not angry with us. In this approach, researchers assume the null hypothesis and accept alternative explanations only if the null hypothesis can be rejected with a very high degree of certainty, usually at least 95 percent. In other words, if a particular event could be shown to have been statistically possible within the counterfactual framework, an abundance of caution led researchers to avoid identifying a direct link between an extreme weather event and anthropogenic climate change.[43]

In recent years, however, a few climate scientists have employed a different approach to extreme weather attribution studies, one rooted in Bayesian statistics, a form of analysis that, perhaps ironically, has historical connections to eighteenth-century arguments in defense of Christian miracles.[44] Named for Thomas Bayes (1702–1761), an English nonconformist minister and mathematician, Bayesian inferences draw on prior knowledge and beliefs when making probabilistic assessments. Rather than measuring the probability that climate change "caused" a storm or heat wave, the Bayesian approach uses prior knowledge and understanding to examine how climate change shaped extreme meteorological events as they unfolded. These data points can be adjusted as new information becomes available, a process sometimes referred to as "updating your priors." In terms of climate change and extreme events, rather than beginning with the fixed model of the frequentists, a Bayesian approach employs known statistical priors such as documented evidence demonstrating that climate change may alter the distribution of natural variables. For instance, we know that increased sea-surface temperatures (which react robustly to greenhouse gas emissions) pump more moisture into the air; this additional moisture, in turn, can elevate rainfall totals during hurricanes while increasing possibility of flooding, property damage, and death. We know, too, that increased atmospheric warmth can deepen the intensity of heat waves.[45] Of course, natural variability in atmospheric circulation systems still occurs, with or without the excessive burning of fossil fuels. But we know for certain that an oversaturated and overheated atmosphere magnifies the intensity of otherwise naturally occurring meteorological events.

With these points in mind, researchers can reframe the question of attribution. As Kevin Trenberth of the National Center for Atmospheric Research writes, the answer to the question of whether an event is caused by climate change is that "the question is poorly posed and has no satisfactory answer. The answer is that *all weather events are affected by climate change because the environment in which they occur is warmer and moister*" (emphasis original).[46] Like an all-powerful but unseen God who sends a drought to punish his people, climate change lurks wherever we stand, and potentially disrupts environments across the globe. By these measures, the counterfactual null hypothesis no longer holds weight, since we cannot legitimately posit an atmosphere unaffected by sustained greenhouse gas emissions. As NASA scientist James Hansen and his colleagues note, we should not ask for the odds on a roll of the climate dice; instead, consider the extent to which the dice are loaded.[47] Rather than asking whether climate change caused an extreme event, a far better question to pose would be, to what extent did climate change alter its form? When a flood runs through city streets, do not ask whether climate change caused the flood. Instead, examine the extent to which ocean waters warmed by greenhouse gas emissions had strengthened it.[48] That's a different way to frame the matter of attribution than asking whether climate change *caused* any particular effect.

While I am not a statistician, I confess that the humanities scholar in me is drawn to the Bayesian method, as it opens up narrative possibilities by looking to how extreme events develop across time and space. After all, a deadly heat wave is more than an event. It is a potential human tragedy with a story—with all its priors—that needs telling. It speaks to the visceral quality of climate change in the here and now. It also raises important ethical questions based on the collected prior knowledge we possess. As Michael Mann and his colleagues have argued, the fear of overstating evidence that drives the more cautious null-hypothesis/frequentist approach may ultimately prevent policy makers from taking the bold steps that may be necessary to protect lives as climate change continues to wreak havoc on populations across the planet.[49] Our prior knowledge can help inform action in ways that a mere statement of odds cannot.

The frequentist and Bayesian approaches both ultimately bring me back to Increase Mather's reflections on the causes of extreme weather. By separating the normal workings of the cosmos from distinct providential interventions, he carved out an interpretive space in which explanation and forewarning overlapped. A disastrous storm could be read as the colonists' just desserts while also signaling worse to come. We now must read our strongest storms in a similar vein. Science is getting better and better at attributing current storms to climate change *and* projecting into the future. In light of such science, we must be prepared to acknowledge that each new weather disaster represents punishment and warning. Our past actions have come home to roost. Portents surround us. Mather and his brethren ultimately did not know whether reforming their sinful ways would lessen God's anger, the Lord's aims being ultimately unknown. They could only tremble and hope. We, on the other hand, possess the requisite knowledge to reduce future punishment. At least that much has changed between Mather's time and ours.

## TEACHING SUGGESTIONS

This essay brings together ideas one might encounter in courses devoted to early American literature or the environmental humanities. In either case, there will likely be some necessary catching up to help students from multiple disciplinary backgrounds. Students in early American literature courses, for example, may be familiar with the Doctrine of Providence but less so with recent discussions of climate change and extreme event attribution. Similarly, courses focused on climate change typically leave out historical documents on the supernatural. One way to bring the issues in the essay into sharper focus for the classroom would be to pair readings in which students examine representative examples of providential weather narratives with contemporary discussions of weather attribution. One example would place Increase Mather's focus on the word "cause" in dialogue with, say, Kevin Trenberth's analysis of the "causes" of contemporary weather events. Such an approach has philosophical implications—just exactly do we mean by "cause"?—while also raising questions of moral culpability, an increasingly pertinent topic in a world shaped by anthropogenic climate change. Similarly, focusing on the notion of natural weather, as opposed to our anthropogenically fueled climate, can generate fruitful discussions of the word "natural" or the distinction between weather and climate. My recent experience in the classroom suggests students are eager for such conversation.

For the last several years I have been part of a team-taught interdisciplinary course on climate change. One of our primary goals is to help students cultivate a climatological story of place. We want them to understand how the locations they inhabit, whether at home or in college, are imbued with intertwining historical, personal, and ecological narratives. These narratives, they soon discover, will likely be altered by climate change. It is in this spirit that I focus on the Boston area, where I live and work, as a springboard for my scholarship. The local climate of eastern New England gives rise to meteorological patterns arising from the Atlantic as well as from storms that gather from points farther west. It is a place with a deep sense of its own history. Factoring those elements into discussions of climate change discourse can help students develop a richer connection to their city and its environs.

## NOTES

1. Karen Ordahl Kupperman, "The Puzzle of American Climate in the Early Colonial Period," *American Historical Review* 87, no. 5 (December 1982): 1262–1289. For histories of colonial New England weather writing, see Peter R. Eisenstadt, "The Weather and Weather Forecasting in Colonial America" (PhD diss., New York University, 1990); and Daryl Sasser, "The Weather and Theology: The Influence of the Natural World on Religious Thought in Puritan New England" (PhD diss., Union Theological Seminary, 2010).
2. Anya Zilberstein, *A Temperate Empire: Making Climate Change in Early America* (Oxford: Oxford University Press, 2016).
3. There is a large body of literature on the Doctrine of Providence. My understanding in this paragraph and throughout the essay draws from David Hall, *Days of Wonder, Days of Judgment: Popular Religious Belief in Early New England* (Cambridge, MA: Harvard University Press, 1989); and Michael P. Winship, *Seers of God: Puritan Providentialism in the Restoration and Early Enlightenment* (Baltimore, MD: Johns Hopkins University Press, 1996).
4. Michael Winship, "Prodigies, Puritanism, and the Perils of Natural Philosophy: The Example of Cotton Mather," *William and Mary Quarterly* 51, no.1 (January 1994): 92–105; Eisenstadt, "Weather and Weather Forecasting," 146–147.
5. Richard Weisman, *Witchcraft, Magic, and Religion in Seventeenth-Century Massachusetts* (Amherst: University of Massachusetts Press, 1984), 31.
6. Thomas Wickman, "The Great Snow of 1717: Settler Landscapes, Deep Snow Cover, and Winter's Environmental History," *Northeast Naturalist* 24, no. 7 (2017): H81–H114.
7. Eliphalet Adams, *A Discourse Occasioned by the Late Distressing Storm* (New London, 1717), 26.

8. Adams, *A Discourse*, 4.
9. Adams, *A Discourse*, 9.
10. Adams, *A Discourse*, 27, emphasis in original. All subsequent emphasis and capitalization in original unless otherwise noted.
11. Adams, *A Discourse*, 28.
12. Adams, *A Discourse*, 26.
13. E. J. Graff, "Boston's Winter From Hell," *New York Times*, February 20, 2015, A27.
14. Chris Mooney, "What the Massive Snowfall in Boston Tells Us about Global Warming," *Washington Post*, February 10, 2015, https://www.washingtonpost.com/news/energy-environment/wp/2015/02/10/what-the-massive-snowfall-in-boston-tells-us-about-global-warming/.

    For the science of snowstorm intensity and climate change, see Andrea Alfano, "Nor'easters May Become More Intense with Climate Change," *Scientific American*, January 26, 2015, http://www.scientificamerican.com/article/nor-easters-may-become-more-intense-with-climate-change/.
15. Susan Joy Hassol et al., "(Un)Natural Disasters: Communicating Linkages between Extreme Events and Climate Change," *Bulletin of the World Meteorological Organization* 65, no. 2 (2016): 2–9.
16. Intergovernmental Panel on Climate Change, *Second Assessment, Climate Change* (1995), 5. https://archive.ipcc.ch/pdf/climate-changes-1995/ipcc-2nd-assessment/2nd-assessment-en.pdf.
17. For the history of attribution studies, see National Academies of Sciences, Engineering, and Medicine, *Attribution of Extreme Weather Events in the Context of Climate Change* (Washington: National Academies Press, 2016); Special Supplements to the *Bulletin of the American Meteorological Society*, "Explaining Extreme Events from a Climate Perspective," a series begun in 2012.
18. Gregory A. Zielinski and Barry D. Keim, *New England Weather/New England Climate* (Hanover, NH: University Press of New England, 2003), 5–6.
19. Benjamin I. Cook, *Drought: An Interdisciplinary Perspective* (New York: Columbia University Press, 2019), vii.
20. Michael Wigglesworth, "God's Controversy with New England, Written at the Time of the Great Drought Anno, 1662, By a Lover of New England's Prosperity," *Electronic Texts in American Studies*, ed. Reiner Smolinksi, accessed October 1, 2020. https://digitalcommons.unl.edu/cgi/viewcontent.cgi?article=1036&context=etas.
21. Scott Slovic, "Epistemologies and Politics in American Nature Writing: Embedded Rhetoric and Discrete Rhetoric," in *Green Culture: Environmental Rhetoric in Contemporary America*, ed. Carl George Herndl and Stuart C. Brown (Madison: University of Wisconsin Press, 1996), 98–100.
22. For Mather's participation in the new science, see Michael G. Hall, *The Last American Puritan: The Life of Increase Mather, 1639–1723* (Middletown, CT: Wesleyan University Press, 1988), 158–173.
23. Increase Mather, *The Doctrine of Divine Providences Opened and Applied* (Boston: 1684), 45, 46.
24. Winship, *Seers of God*, 66; Sasser, "Weather and Theology," 79–84.
25. Vladimir Janković, *Reading the Skies: A Cultural History of English Weather, 1650–1820* (Chicago: University of Chicago Press, 2000), 3, 78–79.
26. Increase Mather, *An Essay for the Recording of Illustrious Providences* (Boston, 1684), Preface, n.p.
27. Hall, *Last American Puritan*, 173.
28. Sasser, "Weather and Theology," 85.
29. For discussion of the 1703 storm, including Queen Anne's reaction, and its modern reconstructions, see Robert Markley, "'Casualties and Disasters': Defoe and the Interpretation of Climatic Instability," *Journal of Early Modern Cultural Studies* 8, no. 2 (Fall/Winter

2008): 102–124; and Richard Hamblyn's introduction to Defoe's *The Storm* (London: Penguin Books, 2003), x–xii.

30. The Boston diarist Samuel Sewall reports hearing about the storm during a meeting with the Royal Governor on March 30, 1704. *Diary of Samuel Sewall, 1674–1729*, vol. 2 (New York: Farrar, Straus, and Giroux, 1973), 97.
31. Markley, "Casualties and Disasters," 114. See also Jan Golinski, "The Weather in Eighteenth-Century Britain," in *Weather, Culture, Climate*, ed. Sarah Strauss and Benjamin S. Orlove (Oxford: Berg, 2003), 19–21.
32. Increase Mather, *The Voice of God in Stormy Winds. Considered, in Two Sermons, Occasioned by the Dreadful and Unparallel'd Storm, in the European Nations. Novemb. 27th. 1703* (Boston, 1704), 11.
33. Mather, *Voice of God*, 16.
34. Mather, *Voice of God*, 7.
35. Mather, *Voice of God*, 52, 53.
36. Mather, *Voice of God*, 42.
37. Mather, *Voice of God*, 14.
38. Andreas Malm, *Fossil Capital: The Rise of Steam Power and the Roots of Global Warming* (London: Verso, 2016), 28–32, 264–272.
39. Rob Nixon, *Slow Violence and the Environmentalism of the Poor* (Cambridge, MA: Harvard University Press, 2011), 2–3.
40. Peter A. Stott, D. A. Stone, and M. R. Allen, "Human Contribution to the European Heat Wave in 2003," *Nature* 432 (December 2004): 610–614.
41. Press release from the American Meteorological Society, December 13, 2017, https://www.ametsoc.org/ams/index.cfm/about-ams/news/news-releases/human-influence-on-climate-led-to-several-major-weather-extremes-in-2016/. For the full report, see *Bulletin of the American Meteorological Society* 99, no. 1 (January 2018).
42. National Academy of Science, *Attribution of Extreme Weather Events in the Context of Climate Change* (Washington, DC: National Academies Press, 2016), 35.
43. Michael E. Mann, Elisabeth A. Lloyd, and Naomi Oreskes, "Assessing Climate Change Impacts on Extreme Weather Events: The Case for an Alternative (Bayesian) Approach, *Climate Change* 144 (2017): 131–142. See also Kevin E. Trenberth, John T. Fasullo, and Thomas G. Shepherd, "Attribution of Climate Extreme Events," *Nature Climate Change* 5 (August 2015): 725–730.
44. In 1768, Bayes's younger colleague, Richard Price, used Bayes's unpublished theorem to critique David Hume's argument that belief in religious miracles or other violations of natural law could not be justified by appeals to cause-and-effect arguments or the testimony of others. Such appeals, Hume argued, could not be traced to direct experience or the long-time observational record. In response, Price drew on Bayes's unpublished essay to assert that by itself the prior uniform observation of the natural laws does not amount to evidence against the occurrence of unusual events. See Stephen M. Stigler, "Richard Price: The First Bayesian," *Statistical Science* 33, no. 1 (2018): 117–125.
45. Mann, Lloyd, and Oreskes, "Assessing Climate Change Impacts," 135.
46. Kevin E. Trenberth, "Framing the Way to Relate Climate Extremes to Climate Change," *Climatic Change* 115 (2012): 289; emphasis original.
47. James Hansen, Makiko Sato, and Reto Ruedy, "Perception of Climate Change," *PNAS* 109, no. 37 (September 2012): E2415–E2423.
48. My understanding of how these questions can be framed is drawn from Trenberth, Fasullo, and Shepherd, "Attribution of Climate Extreme Events," 726.
49. Mann, Lloyd, and Oreskes, "Assessing Climate Change Impacts," 139–140.

# Part Two

## EIGHTEENTH CENTURY + NEW MATERIALISMS

4

# PHILLIS WHEATLEY PETERS'S NIOBEAN SOUNDSCAPES

Shelby Johnson

EIGHTEENTH-CENTURY AFRICAN AMERICAN narratives archive ruptures from kin and world. These fissures often emerge in fleeting moments of repetition or evasion—a stutter of grammar, an unruly flare of unspeakable emotion, a disorienting temporal distortion, a story that refuses closure. In this essay, I trace how Phillis Wheatley Peters's epyllion "NIOBE in Distress for her children slain by APOLLO" represents the disrupted affiliations experienced by African-descended people through an evanescent expression of grief that echoes across a larger environmental order.[1] In Ovid's *Metamorphoses*, Wheatley Peters's immediate source for the poem, Niobe famously expresses pride in her motherhood, and the gods punish her temerity by slaying her fourteen children. Overwhelmed by their sudden deaths, Niobe transforms into stone and becomes part of the landscape, yet continually weeps for her lost sons and daughters.[2] In her translation of Ovid included in *Poems on Various Subjects, Religious and Moral* (1773), Wheatley Peters aligns Niobe with Black experiences of loss by altering the myth's conclusion.[3] She refuses to end with a Niobe converted to stone, and instead closes with her anguished speech to a natural world that in turn echoes her pain: "'Ye heav'nly pow'rs, ah spare me one,' she cry'd, / 'Ah! spare me one,' the vocal hills reply'd." This lyric decision likely was inspired in part by Richard Wilson's painting *The Destruction of the Children of Niobe* (1760), which depicts Niobe *just prior* to her ossification, pleading for her children. An editor, however, interpolated lines to give Wheatley Peters's poem the "correct" mythic conclusion: "A marble statue now the queen appears / But from the marble steal the silent tears."[4] The last twelve lines of the published poem are thus not her translation.[5] Katy Chiles observes that "this final (and unattributable) stanza" illuminates "the vexed status of the poem's miscegenational authorial collaboration" in the infamous "Attestation" that preceded *Poems on Various Subjects*.[6] Niobe, moreover, represents the fraught status of Black authorship itself: "[The] figure of Niobe, on the verge of metamorphosing into a frozen, white, ever-crying melancholic statue becomes an uncanny doppelgänger for the productive, blackened, writing melancholic Wheatley."[7] Yet Niobe's *living* presence in Wheatley Peters's preferred ending counteracts the ways enslavement became, in Kathryn Yusoff's words, "a geologic axiom of the inhuman in which nonbeing was made, reproduced, and circulated."[8] In this essay, I contend that "Niobe in Distress" opposes colonial regimes of nonbeing that rendered Wheatley Peters's body and body of print illegible, while the lyric's soundscapes, which depend on an earthly environment to amplify human and more-than-human aural vibrations and echoes, also reveal patterns of life elided under slavery. Indeed, Wheatley Peters's poem refuses the reduction of blackness and Black creative reproduction (both maternal and poetic) to thinghood within colonial extractive economies.

Scholars of Wheatley Peters have focused on the poem's representation of matters of artistic and maternal life-giving creativity, with Nicole Spigner, Jennifer Thorn, and Devona Mallory arguing that she centers Niobe's unvalued motherhood as a way to resist the commodification of African women's sexuality under slavery.[9] This body of scholarship intersects with broader critical conversations on the exploitation and erasure of Black maternity under slavery, as in work by Jennifer L. Morgan and Kerry Sinanan.[10] Literary critics have turned to her Niobe—and her furious posture in defiance to Olympian tyranny—as an important intervention in lyric representations of Black women's rebellion.[11] Building on this critical genealogy, I maintain that Wheatley Peters's elegiac praxis, derived from her engagements with eighteenth-century landscape aesthetics and the realities of enslaved alienation from kin, has two implications for reading the poem from the perspective of the environmental humanities, and from new materialism in particular.[12] Recent work in new materialism has been invigorated by speculations on interdependence, exigent contingency, and agency in Indigenous and African cosmologies—especially their attentiveness to environmental entities as different forms of *life*.[13] According to these threads, Ovid's myth assigns agency to more-than-human stone in the tears that flow across Niobe's lithic surface.[14] Moreover, "Niobe in Distress" suggests how animate environments, and their organizations of material extension, intention, and duration, can share Niobe's grief and thus assemble a form of communion that challenges what Elizabeth Povinelli names a settler "geontology" that subtends colonial biopolitical subjugations of life and nonlife.[15] Instead of a mere instance of the pathetic fallacy, Wheatley Peters's poem renders a metaphysics of plural existence and a fragile, interconnected expressivity between Niobe and the natural world, a cooperative mourning repertoire that is reciprocal and emphatically *earth-bound*, rather than derived from Olympian divinities. Indeed, their call and response together form an "intra-agency," a "mutual constitution of entangled agencies" that Karen Barad argues conditions life itself.[16]

Yet Wheatley Peters's hesitation in portraying Niobe's lithic transformation also suggests that ecocritical and new materialist readings of her poetry (and her representations of human and nonhuman alliances in "Niobe in Distress" specifically) are fraught from the standpoint of Black studies.[17] While ecocritical research has traced the epistemological, ontological and political possibilities that can arise when nonhuman life is centered, critical race theory also maintains the "human" that must be repudiated is the one "overrepresented," in Sylvia Wynter's words, with mastery, exceptionalism, and possession.[18] Wheatley Peters's poetry became evidence in contentious public debates over the role of climate in the production of the (racialized) human. The Comte de Buffon and Abbe Raynal, as David Waldstreicher surveys, contended "that the New World had led to a natural degeneration of mammalian life—a degeneration that they had seen in the practices of the colonizers, not least their shocking revival of slavery."[19] The implications of this nascent environmental science infuriated the likes of Thomas Jefferson, who argued that Wheatley Peters never possessed the intellectual or affective resources to write poetry: "The compositions published under her name are below the dignity of criticism."[20] For Jefferson, race—regardless of climate or geography—determined her artistic capability. Wheatley Peters's refusal to end with Niobe entombed in stone contests colonial environmental and racial theories, given that Niobe metonymically stands in for both African motherhood and Black authorship, positions antithetical to colonial modes of human-being. "Niobe in Distress" intercedes in conversations scholars have been engaged in regarding the ways African bodies became overdetermined by their violent identification *with* and dispersals *into* early hemispheric land and waterscapes.[21] Christina Sharpe, for one, lingers with the tangible persistence of Black flesh in molecular intimacies: "They, like us, are alive in hydrogen, in oxygen; in carbon, in phosphorous, and iron; in sodium and chlorine. This is what we know about those Africans thrown, jumped, dumped overboard in the Middle Passage; they are with us still."[22] And if Wheatley Peters renders Niobe as alive, rather than as stone, how shall she be with us still?

By turning to soundscapes as a material trace of Niobe's presence in the poem, I contend that Wheatley Peters's Niobe represents a Black "being human" that exceeds colonial commodification of African women and their reproductive bodies. Read this way, the poem "create[s]" what Tiffany Lethabo King calls "a new ecotonal site of possibility for Black embodiment and place-making."[23] "Niobe in Distress" portrays vibrant relationalities between African persons and environments where, instead of suspended in stone, Niobe, miraculously, *speaks* to a responsive world.[24] The poem registers and resists colonial parameters of a "geologic life"—Yusoff's term for "the corporeality of geology as a material embodiment and a systematic framing of materiality," which has "geopolitical and biopolitical consequences for the possibilities of being and nonbeing."[25] In what follows, I consider how Wheatley Peters foregrounds resistant life in Niobe's "blaspheming breath" and the echoes of the "vocal hills," emphases developed at the intersections of elegiac form and landscape aesthetics.[26] I then turn to the poem's evocation of atmosphere to argue that her stress on tremulous soundscapes, rather than inert geologic existences, preserves a sense of ecological contingency and animacy, while opposing the exploitation of enslaved women and their reproductive labor as a resource—like stone—to be extracted.[27] Niobe's living breath and the echoing hills, I conclude in a reflection on the poem's final lines and their pedagogical significance, resonate with histories of Black resistance to colonial presumptions of sovereignty over human and earthbound fellowships. Wheatley Peters offers imaginative possibilities for marshaling contemporary African American dissent from settler geographies in the long afterlives of slavery.

## ELEGIAC ENVIRONMENTS

At first glance, it may seem counterintuitive to read "Niobe in Distress" in the context of new materialism. Critical conversations on the politics of animate matter have often focused on contemporary social and environmental worlds—understandable, given the urgency of our current climate crisis.[28] However, turning to new materialist scholarship for its historicist engagement with more-than-human agency enables us to recognize the poem's critique of Enlightenment hierarchies of personhood, wherein white European men existed at the apex.[29] As both a Black person and a woman, Wheatley Peters was insufficiently "human" herself, according to colonial racial and gender hierarchies, with her Niobe similarly representing an embodiment that King calls a "fleshly matter that exists outside the realm of the body, and, thus, humanity."[30] As a poet whose aesthetic praxis draws from the classics, with their ludic transformations of humans into animals or plants and vice versa, Wheatley Peters intercedes in eighteenth-century efforts to attach personhood to environments and things.[31] Indeed, the literary decision that most defines the poem's ecological perspective—the responsive agency expressed by the echoing hills—is a function of its reconfiguration of elegiac elements. While not precisely an elegy itself, "Niobe in Distress" incorporates a formal quality—the pathetic fallacy—that Mary Jacobus argues is the elegy's most emblematic feature: "The mourning of nature with the mourner . . . is the figure of elegiac emotion par excellence, a call anterior even to dialogue: the figure that gives affect."[32] The natural world that joins Niobe's grief is not only anterior to dialogue, but *assembles* the conditions that enable speech, for no aural reality would exist without material substances to mediate soundwaves—and artists to originate sound itself.

Wheatley Peters's attention to the material dynamics of sound suggests new possibilities for interpreting the formal functions of grief. She deployed no genre more extensively than the elegy, an important aesthetic decision since, as Eric Ashley Hairston argues, the elegy "functioned as more than a funeral poem for classical writers," but was a vehicle for "routinely [expressing] political positions."[33] Wheatley Peters's elegies frequently take as their subject the untimely loss of young children and, occasionally, parents. Many were the children of friends and family to John and Susanna Wheatley, indicating that the premature deaths that fracture such close colonial family networks provide her elegies' immediate exigence.[34] Max Cavitch observes that Wheatley Peters's elegies foreground mutually intelligible grief, or sorrows visible to a colonial reading public, while also grappling with "the unmitigated alienation of slavery and the ungrievable—because largely unknown—losses she sustained as a stolen child."[35] One of the most harrowing elegiac moments in her oeuvre is a comparison between the womb and a (slave) ship's hold in her elegy for Dr. Samuel Marshall, survived by his wife and infant son: "The babe unborn in the dark tomb is tost / And seems in anguish for its father lost."[36] Cavitch argues that this image of a womb that resembles a slave ship tossed on the tempestuous sea is the moment where Wheatley Peters comes closest to representing her terrifying journey onboard the slave ship *Phillis* in 1761—in a grim irony, John and Susannah Wheatley used the ship to give Wheatley Peters her name.[37] The image suggests "both the imaginative consignment of the Marshall child to a ship hold and Wheatley's recognition, in the image of a life that has no history prior to captivity, of her own deracination."[38] Mallory similarly contends that Wheatley Peters's elegies express sorrow at her separation from her mother and insist that women be "in charge of mourning the dead."[39] Wheatley Peters' efforts to memorialize the dead against the inhuman forces of slavery are reflected in Eve Ewing's harrowing lyric on her life: "How many iambs to be a real human girl? / . . . / If I know of Ovid may I keep my children?"[40] Across these historical and aesthetic entanglements, Wheatley Peters's elegiac orientation constitutes an environmental ethos responsive to bleak transatlantic crossings and to the affective resilience required to mourn.

For "Niobe in Distress," Wheatley Peters also drew on source materials in classical literature and art to illuminate conflicts centered on forms of belonging to the earth, as mythic figures struggle over competing claims to territory, children, and mourning rituals.[41] In Ovid's *Metamorphoses*, for instance, Niobe tests the gods' patience when she refuses to heed the prophet Manto, who urges her to revere the goddess Latona and her son and daughter, Apollo and Diana. Niobe declares that because she is the mother of fourteen children (far more than Latona's two), she deserves reverence as a divine figure. Enraged by Niobe's hubris, Latona (through the mediation of Apollo and Diana's arrows) strikes down each of Niobe's sons and daughters. *Metamorphoses* entrenches much of the event's terror in how Niobe becomes a spectator to the offspring of the gods massacring her own children. In Ovid's telling, moreover, Niobe's petrification literalizes her barrenness: as a stone, she can neither bear more children nor vocalize her anguish at their loss.[42] Instead, by becoming coextensive with the earth, Niobe offers a devastating inversion of death and burial, where both her body and her spirit remain aboveground, a constitutive—and yet always silent—part of the landscape.[43]

Ovid's version of Niobe is itself a translation and alteration of various Greek sources. Earlier iterations of Niobe's story sometimes emphasized that the trouble between Niobe and Latona began with another instance of divine cruelty against motherhood: Jupiter has raped Latona, and Juno, in her infinite capacity for jealousy, has prohibited Latona from giving birth to her children on solid ground. The Greek mainland and all islands in the sea were thus forbidden to her, until Latona finally found the obscure floating island of Delos. In "Niobe in Distress," it is worth exploring how Wheatley Peters remains attentive to Ovid and other classical sources' colonialist imaginary of reproductive politics and earthly belonging. As classicist Karl Kerényi notes, Niobe was likely a goddess who originated in Asia Minor, and was imported into the pantheon through Greek colonial expansion.[44] Kerényi further observes that the "throne of the Olympian Zeus, the Zeus of Phidias, was decorated" with images of Niobe and her dead children, suggesting that her punishment was central "to the construction of a divine order in classical antiquity."[45] As a figure who similarly resists a politico-theological order of racial hierarchy in the colonial Americas, Wheatley Peters's rebellious Niobe inhabits the position of African American mothers, whose carnal flesh indexes the libidinal economies of fungibility, vulnerability, and violence that shaped reproduction in the transatlantic world. Niobe's presence archives what Hortense Spillers calls the "zero degree of social conceptualization" for Black women's flesh that accompanies colonial projects of invasion, genocide, and enslavement.[46]

In Wheatley Peters's poem, moreover, Niobe contests the rule of the Olympians out of her maternal authority, insisting that because she holds terrestrial power as a queen and mother, she is entitled to just as much, if not more, honor as Latona. Niobe's reproductive flourishing also informs an ecological perspective derived from her implicit condemnation of Jupiter's punishment of Latona, sentenced to give birth on a world that is barren and unmoored:

> What, shall a *Titaness* be deify'd
> To whom the spacious earth a couch deny'd?
> Nor heav'n, nor earth, nor sea receiv'd your queen,
> 'Till pitying *Delos* took the wand'rer in.[47]

Niobe taunts Latona for having no kingdom on the "spacious earth," and no place to bear and raise Apollo and Diana—which, while it could constitute an instance of intra-female conflict, also stresses the cruelty of Jupiter's own retribution against Latona. Wheatley Peters is attuned to Latona's position as a pregnant stranger in need of shelter and portrays the island of Delos welcoming her as another mother: "'Till pitying *Delos* took the wand'rer in." As an early iteration of a personified natural world in the poem, the floating island of Delos exemplifies a praxis of hospitality opposed to Olympian acts of sexual predation. Throughout Wheatley Peters's elegiac poems, and especially in "Niobe in Distress," her aquatic and land environments alternate between nurturing—"pitying *Delos*"—or tempestuous—"the babe unborn in the dark storm is tost." Without a secure anchor, these settings reveal the raw material circumstances of Wheatley Peters's own estrangement from kin and natality. As later stanzas of "Niobe in Distress" unfold, her alterations to Ovid transform the political and ontological possibilities embodied by African insurgency into an indictment of colonial categories of Black nonbeing.

## THE ATMOSPHERES OF BLACK LIFE

Across "Niobe in Distress," Wheatley Peters intervenes in eighteenth-century visual aesthetics of resistance, while also revealing terrains of antiblack violence. Recently, Sharpe has turned to the weather to name a condition of im/possibility that permeates living conditions for African-descended people: "In what I am calling the weather, antiblackness is pervasive *as* climate. The weather necessitates changeability and improvisation; it is the atmospheric condition of time and place; it produces new ecologies."[48] Antiblack atmospheres also require that Black subjects gather affective resources for survival: resilience, defiance, opacity. Yet as Audre Lorde observes in *Sister, Outsider* (1984), in a passage that implicitly grapples with the figure of Niobe in Black studies, this subjective orientation can impede *intersubjective* vulnerability: "In order to withstand the weather, we had to become stone, and now we bruise ourselves upon the other who is closest."[49] I read Sharpe's emphasis on the political requirements of "changeability and improvisation" alongside Lorde's meditation and remanding of "becoming stone" as a hermeneutic for tracing how "Niobe in Distress" protests political ecologies of antiblack violence. Wheatley Peters's Niobe, in repudiating an impetus to become stone, is neither a figure of terrible hubris nor a plaintive victim of divine punishment. Rather, the poem invests meaning in what Sara Ahmed calls a "willful ecology"—by refusing the silence of stone, Niobe embodies an obdurateness capable of "withstand[ing] the weather," but her resilience is not a solitary experiment in human-being but *shared* in concert with the nonhuman "vocal hills."[50] Wheatley Peters's Niobe thereby extends possibilities for resistance merely hinted at in Richard Wilson's 1760 painting *The Destruction of the Children of Niobe* (figure 4.1), which she either encountered through circulated engravings or saw on exhibition while in England.[51]

The composition of Wilson's painting emphasizes the conflicts over maternal protest and earthly belonging that also engaged Wheatley Peters. A closer look at the painting reveals Niobe as a figure standing near the lowest edge of the painting, surrounded by several of her dead and dying sons and daughters. According to Karen Lerner Dovell, "in contrast to other depictions of the Niobe myth," Wilson positions Niobe and her children as a way to "[call] attention to Niobe's loss, rather than to her alleged sin."[52] Apollo appears along the left-hand border of the

Figure 4.1 Richard Wilson, *The Destruction of the Children of Niobe*, 1760. Courtesy of the Yale Center for British Art.

painting, and like Niobe, is almost out of frame. He nevertheless remains elevated, suggesting his central role as an instrument and mediator of sovereign punishment. James Edward Ford, moreover, observes that the divine figures' placement in the painting underscores their political significance in the context of hemispheric slavery's inhuman momentum: "The placement of Apollo and Diana/Latona in Wilson's painting sustains the possibility that instead of gods we are viewing would-be settlers who are ambushing inhabitants to make the latter mere 'things.'"[53] The scene's chiaroscuro also portrays several of the figures in deep shadow, leaving only Apollo and Niobe, grasping her final child, clearly visible. Through the figures' position and lighting, Wilson (a Welsh artist famous for his landscapes) decenters Niobe in favor of the scenery, including the rocky elevation on which Apollo stands, the shoreline abutting the horizon, and the large tree dominating the foreground. As the eye moves from left to right in viewing the painting, Niobe and her children recede from prominence and appear overwhelmed by the Olympian figures who overshadow portions of the natural environment.

The painting's emphasis on the scale of the figures' earthly surroundings—what Wheatley Peters calls "the spacious earth" in her poem—attempts an aesthetic overreaching: it halts the myth's progression and Niobe remains forever fixed in grief—*just* prior to her chthonic conversion, where she will become part of the rocky elevation from which Apollo's punishment looms. As John Berger reminds us, no painting unfolds in time, so while the painting fixes action at a single moment (Apollo's imminent destruction of Niobe's youngest daughter), its ominous ecology gestures toward the myth's conclusion.[54] Yet the painting's composition also depicts Niobe as *alive* and interceding for her daughter, and thus participates in suspending her death. *The Destruction of the Children of Niobe* strives to evade the myth's impending conclusion within the painting's static composition, what Wheatley Peters elegantly describes as the "tuneful goddess" teaching "the painter in his works to *live.*"[55] Her praise of Wilson anticipates another poem, one dedicated to a fellow African American artist, Scipio Moorhead. In "To S.M. a young *African* painter, on seeing his Works," Wheatley Peters affirms a creative faculty that exists and flourishes between Black artists and Black viewers of Black art:

> To show the lab'ring bosom's deep intent,
> And thought in living characters to paint,
> When first thy pencil did those beauties give,
> And breathing figures learn from thee to live.[56]

Wheatley Peters claims that Moorhead's artistic skill in teaching "breathing figures to live," and her own encounter with works of art that constitute a "new creation rushing on my sight," render Black art as collaborative acts that in turn animate Black *life.* That Wheatley Peters similarly praises Wilson for his attention to Niobe's vitality anticipates the radical revisions she will make to the myth in later stanzas to emphasize the animate qualities of African aesthetics. As Spigner argues, her Niobe emerges as "a woman whose very claim to her children *is the act* of rebellion. She no longer stands as a memorial to a mother's grief. Like the Niobe in Wilson's painting, she emerges reconstituted in flesh and blood, her grief warm, breathing, and palpable."[57]

More overtly, Wheatley Peters pointedly revises the story found in the *Metamorphoses* in order to depict Niobe's anger—what Latona calls her "blaspheming breath"—as the animating principle of Black life and art.[58] As Hairston explains, "Niobe in Distress" is likely one of the "first translations of any kind by an African" in colonial America, and the significance of Wheatley Peters's translation lies in the revisions she made to Ovid's story: "She alters the Ovidian narrative to add her own signature."[59] As translator, Wheatley Peters interpolates four lines near the end of the poem where Niobe, in a tone that is neither resigned nor abject, furiously calls for Jove's vengeance to be visited on those who murdered her children:

> Why is such privilege to them allow'd?
> Why thus insulted by the *Delian* god?
> Dwells there such mischief in the pow'rs above?
> Why sleeps the vengeance of immortal *Jove*?[60]

Niobe forcefully insists on the innocence of her children, themselves blameless of Niobe's supposed act of overreaching, while also contesting Latona's authority to render judgment, and not Jove. This stanza, according to Thorn, intertwines Niobe's sorrow with an anger against what she feels is the unjust punishment of the gods: "In Wheatley's 'Niobe,' what women are, first and foremost, is reproductive and angry."[61] Niobe's complaint, expressed in a series of blistering rhetorical questions, registers her rage at the generational devastations enacted by Apollo and Diana, and a lineal wreckage sanctioned by Jove's own passivity and silence—his accountability to justice, or "vengeance," as Niobe scornfully calls it, "sleeps." This regime of capricious sovereignty recurs throughout Greek and Latin mythology, where assertions of kinship are always precarious under fickle gods, but Wheatley Peters powerfully recasts Niobe as a mother who reclaims her children. "Niobe in Distress" thus protests against politico-theological acts of colonial violence throughout the Atlantic world, where the ruinous momentum of slavery awakens and suffuses the atmospheres in which African women live.

## "HER BLASPHEMING BREATH": PROTEST AND PEDAGOGY

Ultimately, Wheatley Peters's most telling alteration in the poem is her evasion of Niobe's final transformation into stone by ending with the following stanza:

> One only daughter lives, and she the least;
> The queen close clasp'd the daughter to her breast:
> "Ye heav'nly pow'rs, ah spare me one," she cry'd,
> "Ah! spare me one," the vocal hills reply'd:
> In vain she begs, the Fates her suit deny,
> In her embrace she sees her daughter die.[62]

Sometime later, a final stanza was added to the poem, which more conventionally ends with Niobe's familiar fate: "A marble statue now the queen appears / But from the marble steal the silent tears."[63] If we take lines 207–212 as Wheatley Peters's preferred ending, however, then she makes an evocative aesthetic decision, one that highlights something Ovid's rendering of Niobe does not address—what happens to the children's bodily remains. As the *Metamorphoses* frames events, enough time passes after the deaths of her sons that Niobe and her daughters can in their grief attire themselves in black, loosen their hair, and congregate around their brothers' funeral biers as a prelude to burial.[64] It is at this moment that Diana massacres the girls, disregarding Niobe's pleas for the youngest to be spared. Ovid last describes Niobe's sons and daughters as unburied, exposed to the elements and always before Niobe's sight: "She sat bereft / Amid her sons, her daughters and her husband / All lifeless corpses, rigid in ruin."[65]

Ovid's stanza concludes with Niobe repeating the "rigid ruin" of her family's bodily remains by turning to stone. Yet Wheatley Peters's translation alters the repetition of Niobe's living rigor mortis, choosing instead to render a performative reiteration through the echo the "vocal hills reply'd" to her sorrow. John Shields argues that Wheatley's depiction of nature's echo "increases the intensity of her punishment," as the landscape merely repeats her "vain" plea, instead of responding to it.[66] Lucy K. Hayden, however, contends that nature's echo "intensifies the pathos" of Niobe's loss by reiterating her sorrow.[67] Building on Hayden, I do not think "Niobe in Distress" concludes with simple pathos, with the unanswered call of her terrible grief—another silence in the archives of slavery. By representing Niobe's spoken grief echoed by "the vocal hills," Wheatley Peters goes beyond accentuating the intensity of Niobe's punishment and strategically reverses the work of mourning Ovid offered. Evading Niobe's lithic transformation, she refuses to strip Niobe of her capacity to speak. Moreover, the echo of the "vocal hills" distributes Niobe's agony to include the earthly order in Niobe's contestation against Olympus. Wheatley Peters assembles a form of life that refuses colonial legal consignments of African maternal bodies to nonbeing *and* settler definitions of the earth as inert matter. This form of life is necessarily aural and collaborative, sounding between Niobe and the earth, and thus exemplifies what Katherine McKittrick, Frances H. O'Shaughnessy, and Kendall Witaszek have recently named "rhythm." In an African-descended tradition of "bios-mythois," rhythms are "repeated or patterned sounds" that do "not privilege singular ways of being but rather [insist], in advance, that collaborative engagement is necessary to who and what we are."[68] Above all, the materialities subtending the call and response between Niobe and her surround relinquish a colonial and patriarchal configuration of estranged reproduction for a vision of Black and ecological mutual existence—of beings who mourn and breathe *together*.

Given Wheatley Peters's interventions in racial governance and environmental dwelling, I believe "Niobe in Distress" necessitates pedagogical approaches responsive to the Black Lives Matter movement for the ways it directs our attention to the ongoing devaluation of Black life, the state-sanctioned murder of Black people, and the grief of Black mothers. The poem's elegiac tropes foreground questions of protest and mourning—and resonate with many of my students' experiences with contemporary activism. Using our discussions of elegy and form as a point of departure, I ask students to consider the relationship between Niobe's "blaspheming breath" and her deep pain: How can forms of public grief authorize dissent? In what ways may breathing inspire idioms of political insurgency? Might something like Niobe's furious anguish give rise to radical resistance? Can breathing *be* blasphemy? Students often note the ways Niobe echoes the fortitude of BLM founders such as Alicia Garza, Opal Tometi, and other activists and mothers in the movement, who urge us to remember the last words of Eric Garner and George

Floyd: "I can't breathe." In a proposal that powerfully recalls Niobe's resolute pose, Sharpe suggests that a "group noun for young (and middle aged and old) Black folks in the street is a breathing."[69] In previous semesters, I have taught "Niobe in Distress" as a representation of (and resistance to) the biopolitical commodification of African women's reproduction and maternity.[70] Yet Sharpe's definition of Black collectivity as a "breathing" underlines that Black survivance is historically conditioned by colonial ecologies of nonbeing, an insight that feels even more vital as communities of color respond to acute health disparities exacerbated by COVID-19 and its attendant stresses on the respiratory system—disparities also intensified by long histories of exposure to environmental toxicities. Taken together, then, these historical pressures urge us to imagine our pedagogic rhythms anew in the wake of global pandemic and police violence.

Both then and now, breath hovers and flows between spiritual and material registers, a metonym for soul or spirit, but also for fleshly life. From this perspective, Niobe's "blaspheming breath" participates in what Edward E. Baptist calls a "vernacular history of being stolen hidden in the breath of captives."[71] This vernacular history exists in imaginative and substantial plots at odds with settler ecologies of stolen land and life, and flourishes in places like the gardens planted by Eric Garner, as Ross Gay lyrically evokes: "he put gently into the earth / some plants which, most likely, / . . . / [make] it easier / for us to breathe."[72] In "BLM 2020: Breathing, Resistance, and the War against Enslavement," moreover, Kerry Sinanan traces how scholarship on Black rebellion persistently recurs to imaginaries of both breath and geology—subsoil, bedrock, volcanic eruption—to foreground revolt as a counter-ecology.[73] At first glance, breath and stone might appear to be counterintuitive sites from which to think with resistance, for the former seems insubstantial and invisible, while the latter figures forth in all its fixed materiality, and yet philosophers and literary critics have argued otherwise. Jeffrey Cohen's survey of lithic existence in *Stone: An Ecology of the Inhuman* (2015) urges a reconceiving of stone as not truly inert—or "worldless," as Martin Heidegger famously designated—but as vibrant, always in motion.[74] Indeed, Cohen argues that stone is a geologic substance against which "a limit-breaching intimacy persistently unfolds."[75] In the arena of transatlantic slavery, images of stone-in-motion, such as volcanic eruption and flame (also, sometimes, a metaphor for breath), frequently appear to describe the "limit-breaching intimac[ies]" of insurgencies like the Haitian Revolution. As CLR James argues in *The Black Jacobins* (1938, 1963): "In a revolution, when the ceaseless slow accumulation of centuries bursts into volcanic eruption, the meteoric flares and flights above are a meaningless chaos and lend themselves to infinite caprice and romanticism unless the observer sees them always as projections of the sub-soil from which they came."[76] As with "Niobe in Distress," this critical genealogy and unruly praxis in Black studies represent, in Sinanan's words, a "bedrock of breathing resistance" against the inhuman forces of colonialism.[77] Wheatley Peters's Niobe constitutes a perhaps less noticed, but no less significant, rebellious posture, a subversive touchpoint in a historical continuum that includes the Haitian Revolution and the BLM movement. Above all, when Wheatley Peters renders the embodied and environmental terrain of Black motherhood, she strives to retrieve collective forms of life—of breathing(s)—that endure within the rhythmic echoes that roll across the "the vocal hills."

## NOTES

1. By using Phillis Wheatley Peters's married name to John Peters, I follow the practice illustrated by Honorée Fanonne Jeffers's *The Age of Phillis* (Middletown, CT: Wesleyan University Press, 2020), exemplified by poetic practice on pages 119–132. Jeffers eloquently argues that "Peters" was the only name the poet chose for herself, given that "Phillis" was a relic of the slave ship *Phillis,* which transported her to Boston and "Wheatley" the name of her enslavers.
2. Ovid, *Metamorphoses*, trans. A. D. Melville (Oxford: Oxford World Classics, 2009), 83–84.
3. For Wheatley Peters's classicism, see Lucy Hayden, "Classical Tidings from the Afric Muse: Phillis Wheatley's Use of Greek and Roman Mythology," *College Language Association Journal* 35, no. 4 (1992): 432–447; John Shields, "Phillis Wheatley's Use of Classicism," *American Literature* 52 (1980): 97–111, and *The American Aeneas: Classical Origins of the American Self* (Knoxville: University of Tennessee Press, 2004); Tracey Walters, *African American Literature and the Classicist Tradition* (London: Palgrave Macmillan, 2007); Emily Greenwood, "The Politics of Classicism in the Poetry of Phillis Wheatley," in *Ancient Slavery and Abolition*, ed. Edith Hall, Richard Alston, and Justine McConnell (Oxford: Oxford University Press, 2011), 153–180. Much of this work on Wheatley is occurring within reassessments of African American classicism. See, for instance, Algernon Austin, *Achieving Blackness* (New York: New York University Press, 2006); William W. Cook and James Tatum, *African American Writers and Classical Tradition* (Chicago: University of Chicago Press, 2011); Margaret Malamud, *African Americans and the Classics* (London: I.B. Tauris, 2016); John Levi Barnard, *Empire of Ruin* (Oxford: Oxford University Press, 2018).
4. Wheatley Peters, "Niobe in Distress," lines 209–210.
5. A footnote observes, "This Verse to the End is the Work of another Hand" (59).
6. The "Attestation" states, "WE whose Names are under-written, do assure the World, that the POEMS specified in the following Page, were (as we verily believe) written by PHILLIS," followed by eighteen names of prominent men in Boston (8). See Henry Louis Gates Jr., *The Trials of Phillis Wheatley* (New York: Civitas Books, 2010) for a discussion of the "Attestation."
7. Katy Chiles, *Transformable Race* (Oxford: Oxford University Press, 2014), 62.
8. Kathryn Yusoff, *A Billion Black Anthropocenes or None* (Minneapolis: University of Minnesota Press, 2018), 5.
9. See Nicole A. Spigner, "Phillis Wheatley's Niobean Poetics," in *Brill's Companion to Classics in the Early Americas*, ed. Matthew E. Duquès, Adam Goldwyn, and Maya Feile Tomes (Leiden: Brill, 2021), 320–342; Devona Mallory, "I Remember Mama: Honoring the Goddess-Mother while Denouncing the Slaveowner-God in Phillis Wheatley's Poetry," in *New Essays on Phillis Wheatley*, ed. John C. Shields (Knoxville: University of Tennessee Press, 2011), 19–34; and Jennifer Thorn, "'All Beautiful in Woe': Gender, Nation, and Phillis Wheatley's Niobe," *Studies in Eighteenth-Century Culture* 37, no. 1 (2008): 233–258.
10. See Jennifer L. Morgan, *Laboring Women: Reproduction and Gender in New World Slavery* (Philadelphia: University of Pennsylvania Press, 2004); and Kerry Sinanan, "Lost Mothers in the Caribbean Plantation and Contemporary Black Maternal and Infant Mortality," in

*Caribbean Literature in Transition*, Vol. 1: 1800–1920, ed. Evelyn O'Callaghan and Tim Watson (Cambridge, UK: Cambridge University Press, 2021), 390–408.

11. For Wheatley Peters's resistant ecologies, see John C. Shields, "Phillis Wheatley's Subversive Pastoral," *Eighteenth-Century Studies* 27, no. 4 (1994): 631–647. For a more recent account, see James Edward Ford's brilliant reading that "Niobe in Distress" expresses a political theology that contests state violence, extending philosophies on violence in Kant and Benjamin. See Ford, "An African Diasporic Critique of Violence," in *Systems of Life*, ed. Richard A. Barney and Walter Montag (New York: Fordham University Press, 2018), 56–80.
12. I route my argument through new materialism, rather than object-oriented ontology, because this scholarship foregrounds African and Native ontologies, enabling a reading of the vexed mutualities and entanglements between human flesh and ecologies in Wheatley Peters's poetry.
13. It is tempting to speculate on Wheatley Peters's ideas about animate matter, given her likely exposure to Senegalese and Akan spiritualities, as well as North African Islamic thought. John C. Shields in *Phillis Wheatley's Poetics of Liberation* argues that Wheatley's spiritual commitments are a highly personalized and syncretistic blend of animism, Islam, classicism, and Christianity (77, 132, 148). Other scholars, such as Regina Jennings and Will Harris, have also explored Wheatley's possible adherence to West African sun worship and Islam. See Shields, *Phillis Wheatley's Poetics of Liberation* (Knoxville: University of Tennessee Press, 2008); Regina Jennings, "African Sun Imagery in the Poetry of Phillis Wheatley," *Pennsylvania English* 22, no. 1–2 (2000): 68–76; and Will Harris, "Phillis Wheatley: A Muslim Connection," *African American Review* 481, no. 2 (2015): 1–15.
14. For nonhuman agency, see Jane Bennett, *Vibrant Matter* (Durham, NC: Duke University Press, 2010), 3–5; Stacy Alaimo, *Bodily Natures* (Bloomington: University of Indiana Press, 2010), 129–147; Donna Haraway, *Staying with the Trouble* (Durham, NC: Duke University Press, 2016), 88–107; Karen Barad, *Meeting the Universe Halfway* (Durham, NC: Duke University Press, 2007), 33–35.
15. Elizabeth Povinelli, *Geontologies* (Durham, NC: Duke University Press, 2016), 24–28. For work in Native studies that explores nonhuman personhood, see Mario Blaser, *Storytelling Globalization from the Chaco and Beyond* (Durham, NC: Duke University Press, 2010), 23–25; Marisol de la Cardena, *Earth Beings* (Durham, NC: Duke University Press, 2015), 31–32; Eduardo Kohn, *How Forests Think* (Berkeley: University of California Press, 2013), 118–119. See also Blaser and de la Cadena's edited collection, *A World of Many Worlds* (Durham, NC: Duke University Press, 2018).
16. Barad, *Meeting the Universe Halfway*, 33. For "call and response" and nonhuman voices, see Cynthia Willett, *Interspecies Ethics* (New York: Columbia University Press, 2014), 9–17.
17. Stephen Best and Saidiya Hartman, "Fugitive Justice," *Representations* 92, no.1 (2005): 1–15, 9; Joshua Bennett, *Being Property Once Myself* (Cambridge, MA: Beacon, 2020), 12–13.
18. Sylvia Wynter and Katherine McKittrick, "Unparalleled Catastrophe for Our Species?: Or, to Give Humanness a Different Future: Conversations," in *On Being Human as Praxis* (Durham, NC: Duke University Press, 2014), 9–24. King and Yusoff also attend to racial capitalism and anthropocentric climate change as *co-constitutive*, and critique the settler and antiblack discourses that structured colonized subjectivity and landscapes. See Tiffany Lethabo King, *The Black Shoals: Offshore Formations of Black and Native Studies* (Durham, NC: Duke University Press, 2019), 13–21, and Yusoff, *A Billion Black Anthropocenes*, 7–11 and 52–57. See also Alexander Weheliye in *Habeas Viscus* (Durham, NC: Duke University Press, 2014), 1–16; Bennett, *Being Property Once Myself*, 12–13.
19. David Waldstreicher, "Ancients, Moderns, and Africans: Phillis Wheatley and the Politics of Empire and Slavery in the American Revolution," *Journal of the Early Republic* 37, no. 4 (2017): 729.
20. Thomas Jefferson, *Notes on the State of Virginia*, ed. David Waldstreicher (Boston: Beacon, 2002), 178.

21. See Monique Allewaert, *Ariel's Ecology* (Minneapolis: University of Minnesota Press, 2013), 51–63. See also Paul Outka, *Race and Nature from Transcendentalism to the Harlem Renaissance* (London: Palgrave Macmillan, 2008); and Leilani Nishime and Kim D. Hester Williams, eds., *Racial Ecologies* (Seattle: University of Washington Press, 2018).
22. Christina Sharpe, *In the Wake: On Blackness and Being* (Durham, NC: Duke University Press, 2016), 19.
23. Tiffany Lethabo King, *Black Shoals*, 114.
24. June Jordan, "The Difficult Miracle of Black Poetry in America: Something Like a Sonnet for Phillis Wheatley," *Some of Us Did Not Die: New and Selected Essays by June Jordan* (London: Penguin Books, 2002), 174–186.
25. Yusoff, *A Billion Black Anthropocenes*, 3.
26. Wheatley Peters, "Niobe in Distress," line 99.
27. I am influenced here by Jennifer Morgan, who argues, "As slavery took root in North America and the Caribbean colonies, the interplay between inherited beliefs about gender, race, and civility coalesced to shape slaveowners' implicit expectation that their wealth and, indeed, that of entire colonial empires, derived from the reproductive potential of African women." See Morgan, *Laboring Women* 8.
28. Scholars are turning to the early modern period as an origin for the current climate crisis. The Little Ice Age (1300–1800) and its era of global cooling was exacerbated by colonial expansion and Native genocide in the Americas. See Mark S. Maslin and Simon L. Lewis, "A Transparent Framework for Defining the Anthropocene Epoch," *Nature* 519 (2015): 171–180; and Dana Luciano's response, "The Inhuman Anthropocene," *Avidly*, March 22, 2015.
29. On the question of objects and agency in the eighteenth century, scholars have turned to commodities and "it narratives" as instances of more-than-human personhood. See Jonathan Lamb, *The Things Things Say* (Princeton, NJ: Princeton University Press, 2011), and Elizabeth Kowaleski Wallace, "The Things Things Don't Say: *The Rape of the Lock*, Vitalism, and New Materialism," *Eighteenth Century: Theory and Interpretation* 59, no. 1 (2018): 105–122.
30. King, *Black Shoals*, 53.
31. Sylvia Wynter argues that in its transition from medieval Christendom to the colonial West, Europe "overrepresented" what it meant to be human, imaging its own narrow, culturally specific definitions of human-being. Europeans believed the category of "human" applied to only themselves, an overdetermination Wynter calls Man, and imagined Man to be individualistic, autonomous, rational, economical—all values Europeans came to prize concurrent with the rise of capitalism. See Wynter and McKittrick, "Unparalleled Catastrophe," 45–54.
32. Mary Jacobus, "'Distressful Gift': Talking to the Dead," *South Atlantic Quarterly* 106, no. 2 (2007): 400.
33. Eric Ashley Hairston, "The Trojan Horse: Memory, Transformation, and Afric Ambition in *Poems on Various Subjects, Religious and Moral*," *New Essays on Phillis Wheatley*, ed. John C. Shields (Knoxville: University of Tennessee Press, 2011), 87.
34. For Wheatley Peters's elegies and their colonial religious contexts, see Paula Bennett, "Phillis Wheatley's Vocation and the Paradox of the 'Afric Muse,'" *PMLA* 133, no. 1 (1998): 64–76; Gordon Thompson, "Methodism and the Consolation of Heavenly Bliss in Phillis Wheatley's Funeral Elegies," *CLA* 48 (2004): 34–50; Jennifer Thorn, "Phillis Wheatley's Ghosts: The Racial Melancholy of New England Protestants," *Eighteenth Century: Theory and Interpretation* 50, no. 1 (2009): 73–99; Max Cavitch, *American Elegy* (Minneapolis: University of Minnesota Press, 2007), chap. 5; Caroline Wigginton, *In the Neighborhood* (Amherst: University of Massachusetts Press, 2016), chap. 3; and Antonio Bly, "'On *Death's* Domain Intent I Fix My Eye': Text, Context, and Subtext in Phillis Wheatley's Elegies," *Early American Literature* 53, no. 2 (2018): 317–341.
35. Cavitch, *American Elegy*, 187.
36. Wheatley Peters, "Niobe in Distress," lines 21–22.

37. Cavitch, *American Elegy*, 189. For the womb/ship relation, see Fred Moten and Stefano Harney, *The Undercommons* (New York: Autonomedia, 2013), 93–94, and Sharpe, *In the Wake*, chapter 3.
38. Cavitch, *American Elegy*, 189.
39. Mallory, "I Remember Mama," 32.
40. Eve L. Ewing, "Eve Ewing on Phillis Wheatley," *New York Times: The 1619 Project*, August 18, 2019, p. 42.
41. For an overview of eighteenth-century adaptations of Ovid, see James Horowitz, "Ovid in Restoration and Eighteenth-Century England," in *A Handbook to the Reception of Ovid*, ed. John F. Miller and Carole E. Newlands (New York: Wiley Blackwell, 2014), 355–370.
42. Ovid, *Metamorphoses*, 82–83.
43. Wheatley Peters's revision of Niobe participates in adaptations of classical women (like Medusa) whose aberrant sexualities inspire revulsion, fascination, and fear. Niobe could be a feminist iteration of the "Black Prometheus," a figure of enslaved revolt that Jared Hickman traces in *The Black Prometheus* (Oxford: Oxford University Press, 2016), 3.
44. Karl Kerényi, *Goddesses of Sun and Moon*, trans. Murray Stein (Dallas: Spring Publications, 1979), 61–62.
45. Kerényi, *Goddesses*, 67. See also Karen Lerner Dovell, "The Interaction of the Classical Traditions of Literature and Politics in the Work of Phillis Wheatley," in *New Essays on Phillis Wheatley*, ed. John C. Shields (Knoxville: University of Tennessee Press, 2011), 44.
46. Hortense Spillers, "Mama's Baby, Papa's Maybe: An American Grammar," in *Black, White, and in Color* (Chicago: University of Chicago Press, 2003), 206.
47. Wheatley Peters, "Niobe in Distress," lines 74–78.
48. Sharpe, *In the Wake*, 106.
49. Audre Lorde, *Sister, Outsider* (New York: Crossing Press, 2007), 160.
50. Sara Ahmed, "Willful Stones," *Feminist Killjoys*, January 26, 2016, https://feministkilljoys.com/2016/01/29/willful-stones/.
51. Vincent Carretta, *Phillis Wheatley* (Athens: University of Georgia Press, 2011), 105.
52. Dovell, "Interaction of the Classical Traditions," 44.
53. Ford, "African Diasporic Critique of Violence," 72.
54. John Berger, "The Changing View of Man in Portrait," in *Selected Essays and Articles* (London: Pelican, 1972), 40.
55. Wheatley Peters, "Niobe in Distress," lines 2, 4 (emphasis added).
56. Phillis Wheatley Peters, "To S.M. a young *African* painter, on seeing his Works," in *Phillis Wheatley*, ed. Vincent Carretta (London: Penguin Books, 2001), lines 1–4.
57. Spigner, "Phillis Wheatley's Niobean Poetics," 331.
58. Wheatley Peters, "Niobe in Distress," line 99.
59. Hairston, "The Trojan Horse," 85.
60. Wheatley Peters, "Niobe in Distress," lines 165–168.
61. Thorn, "'All Beautiful in Woe,'" 249.
62. Wheatley Peters, "Niobe in Distress," lines 207–212.
63. Wheatley Peters, "Niobe in Distress," lines 209–210.
64. Ovid, *Metamorphoses*, 83.
65. Ovid, *Metamorphoses*, 83–84.
66. Shields, "Phillis Wheatley's Use of Classicism," 110.
67. Hayden, "Classical Tidings," 438.
68. Katherine McKittrick, Frances H. O'Shaughnessy, and Kendall Witaszek, "Rhythm, or On Sylvia Wynter's Science of the Word," *American Quarterly* 70, no. 4 (2018): 870.
69. Christina Sharpe, Twitter, June 6, 2020, https://twitter.com/hystericalblkns/status/1269331358704259072.
70. My students and I engaged with the horrifying gynecological research of J. Marion Sims, who conducted experiments on the bodies of Black women without anesthesia. See Deirdre Cooper Owens, *Medical Bondage*, 2nd ed. (Athens: University of Georgia Press, 2018).

71. Edward E. Baptist, *The Half Has Never Been Told* (New York: Basic Books, 2016), 191.
72. Ross Gay, "A Small Needful Fact." https://poets.org/poem/small-needful-fact/ (accessed February 27, 2021). Wynter connects the rise of the plantation to the novel as a genre that has elided but could not erase what she calls people of color's "secret histories." "Plot" offers a way to trace worlds that colonialism occludes, that persist in gardens, ecologies, and lived experience. See Sylvia Wynter, "Novel and History, Plot and Plantation," *Savacou* 5 (June 1971): 95–102.
73. Kerry Sinanan, "BLM 2020: Breathing, Resistance, and the War against Enslavement," *Age of Revolutions*, June 10, 2020, https://ageofrevolutions.com/2020/06/10/blm-2020-breathing-resistance-and-the-war-against-enslavement/.
74. Martin Heidegger, "The Origin of the Work of Art," in *Poetry, Language, Thought*, trans. Alfred Hofstadter (New York: Harper, 1975), 43.
75. Jeffery Jerome Cohen, *Stone: An Ecology of the Inhuman* (Minneapolis: University of Minnesota Press, 2015), 1.
76. CLR James, "Preface to the First Edition," *The Black Jacobins* (London: Vintage, 1938), xi. Also cited in Sinanan, "BLM 2020." We might also recall the searing description of the slave ship *San Dominick* in Herman Melville's *Benito Cereno* as "the slumbering volcano." See Melville, *Benito Cereno*, ed. Brian Yothers (Peterborough: Broadview, 2019), 74.
77. Sinanan, "BLM 2020."

5

# SYPHILIS AND NATURAL HISTORY

## The Ethical Limits of Human Mastery

MARIAH CRILLEY

IN EARLY 2020, as the coronavirus pandemic raged and people across the world succumbed to myriad effects of viral spread, reporters pressed Dr. Anthony Fauci, the director of the National Institutes of Allergy and Infectious Disease (NIAID): When would the country "reopen"? "The virus," Fauci responded, would "decide."[1] In 1707, Hans Sloane, future president of the Royal Society, chronicled another pandemic in his natural history of the West Indies: syphilis "spread itself through *Italy*, and *France*" so quickly that soon "it had possess'd many People in all the Provinces of *Europe*."[2]

Though divided by 311 years and different diseases, both Fauci and Sloane make disease the subject of their sentences, assigning it syntactic agency. The disease acts—"spreads," "possesses," and even "decides." Why? This cannot be mere rhetorical flourish or pathetic fallacy. Both men's professional achievements demonstrate that they understand their necessary rhetorical conventions: in modern science, objects are acted upon, not acting, and in the natural history, the observing human subject should ground the sentence and its action. The strange, shared syntax also cannot be rhetorical manipulation for personal or political ends alone. If Fauci wanted to demonstrate that reopening was out of his control, he would need to assign the virus agency consistently. Sloane would need to be similarly consistent if he wanted to minimize colonization's and early science's responsibility for transmitting syphilis across Europe. Nor is the syntax simply a by-product of historical context. Certainly, the eighteenth century considered the environment powerful, determining health and identity. Just as certain, anthropogenic climate changes means that we are witnessing an environment that is more visible, palpable, and destructive than ever before, as Chapters 1, 2, and 3 have addressed. Yet even a shared sense of a powerful and affecting environment cannot explain why both men make disease the subject of their sentences, especially when such syntax deviates so significantly from their training.

What if we are missing a simpler, though undeniably more radical, explanation? What if we take their syntax seriously? What if we recognize the coronavirus and syphilis as agents? To do so does not require proving that early Atlantics believed in microbes, an anachronistic theory of disease. Nor does it require proving that directors of the NIAID truly believe that viruses make choices. Rather, by redefining agency through new materialist philosophy, we can recognize and parse how nonhumans like microbes work on our bodies, selves, and even language, whether we name this work "agency" or not.

The new materialisms theorize nonhuman agency and being. Agential realists like Jane Bennett and Karen Barad analyze the intersection of human and nonhuman agency, while object-oriented ontologists such as Graham Harman examine the individual and inaccessible being of each nonhuman.[3] Though one emphasizes relation and the other irreducibility, both perspectives strive to combat the dangerous solipsism of the Anthropocene by centering and reimagining nonhumans. Our current pandemic, and the emerging "pandemic era" that it heralds, illustrates the vital necessity of uniting the new materialisms with the broader environmental humanities—of the need to account for the increasingly intimate and dangerous agency of nonhumans, crystallized in the universal yet idiosyncratic experience of pandemic illness.[4]

As I approach it, the environmental humanities, therefore, should and must intersect with the medical humanities, especially given the environmental corollaries and manifestations of the viral agents I address here. Disease may be a lively and irreducible agent, but its agency is interactive, affecting human life in painful, deadly, and sometimes generative ways. As the medical humanities emphasizes, illness is exceptionally complex: it can be global but also individual, can connect us to others in networks of shared experience or painfully isolate us, can generate a profound intimacy with our own bodies or alienate us from them, can enable new insights and expressions or disable bodies and imaginations. Illness represents one of our most powerful interactions with nonhuman agency and our own material bodies. Thus, the new materialisms provide a complex paradigm beyond science and medicine to theorize our materiality without resorting to normative definitions of human or nonhuman embodiment.

To reimagine nonhuman agency, illness, and our pandemic era, and posit an alternative future, we must apply the new materialisms to the eighteenth century, to the period that engineered the systems of capital, globalization, and science that led us here. To do so, this essay proposes a method for reading for and listening to disease's agency. Such a method is necessary and useful because to rebuild our relationship with the nonhuman world we must attempt to listen, understand, and integrate it. First, we must closely examine a disease's ontology and etiology, its unique being, evolution, transmission, and interaction with other matter; next, we must similarly examine patterns in genres of writing; then, we can compare ruptures in a genre's patterns against a particular disease's unique being, tracing a disease's presence throughout a text or even an entire genre; and, finally, we must listen to what those ruptures say, what they critique and construct. To read, listen, and respond to disease's textual agency is to begin to build a world for humans and nonhumans alike.

Sloane's natural history, *A Voyage to the Islands of Madera, Barbados, S. Christophers and Jamaica* (1707), grapples with many diseases, but syphilis unsettles the text in distinct and traceable ways. Syphilis' precise nature and origin were contested in the eighteenth century and remain unclear. It went by many names, mimicked other diseases, and evolved oddly over time. In turn, Sloane's natural history should ostensibly conform to important rules for proper style and content. Rhetorical techniques like virtual witnessing established the new genre's epistemological claim to true knowledge. Yet the sections on syphilis refuse to comply with such rhetorical conditioning. They suddenly withhold patients' names, confuse conventional case-study plotting, invoke ornamental citations, and undermine author-centered syntax—rhetorical techniques vital to the natural history. In this consistent inconsistency, we glimpse syphilis "itself," the ontological agent that necessarily escapes human comprehension. Syphilis disrupts the certainty that the epistemology and genre of the natural history promised, demonstrating both the limits of human knowledge and the danger that faith in human knowledge poses. Instead, syphilis posits its own incomprehensibility as a more ethical framework for understanding the complexity of disease—human illness and health disparities, microbial agents and evolutions, global supply chains and pandemics. To recognize the limits of knowledge and certainty, to make agency and action diffuse and unmoored from humanity, and to practice listening to what these agents and their actions reveal is a powerful and necessary tool for living in our pandemic era.

## READING FOR SYPHILIS

In the early modern period, syphilis was a cypher—without name, origin, or even identity. It went by many names, including *lues venerea*, the pox, the great pox, and the French or Spanish pox. Sloane himself never uses the term "syphilis," opting for "pox" instead. Undoubtedly, this linguistic variation resulted from the illness's uncertain origin. While many early writers located syphilis' genesis in the Americas, imported to Europe by Christopher Columbus and his crew, others refused to believe that any disease could be completely "new," citing references to the same or a similar disease in the Old Testament and Hippocratic texts.[5] Ultimately, these debates signaled the difficulty of identifying and defining syphilis. To Europeans, syphilis seemed like a variation on a familiar venereal disease, gonorrhea, but it also acted like another purportedly American ailment, yaws. It could have been a combination of these known diseases, dictated and exacerbated by the torrid climate, or it could have been something new entirely.[6] Sloane himself seems to recognize gonorrhea and yaws as illnesses distinct from syphilis, even though he sometimes uses "gonorrhea" as a symptom of syphilis.[7]

Syphilis's precise ontology, its symptoms and progression, continued to thwart doctors through the nineteenth century. One physician described the disease as an "imitator" that mimicked other diseases like psoriasis, smallpox, measles, epilepsy, and blindness, but lacked any identifiably distinct symptoms of its own. It was "mixed up," and its "development" was "seldom quite regular" and frequently "rapid."[8] Although few symptoms are particular to one disease only, an accurate diagnosis requires a clear progression of identifiable symptoms over time, what the medical humanities and narrative medicine calls "narrative emplotment."[9] Thus, even four centuries after its entrance into the European record, syphilis remained a diagnostic quagmire. The physician was forced to concede, "Whenever the evolution of a disease is irregular, and the use of definite nomenclature thus rendered difficult, we may suspect that the malady is syphilitic."[10] If it was not anything else, it might be syphilis.

Syphilis remains mysterious even now. The bacterium *Treponema pallidum* is sexually transmitted, beginning with a chancre, an ulcer in the genitals, fever, swollen glands, rashes, and other ulcers in the mouth and genitals. These symptoms usually resolve in a few weeks, but the bacterium remains in the body for decades if untreated, producing internal ulcers and eventually leading to visual and auditory impairments, strokes, paresis, and dementia.[11] On the one hand, we do understand syphilis' etiology and symptoms now. We also know how to cure it—with antibiotics, at least for now.[12] On the other hand, much is still unknown: its relationship to other treponemic diseases like pinta, yaws, and bejel remains unclear; its geographic origin is still contested;[13] and, its violent, epidemic entrance into Europe, its nineteenth-century presentation, and the contemporary illness are wildly different, suggesting some uncertain evolution.[14] That the disease remains so unknowable and uncertain so long after its initial description suggests that something about the disease itself, its ontology, is unknowable and uncertain. What might it mean, then, for something about which we know so much to remain absolutely aporic?

Although syphilis may be uncertain and unknowable, new materialisms provide a framework for reading its intersection with human agency, bodies, and bodies of literature. Jane Bennett and Karen Barad argue that agency is not intentional, causal, rational, or effective. Nor is it unique to humanity. Rather, agency is the simple but diffuse capacity to affect—to impact, inform, and transform other bodies. Barad, working with quantum physics, argues that agency derives from inseparability: "Individually determinate entities do not exist. . . . Rather, determinate entities emerge from their intra-action."[15] No thing is inseparable from any other thing, even though we can and must separate them for ontological and epistemological clarity. Agency, therefore, is not a possession; it is an always-in-tandem, though not necessarily harmonious, action. Diseases end lives, challenge identities, decimate cultures and economies, foster new cultures, and change history. They act—through, with, and against us. It is not that *Treponema pallidum*, the bacterium that causes syphilis, does not exist without me, nor that I do not exist without it, but that syphilis does not exist but for this intra-action. This intra-action, in turn, gives shape and meaning to discrete categories like microbe and human. Neither disease's nor humanity's agency is self-evident. We intra-act, intersect, coproduce, and cocreate.

Yet for object-oriented ontology (OOO), the sensual qualities of an object, perceived by humans, are not the object itself. The object is not fundamentally inseparable, as Barad argues, but fundamentally inaccessible. For Graham Harman, the "real" object is "withdrawn": its "reality is free of all relations."[16] Syphilis names the sensual qualities of a withdrawn *Treponema pallidum*, what we perceive and experience of the microbe, but that is not necessarily the microbe's reality. Thus, agential realism offers a critical paradigm for parsing how nonhumans and humans intersect, biologically, culturally, and textually, while OOO reminds us that disease is also withdrawn and inaccessible—above, below, or beyond human symptoms and illness, eluding both understanding and narration.

Nonetheless, such agency writes our bodies and our bodies of literature. According to Julian Yates, nonhumans represent. For example, an orange "is a thoroughly rhetorical entity."[17] It is "a representation of what the plant assumes that we sighted, smelling, tasting animals might desire."[18] Similarly, *Treponema pallidum* represents itself and writes the human body with sores, ulcer, and fever. Yet Yates extends further, arguing that nonhumans can write and rewrite human texts: a literary inscription of an orange "exists not as a moment of capture so much as a ceding of our writing system to others."[19] To Yates, only some writers can "cede" authorial control; however, the deeply personal yet universal experience of illness asserts that ceding control is inevitable. If intra-action occurs physiologically to produce illness without human permission, then it can also occur textually. An author need not be infected with a microbe for their work to be affected; rather, the microbe's ontology could infect the culture and, thereby, the human and their writing.

Diseases affected and inscribed human bodies and bodies of literature. More pointedly, a disease's idiosyncratic ontology—for example, the unnamed, ahistorical, inaccessible "it" of syphilis—generated a traceable pattern of textual symptoms. In Sloane's *Voyage*, the passages on syphilis fail to meet generic and epistemological conventions, but they do so consistently, skirting the authentication protocols like naming and plotting necessary to establish certain truth. This consistent inconsistency inscribes syphilis's own ontological uncertainty onto the pages of the natural history, challenging its epistemological confidence and positing radical ambivalence in its place.

## REPRESENTING THE THING ITSELF: THE NATURAL HISTORY, ROYAL SOCIETY, AND SLOANE'S *VOYAGE TO JAMAICA*

Before modern medicine and science were professionalized in the nineteenth century, the Royal Society pioneered a new kind of "science" based on observing, experimenting, and transcribing "all the works of *Nature*."[20] They rejected the ancient model of natural knowledge, which privileged deductive reasoning through classical citation. Instead, the Royal Society prioritized direct experience through close observation of natural phenomena in the laboratory (natural philosophy) and in the wider world (natural history). Sloane was one of the earliest trained natural historians to observe the Americas directly and, thus, dedicated to its epistemology of direct experience: "the Knowledge of Natural-History, being observations of Matters of Fact, is more certain than most Others, and in my Slender Opinion, less subject to Mistakes than Reasonings, Hypotheses, and Deductions are."[21] The natural history was not simply a method for creating knowledge; rather, it was knowledge. As Michel Foucault explains, natural historians believed that they could "reduce the distance between" things and language "so as to bring language as close as possible to the observing gaze, and the things observed as close as possible to words."[22] Eschewing mere representation, the natural history aimed to replicate the thing itself on the page and thereby generate knowledge.

This epistemology, however, was not self-evident. It required work, what Steven Shapin and Simon Schaffer have called a "literary technology" that would ensure the least depreciation of direct experience to paper.[23] To accomplish this task, the new epistemology needed "virtual" rather than live witnesses to corroborate and authenticate direct experience. These witnesses would replace ancient scholars and citations, authenticating experiments and experience, but they did not need to be "eye" witnesses. Rather, the Royal Society codified "virtual witnessing" through the publication and circulation of its journal, *Transactions*.[24] Through this "literary technology," a network of like-minded scholars corroborated distant experiments and observations, creating "matters of fact" without actually seeing a chemical reaction or stepping foot in the Caribbean themselves. Thus, these virtual witnesses functioned as an epistemological safety valve on the new method of direct experience—if and only if the writer could inscribe that experience faithfully on the page.

To authenticate, the witnesses needed to believe the writer; to be believable, the writer needed to follow the Royal Society's stated and unstated literary conventions. First, the writer needed to eliminate bombast and strive for linguistic simplicity. As the Society's earliest statute on the topic dictates, "the matter of fact shall be barely stated, without any prefaces, apologies, or rhetorical flourishes."[25] Instead, the writer should strive for "primitive purity, and shortness when men deliver'd so many *things*, almost in an equal number of *words*."[26] Plain language and short, precise sentences were not just preferable but epistemologically necessary. Introductions, digressions, colorful language, and ornamentation were too "bewitching to consist with *right practice*," endangering the natural historian's ability to write the thing itself, the witnesses' ability to authenticate, and, thus, the new science.[27]

Second, the natural historian and philosopher needed to center the self in order to establish the credibility necessary for virtual witnessing. Unlike modern science, the natural history was guided by the paradigm of "truth-to-nature," which recognized that the naturalist played a critical role in selecting, comparing, and generalizing truth from nature.[28] While the Society did not issue any explicit rules on centering the self in writing, Dwight Atkinson's discourse analysis of the *Transactions* reveals that the early Society privileged writing that was author-centered in content and style.[29] Typically, authors were modest about their own abilities and deferential to other society members. For example, Sloane champions the natural history's "more certain" method, quoted above, but hedges, admitting that this is only his "Slender Opinion."[30]

Beyond humility, Society members' published writing frequently invited readers into their "affective states and psychological processes."[31] Sloane describes his desire for direct experience as an almost spiritual discontent that haunted him from his "Youth" through his adulthood: "I . . . could not be so easy. . . . These Inclinations remain'd with me sometime after I had settled my self to practice Physic in London, and had had the Honour to be Admitted a Fellow of the College of Physicians, as well as of the Royal Society."[32] Here, Sloane exposes his enduring dissatisfaction despite his "settled" state and many honors, encouraging the reader to experience his own emotions and psyche. As an obvious component of such personal content, member reports overwhelmingly employed first-person and active verbs. By centering the self in content and syntax, the author built the credibility necessary for establishing their work as "matter of fact."

Each of these literary conventions ultimately functioned to support the new science's end goal, representing the thing itself: plain language and author-centered writing demonstrated the writer's credibility; that credibility provided the foundation for virtual witnessing; and virtual witnessing corroborated the direct experience that was vital to the new science's method and epistemology. Without these rhetorical conventions, knowledge itself was at risk.

## FAILING TO REPRESENT THE THING ITSELF: SYPHILIS ELIDES AND REVISES SLOANE'S *VOYAGE*

As a trained natural historian and philosopher, and a future president of the Royal Society, Sloane's most important goal was to transmit nature itself through the page and across the Atlantic. Yet he struggles. He shifts genres, repeats himself, and employs multiple volumes and an unwieldy index—all of which undermine his credibility and the natural history's claims to truth and knowledge.[33] The passages on syphilis, however, fail consistently: case studies never establish witnesses nor describe patient's experiences clearly, and the natural history of syphilis cites ancient sources and decenters the author. In each of these ways, Sloane fails to meet the Royal Society's literary standards and, thus, to establish the knowledge that such conventions ensured. In these moments, syphilis "itself" infects, intra-acts, and revises Sloane's writing. It challenges the natural history's faith that things "themselves" could be represented and known and offers an alternative epistemology.

Most natural histories included observations on disease and many natural historians were practicing physicians, but the genre did not necessarily require formal case studies of particular people and illnesses. Sloane's *Voyage*, however, dedicates an entire section to studies of individual cases of illness. The case study, case narrative, or case history has ancient origins, but it gained popularity in Sloane's era through the work of Thomas Sydenham, the "English Hippocrates." Like the natural history, the genre of the case study codified a version of virtual witnessing through rhetorical conventions. As the Royal Society rejected classical citations for immediate experience, Sydenham rejected the theoretical or scholastic approach to medicine in favor of close attention to individual patients.[34] In turn, the case study functioned like the Royal Society's *Transactions*, circulating observations, experiments, and resources among a network of physicians with a finite pool of patients. As such, it also developed a formulaic organization including "a description of the patient, an account of his or her past patterns in sleep, diet, and exercise, description of therapies, their effects and outcomes—'complete cure,' 'relief,' or death."[35] In his case studies, Sloane opens with the patient's name, symptoms, former treatments, diagnoses, new treatments, and outcome, whether health or death. Additionally, he classifies each case according to illness in the margin of the page, for example, "Of a Tertian" or "Of a Colick." The case study strives for "narrative emplotment," translating the subjective experience of illness into a clear progression of identifiable symptoms, treatment, and outcome.[36] Thus, direct experience, witness naming, clear language, and formulaic plotting were just as important to the new medicine as they were to the natural history: they established the personal credibility required for epistemological certainty.

Yet Sloane's case studies are inconsistent. In particular, the cases of syphilis dramatically shirk these stylistic rules. First, they fail to deliver the patient's names. In every other case in his natural history, Sloane delivers the ill person's first name, providing the authentication that both he and his genre required; however, no case of syphilis receives a name.[37] Instead, he replaces them with blank spaces: "One——who had a *Gonorrhea* often" (here, gonorrhea is used as a symptom of syphilis rather than gonorrhea proper);[38] and, "One——, aged about Fifty."[39]

These blank spaces may aim to protect the sufferers' identities, to mitigate the scorn and shame of venereal disease. Nonetheless, they undermine the vital function of witness naming, subverting Sloane's credibility and virtual witnessing. Ontologically and linguistically ambiguous itself, syphilis seeds uncertainty into the heart of the natural history's drive to name, know, and possess.

Second, the case studies on syphilis refute the clear plotting required of both the natural history and the case study. In one account, an unnamed man, complaining of abdominal pain, seeks Sloane's aid. Sloane prescribes "such things as usually avail in such Cases," but the typical treatments fail, and the man grows worse.[40] From this and the patient's pain in urination, Sloane realizes he has misdiagnosed the man and begins to suspect "the Pox to be the chief of his complicated Diseases, and questioning him very hard about the Matter, he at length confess'd it."[41] The man soon dies. In another case, Sloane argues that a woman with syphilis dies after being poisoned by an enslaved woman, "either designedly to do her good" or "maliciously."[42] Still, even he does not seem satisfied with this answer: "of what, or how," she died, "I know not."[43] While his patients' deaths are not particularly strong evidence of his medical skill, both natural historians and physicians frequently included failures to illustrate their unflinching commitment to truth.

What is strange about these cases, then, is not Sloane's failure to cure, but rather his failure to narrate and emplot that failure clearly and plainly. In the first case, Sloane assumes a kind of bravado, claiming to have interrogated the man "very hard" until he elicited the truth, but that truth is not particularly certain: "the Pox," Sloane's diagnosis, may have been one of many "complicated Diseases."[44] Similarly, though Sloane attempts to blame an enslaved woman for the second patient's death, he still admits that he does not know her intentions or even whether her actions had caused the death at all. Royal Society reports were supposed to be circumspect on anything that was uncertain, but these cases vacillate between a braggadocious certainty and ambivalence. Syphilis, whose origin and evolution remain unclear, generates a plot just convoluted enough to undermine Sloane's personal and epistemological credibility.

Beyond the case studies, Sloane's natural historical observations on syphilis also reject the required rhetorical conventions, devolving into the citations that the new science rejected. In the passage cited at the beginning of this essay, where Sloane explains syphilis' origin and pandemic spread across Europe, he cites multiple sources: "I am of Opinion notwithstanding what these have said, and some other less material Passages in ancient Writers and Histories, and what *Joanne sab Arderne* has written about *Anno* 1360. And likewise what *Stow* says of the Laws of the publick Stews in *Southwark*, that this was a Distemper altogether new in *Europe, Africa*, and *Asia*, before it was brought from the *West-Indies*."[45] Sloane collects not only ornamental but also contradictory citations here. He claims to support the

Columbian origin of syphilis, but before that he cites three sources that disagree with his conclusion and absolutely no sources that support him.[46] On the one hand, these citations may not contradict the new science's empiricism. After all, Sloane could experience the fourteenth-century epidemic only through sources. He also rejects these sources instead of making them the central support of his assertions of truth. On the other hand, Sloane's syntax obscures any clear reading of the sentence. His claim—"I am of Opinion notwithstanding what these have said . . . that this was a Distemper altogether new in *Europe*"—is quite literally split in half by these contradictory sources.[47] That is, these sources are embedded in a complex sentence that resists plain plotting as much as the case studies. Between these sources, their contradictions, and complex syntax, syphilis' origin frustrates natural historical certainty.

Finally, Sloane's natural history of syphilis undermines the author-centered syntax central to the genre's epistemology. While the failures of the case studies and origin story clearly undermine Sloane's credibility, the passage that began this essay most obviously decenters the observer-writer: "*Columbus*, likewise brought into *Europe* in his Ship, and First Voyage, from these places, the Pox, which spread so quickly all over *Europe*, that *Antonio Benivenius*, who was at that time a great Practiser in Physick in *Florence* . . . tells us, that the *Lues Venerea* then beginning in *Spain*, had spread itself through *Italy*, and *France*, and that in the Year 1496, it had possess'd many People in all the Provinces of *Europe*."[48] While Sloane shifts through multiple subjects, "*Lues Venerea*," emphasized by the intensive pronoun "itself," seems to be the most important actor in the sentence. In a proper natural history, Sloane should be the subject of this sentence—to warrant his report, to establish witnesses, and, thus, to generate real knowledge. Instead, syphilis spreads syphilis. As the emphasized subject of the sentence, syphilis undermines Sloane's author-centered syntax and everything else contingent on it. Failing the requisite rhetorical conventions—witness naming, clear plotting, citations, and author-centered syntax—Sloane also fails at the natural history, undermining personal credibility, virtual witnessing, and warranted knowledge.

Yet, when read through new materialist philosophy, Sloane's consistent inconsistencies suggest not Sloane's failure but an alternative agent and writer: syphilis "itself." Together, these rhetorical failures trace syphilis' uncertain and challenging ontology: inconsistent naming, unclear nosology, inscrutable origin and evolution. In these ruptures, we glimpse but never fully access the "it" in syphilis "itself." It is this inaccessible but traceable syphilis that challenges the new science's epistemology and poses a potentially powerful alternative.

The pronoun "itself" illustrates. In addition to functioning intensively, "itself" also acts reflexively, constructing the antecedent as the subject and object of the verb. For example, in the phrase, "I spread myself out," "I" is both doing and receiving the action. As Edward T. Jeremiah argues, reflexive pronouns, thus, create a "conceptual separation" that generates "another level of the subject" that "forever eludes objectification."[49] Some part of "myself," the "I" that is spreading, must be conducting rather than receiving the action. Thus, some part must be beyond objecthood. By splitting the subject, the pronoun effectively undermines the paradigm of the singular, self-contained, human person. Yet Jeremiah's analysis should extend inversely as well: the reflexive pronoun also ruptures the object. If some part of syphilis-as-subject escapes objectification, then some part of syphilis-as-object also evades objectification, undermining the concept of the singular, knowable, controllable thing. This particular moment of reflexivity challenges the objectification of both the human and the nonhuman. Neither person nor thing can be fully transcribed, fully known, fully mastered as the natural history's epistemology and rhetorical conventions promised.

Syphilis "itself," in turn, exposes the danger of the natural history's ideology. Its faith in pure observation and certain knowledge fueled, legitimated, and absolved colonization and chattel slavery.[50] As Mary Louise Pratt argues, the natural history was "simultaneously innocent and imperial," purporting "to do virtually nothing in or to the world" while also making it available for conquest.[51] Cataloging flora, fauna, diseases, and people, the natural history identified the cash crops and potential impediments to global, imperial capitalism, including best practices for enslaving Africans. Christopher Iannini has shown how Sloane struggled to reconcile chattel slavery with his role as a natural historian.[52] Yet the natural history actively supported American colonization, the genocide of Indigenous peoples, and the enslavement of Africans.[53] Moreover, as the precursor to professional science and medicine, the natural history founded medical abuses and health-care disparities, from the infamous Tuskegee experiments to subtle and not-so-subtle eugenicist policies inflicted on people with disabilities to the disproportionate burden of our current coronavirus pandemic on people of color, those with chronic illness, and those at the intersection of these identities. Similarly, natural history's relationship to early capitalism grounded modern climate change, this pandemic, and all those likely to follow. Syphilis demonstrates the danger of the natural history's epistemology: unchecked faith in inexorable knowledge, mastery, and white, able-bodied, human supremacy endangered and continues to endanger humans and nonhumans alike.

Yet syphilis "itself" also posits a more ethical orientation to the world—uncertainty, ambiguity, and ambivalence. Those blank spaces and unclear plots in Sloane's case studies undermine his credibility, but they also reveal just how much the natural history could not affix with the observing eye and transcribing pen—both the utterly inaccessible experience of the microbe and the ineffable and exceptionally idiosyncratic experience of human pain and suffering. Whereas the natural history and its successors assume that complete knowledge and mastery is possible, inevitable, and desirable, syphilis' ontological uncertainty asserts that all knowledge is partial, contingent, intra-active, and difficult. Such uncertainty and contingency would serve as a more ethical framework for responding to human and nonhuman intra-actions, including climate change and illness.

Perceiving syphilis' or any disease's agency should not be radical. It is intuitive. We both feel and know how the coronavirus has shaped our lives, regardless of "intentionality." What's radical is to pretend that this isn't agency. Yet we cannot simply perceive nonhuman agency; we must attend to it. This argument offers a method for listening to such nonhuman agency: closely examine a disease's ontology and etiology; observe generic patterns; trace generic ruptures that parallel ontology and etiology; and, finally, listen carefully to what those ruptures reject and propose. Diseases like syphilis are destructive. They kill and maim and wreak havoc. Yet they, like all life, exist to survive and, therefore, produce. Syphilis rejects and undermines Sloane's text and the epistemology of the natural history, but its ontological being and textual presence also produces and proposes an alternative orientation to the human and nonhuman world. When we search for nonhuman agency, especially the nonhuman agency of an "enemy" like disease, we cannot simply critique. We must listen, and we must coproduce. Our world needs it.

## PEDAGOGICAL IMPLICATIONS: SYSTEMS MAPPING FOR UNCERTAIN COMPLEXITY

Like many, I don't teach in my research field. Instead, I teach in an interdisciplinary studies program that enables a diverse group of students to complete individualized degrees. Many students study business and psychology, or health sciences and sociology, or communications and political science. What unites such disparate students and coursework is a commitment to solving problems that span disciplines and their siloed objects of study, method, and epistemology—problems like the coronavirus pandemic.

One tool that I use to help students solve problems—to understand complexity, appreciate ambiguity, consider stakeholders (including nonhumans), and, ultimately, to build solutions and alternative futures—is systems-mapping. While techniques vary, all systems maps begin with a problem. The thinker then considers and illustrates the many factors—physical, conceptual, human, nonhuman, systemic—that contribute to and result from that problem, using arrows, lines, and feedback loops to visualize the problem fully through time and space. Thus, systems-mapping forces problem-solvers to focus on the problem itself rather than on what their expertise allows them to see and address. It provides a tool for abstract, structural thinking that magnifies rather than reduces complexity.

In my fall 2020 senior capstone class, all students began with the problem of the pandemic, using systems-mapping first to explore the pandemic's complexity and then to narrow to subproblems to research and solve. We began with a full-class brainstorm, free-associating as many factors as we could imagine that had contributed to the rise of the pandemic. We used the white board function on Zoom, but had we been face-to-face, I would have encouraged them to yell out ideas as I scrambled to write them quickly on the board. By the end of the exercise, we had generated an overwhelming, messy morass—the raw material for a true systems map.

Equipped with this brainstorm, I asked students to organize this mess into a logical systems map, using arrows and visual movement to structure time, order, relationship, and feedback among these many factors. In our next class, we worked together in small groups to synthesize these maps. Eventually, we debated, compromised, and collaborated as a class on a full map that visualized the system. Students used this expansive, collaborative map both to imagine the complexity of the pandemic and to identify smaller problems to investigate and intervene in over the course of the semester.

Such systems-mapping enables the ambivalent, uncertain, and yet productive orientation to the world that I argue syphilis proposes in Sloane's natural history. First, the systems map's emphasis on the problem forces problem-solvers to recognize their own limitations and biases, whether disciplinary (like the natural history's methodological faith in empiricism) or otherwise. Second, it instructs thinkers to consider all factors and stakeholders that contribute to a problem, including nonhumans like organizations, ideas, and microbes. Third, it emphasizes complexity and the need to parse complexity without encouraging simplicity, without assuming that that complexity can be fully understood. And, finally, the systems map aims for action—for intervening and hopefully solving the problem it analyzes.

The systems map listens to syphilis and rejects the natural history's model of knowledge production, emphasizing disciplinary and personal humility, nonhuman factors and stakeholders, complexity and ambiguity, and the need for solutions, however partial and incomplete. To build assignments that enable students to value complexity and persist toward solutions is to remind them that they themselves are agents in their learning and communities. Setting course descriptions, learning objectives, and even my own disciplinary expertise aside, that's all I really want as a teacher.

## NOTES

1. Anthony Fauci, "Dr. Fauci: The Virus Decides When the Country Will Open Back Up," *CNN*, April 10, 2020, video, 2:26, https://www.cnn.com/videos/health/2020/04/10/dr-anthony-fauci-coronavirus-response-newday-vpx.cnn.
2. Hans Sloane, *A Voyage to the Islands Madera, Barbados, Nieves, S. Christophers and Jamaica* (London: 1707), ii.
3. Jane Bennett, *Vibrant Matter: A Political Ecology of Things* (Durham, NC: Duke University Press, 2010); Karen Barad, *Meeting the Universe Halfway: Quantum Physics and the Entan-*

*glement of Matter and Meaning* (Durham, NC: Duke University Press, 2007); Graham Harman, *The Quadruple Object* (Alresford, UK: Zero Books, 2011).

4. David M. Morens and Anthony S. Fauci, "Emerging Pandemic Diseases: How We Got to COVID-19," *Cell* 182, no. 5 (2020): 1077.
5. The Columbian origin of syphilis was contested in Sloane's era and continues to be debated now. For an overview of both the historiographical and scientific debate, see Kristin N. Harper et al., "The Origin and Antiquity of Syphilis Revisited: An Appraisal of Old World Pre-Columbian Evidence for Treponemal Infection," *American Journal of Physical Anthropology* 146, no. S53 (2011): 99–133.
6. Kelly Wisecup, *Medical Encounters: Knowledge and Identity in Early American Literature* (Amherst: University of Massachusetts Press, 2013), 173–180.
7. Sloane, *A Voyage*, cxli, ciii, xciii. Sloane refers to patients with gonorrhea and yaws by name, unlike those with syphilis, suggesting that he understood the three diseases as distinct.
8. Jonathan Hutchinson, "An Address on Syphilis as an Imitator," *British Medical Journal* 1, no. 953 (1879): 541.
9. Kathryn Montgomery Hunter, *Doctor's Stories: The Narrative Structure of Medical Knowledge* (Princeton, NJ: Princeton University Press, 1991), 45.
10. Hutchinson, "Address on Syphilis," 541.
11. Dorothy H. Crawford, *Deadly Companions: How Microbes Shaped Our History* (New York: Oxford University Press, 2007), 124–130.
12. Lola V. Stamm, "Global Challenge of Antibiotic-Resistant *Treponema pallidum*," *Antimicrobial Agents and Chemotherapy* 54, no. 2 (2010): 583–589. While penicillin remains effective in treating syphilis, *Treponema pallidum* has evolved resistance to the second-line antibiotic macrolides.
13. Harper et al., "The Origin and Antiquity of Syphilis," 99–133.
14. Early modern syphilis seems to have been significantly more virulent than the modern disease, suggesting some mutation. Moreover, some scientists have suggested that syphilis mutated again between the eighteenth and nineteenth centuries. See E. H. Hare, "The Origin and Spread of Dementia Paralytica," *Journal of Mental Science* 105, no. 440 (1959): 594–626.
15. Barad, *Meeting the Universe Halfway*, 128.
16. Harman, *Quadruple Object*, 47.
17. Julian Yates, "Represent," in *Veer Ecology: A Companion for Ecological Thinking*, ed. Jeffrey Jerome Cohen and Lowell Duckert (Minneapolis: University of Minnesota Press, 2017), 132.
18. Yates, "Represent," 131.
19. Yates, "Represent," 127.
20. Thomas Sprat, *The History of the Royal-Society of London*, 2nd ed. (London: 1702), 61. The natural history, as both a genre and an epistemology, has been studied widely. Notable analyses that emphasize rhetorical conventions and inform this study include Susan Scott Parrish, *American Curiosity: Cultures of Natural History in the Colonial British Atlantic World* (Chapel Hill: University of North Carolina Press, 2006); Christopher Iannini, *Fatal Revolutions: Natural History, West Indian Slavery, and the Routes of American Literature* (Chapel Hill: University of North Carolina Press, 2012); Britt Rusert, *Fugitive Science: Empiricism and Freedom in Early African American Culture* (New York: New York University Press, 2017); and Greta LaFleur, *The Natural History of Sexuality* (Baltimore, MD: Johns Hopkins University Press, 2018).
21. Sloane, *A Voyage*, unpaginated preface.
22. Michel Foucault, *The Order of Things: An Archaeology of the Human Sciences* (New York: Vintage Books, 1994), 132.
23. Steven Shapin and Simon Schaffer, "Seeing and Believing: The Experimental Production of Pneumatic Facts," in *Leviathan and the Air-Pump: Hobbes, Boyle, and the Experimental Life* (Princeton, NJ: Princeton University Press, 1989), 22–79. While Shapin and Schaffer

focus on natural philosophy, the natural history was also built on direct experience and its inscription. Moreover, both fields were codified by the Royal Society.

24. Shapin and Schaffer, "Seeing and Believing."
25. "Statutes of the Royal Society Enacted in 1663," *A History of the Royal Society with Memoirs of the President, Compiled from Authentic Documents*, ed. Charles Richard Weld (London: 1848), 527.
26. Sprat, *The History of the Royal-Society*, 113.
27. Sprat, *The History of the Royal-Society*, 112.
28. Lorraine Daston and Peter Galison, "Truth-to-Nature," in *Objectivity* (Princeton, NJ: Zone Books, 2007), 55–114.
29. Dwight Atkinson, "The *Philosophical Transactions of the Royal Society of London*, 1675–1975: A Sociohistorical Discourse Analysis," *Language in Society* 25, no. 3 (1996): 339.
30. Sloane, *A Voyage*, unpaginated preface.
31. Atkinson, "The *Philosophical Transactions*," 339.
32. Sloane, *A Voyage*, unpaginated preface.
33. Iannini, "Strange Things, Occult Relations: Emblem and Narrative in Hans Sloane's *Voyage to . . . Jamaica*," in *Fatal Revolutions*, 35–74. While Iannini notes many of Sloane's generic oddities, emphasizing how chattel slavery undermined the natural historian's ability to represent "things," his study of Sloane's genre and rhetoric does not extensively analyze the case studies.
34. Peter Anstey, "The Creation of the English Hippocrates," *Medical History* 55, no. 4 (2011): 457–478.
35. Londa L. Schiebinger, *Plants and Empire: Colonial Bioprospecting in the Atlantic World* (Cambridge, MA: Harvard University Press, 2004), 160.
36. Hunter, *Doctor's Stories*, 45.
37. Even yaws (ciii) and gonorrhea (cxli), which were similar and as yet undifferentiated diseases, received names. On medical differentiation, see M. Tampa et al., "Brief History of Syphilis," *Journal of Medicine and Life* 7, no. 1 (2014): 4–10.
38. Sloane, *A Voyage*, xciii.
39. Sloane, *A Voyage*, ci.
40. Sloane, *A Voyage*, ci.
41. Sloane, *A Voyage*, ci.
42. Sloane, *A Voyage*, cxxi.
43. Sloane, *A Voyage*, cxxi.
44. Sloane, *A Voyage*, ci.
45. Sloane, *A Voyage*, ii–iii.
46. One of these three sources is cited in a preceding passage, not quoted here. See note 5 on the Columbian origin hypothesis.
47. Sloane, *A Voyage*, ii–iii.
48. Sloane, *A Voyage*, ii.
49. Edward T. Jeremiah, *The Emergence of Reflexivity in Greek Language and Thought: From Homer to Plate and Beyond* (Leiden: Brill, 2012), 31.
50. See, especially, Beth Fowkes Tobin, *Colonizing Nature: The Tropics in British Arts and Letters, 1760–1820* (Philadelphia: University of Pennsylvania Press, 2004), and Iannini, *Fatal Revolutions*. On the natural history's ideological influence on other popular genres, see Rusert, *Fugitive Science*, and LaFleur, *Natural History of Sexuality*. Finally, on how Creoles, women, Africans, and Indigenous peoples undermined the colonial power of the natural history and medicine, see Parrish, *American Curiosity*; Rusert, *Fugitive Science*; and Wisecup, *Medical Encounters*.
51. Mary Louise Pratt, *Imperial Eyes: Travel Writing and Transculturation* (New York: Routledge, 1992), 33.
52. See note 33.
53. See note 50.

# Part Three

## EIGHTEENTH CENTURY + BLUE HUMANITIES

# 6

# SHORE/LINES

## Drawing Environmental Change on Eighteenth-Century Prince Edward Island

CLAIRE CAMPBELL

## MAKING LANDFALL: AN INTRODUCTION

In early September 2019, Hurricane Dorian careened up the Atlantic coast. Like most fall storms, it had gathered strength as it moved through the Caribbean before arcing north and crashing, slowing as it did so, into the Maritimes. But as Dorian crossed over Prince Edward Island, it caused more destruction than most storms, eroding up to six feet of coastline in places, toppling 80 percent of tree cover in others. Dorian was both historic and foreboding; storms like this, after all, will become only more common with climate change and oceanic warming.[1]

The fall storm season of hurricanes, cyclones, and tropical depressions along the Atlantic seaboard has generated its own catalog of images. There is the satellite view, with its compelling, almost artistic aesthetic, a coherent composition anchored by the storm's eye, with a palette contrast of swirling white against darker blue. There is the much tidier graphing of the predicted storm track, annotated with place names, time stamps, and wind speeds, the full arc of the storm visible at a glance (figure 6.1). And finally, there is the human portrait, on the ground—the hapless meteorologist yelling over the gale into the camera, footage of people or cars sloshing through feet of water, trees bent double by pounding wind and rain. That nearly every storm update requires sampling from all three types of images is an implicit admission that the neat iconography of weather maps cannot convey the deeply physical experience of the storm itself. At the same time, these real-time updates create a biographical narrative of the storm as a singular event, not as one of many with more to come.

But there will be more. Climate change is the environmental issue of our age, and scientists have identified P.E.I. (Prince Edward Island), Canada's smallest province, as one of the country's most important harbingers of that change. The island's dynamic northern shore, which is made up of sandy barrier islands, coastal dunes, and tidal deltas, is of particular concern, but the capital city of

Figure 6.1 Hurricane Dorian made landfall over North Carolina, September 6, 2019. National Oceanic and Atmospheric Administration (NOAA).

Charlottetown is also vulnerable to storm surge and flooding.[2] Seen from above, arterial rivers threaten to pierce the island through, as the soft red sandstone dissolves outward with the tide, bleeding into the ocean. A small island (only about 140 miles long) lying low in an increasingly stormy ocean seems an apt metaphor for our state of vulnerability in an era of climate change. Like most aspects of the climate crisis, though, the study of the island's coastlines has been scientific and statistical in tone, direction, and presentation. What can the environmental humanities bring to this most pressing question? What can historic texts and images tell us about how slivers of shoreline have been used, understood, and changed? Why does it matter what and how we see?

The practice of storm tracking—the presentation of environmental data generally—reveals the need for cartographic literacy: the ability to read maps critically and situate them in historical context. The prospect of questioning a map often comes as a revelation to my students, who know maps only through the ubiquitous Google, where water is a single matte shade of blue and Starbucks are located more precisely than shorelines. For them, as for many, a map is a convenience, and geography a functional fait accompli. But places and their representations are both constructions, the work of human hands intended for human use. A cartographic biography—telling the story of a place through maps that "share distinct sinews of intent and rationale"—is a valuable way of approaching environmental history and the querying spirit of the environmental humanities.[3] Historical maps show history and nature, alike in motion (perhaps ironically so), being made and remade. They invite us to ask, Who was here? What did they think of it? How did they live here? How did they change it?

In thinking through these questions, we might see landscape as a *palimpsest,* a term borrowed from the study of medieval texts. A palimpsest, in the original meaning, refers to a manuscript whose ink has been scraped off so the parchment can be reused. Even after fresh ink is written over the old, the older inscription remains visible. If we consider the natural world as the parchment in this metaphor, then the layers and lives of past occupants are the ink that has been erased yet remains visible: the marks they made on the land, the paths and grooves they wore and carved. (When explaining the palimpsest in class, I ask students if they used the erasable Paper Mate pens in elementary school—the ones whose erasers never entirely worked, so there was still blue ink amid the eraser rubbings. The analogy works every time.)

Prince Edward Island is one of the easier places to read as palimpsest, in part because we want to. At the end of the nineteenth century, the island was tagged as "the Garden of the Gulf" for its small family farms draped over arable soils, a setting unusual in the north Atlantic. Soon thereafter, Lucy Maud Montgomery brought a gentle, pastoral picture of gabled farmhouses, orchards and hedgerows, and red-dirt roads into the global imagination, which Kate Scarth takes up at greater length in chapter 11. Perennial provincial tourism campaigns have sustained this image with the promise of return to such a time and place. The island, thus, occupies something of a paradox in the public imagination, where the nostalgia for rural permanence persists alongside anxiety about rapid coastal change. In fact, this image of the island, with all its presumptions of landward cultivation and rural stability, is an artifact of the long eighteenth century. This particular image of the island begins to take shape in the maps that follow, especially after 1760, as British authorities recast the island as linear, arable, and property.

But this is not the only or even the primary image of the island that emerges from the age of the eighteenth century. Known as Epekwit'k ("lying on the water") or Kjiktúlnu to the Mi'kmaq, Isle St. Jean to the French, Prince Edward Island—renamed only at the very end of the century for the fourth son of George III—sat at the crossroads of maritime traffic and imperial conflict, of Indigenous, European, and colonial interests.[4] Accordingly, for most of the eighteenth century, maps of Prince Edward Island reflected extensive interest in and growing knowledge of inshore and coastal or littoral zones. In this way, they offer an important resource for measuring and marking coastal change over time, and returning our attention to shorelines in our public discussions about sea level rise and climate change. If we have turned away from the water in the road-riven twentieth century, the water is now returning to us.[5]

## THE CARTOGRAPHIC RECORD AND ENVIRONMENTAL HISTORY

In February 1746, the *Gentleman's Magazine* of London quoted the preeminent royal French cartographer Jacques-Nicholas Bellin as saying, "Geography is so necessary to illustrate history that they ought to be inseparably connected."[6] Bellin's choice of word is important, for "illustrate" here means to supply visual representation as well as spatial explanation. Maps are unusual as historical sources for their Janus-like view to historical time, multifaceted form, and efforts to situate human motivation in greater-than-human space. They can allude simultaneously to past, present, and future; they document older interventions and current practices, and declare—and legitimate—agendas and anticipated presence.

As we track chronologies and changes over time, we also have to think about choices of scale and perspective. Meanwhile, we absorb the numerous means of representation at work in these maps as visual, textual, and physical artifacts. Maps speak through symbolic means (e.g., place names, the cartouche) and material record (e.g., land use, property). They are authored by forces of politics, capital, and belief, as well as (if not sometimes more than) science or curiosity. As we identify farmland or harbors, roadways by land or shoals at sea, we see quests for possession, "improvement" and profit, surety and order. As Matthew Edney observes, "As a mimetic picture, the map presents the world to the reader's sight, permitting the reader to view and observe not just the map but, through the map, the world. The reader can, in most instances, take hold of the map as an artifact, turn it whichever way they desire, and physically dominate it. And through the map, they can imagine they dominate the world they depicted. . . . The physical domination of the map promotes the intellectual conviction that one dominates the depicted world."[7] And yet, beneath the increasingly confident claims of the imperial record, we see the persistence of older knowledge and environmental features that require negotiation and adaptation. In short, the map is a rich, complex, but deeply complicated record of human engagement with the natural world. Yes, Monsieur Bellin, geography does indeed illustrate history. But then, history also illuminates geography.

This is a premise of environmental history. Environmental history considers our relationship with the natural world in the past: how nature informed human choices and actions and how these actions, in turn, affected the environment that surrounded them. Tracking such changes seems particularly important in the Anthropocene, a period defined by the indelible mark of human intervention on the physical world. Maps from the nineteenth century onward might seem to be more useful here: they seem more precise, more "accurate," and therefore more suited to comparison and georeferencing (the practice of overlaying and stretching older maps to a baseline map to measure change). But to understand the forces that have given rise to ripped-from-the-headlines environmental crises, such as Hurricane Dorian, we must reach farther back in time.

Environmental history needs to engage more fully with the eighteenth century in order to understand the shape of the Anthropocene and the deeper roots of human thought and action that have given rise to the environmental dilemmas of our day. Older maps defy ArcGIS and its uncritical teleology of Google Maps, requiring instead a historical imagination to reach *into* the text to extract meaning beyond data points. The meaning is there: coastal knowledge was critical for Indigenous and European peoples alike, whether for navigation from sea, harvests on shore, or the siting of settlements on land. But this poses a profound challenge to the very idea of a map as we know it: How are we to represent a changeable, irregular littoral, in flat, fixed form?[8]

Islands, as Adam Grydehøj argues, offered colonizing powers exceptional opportunities to craft *legible geographies*: "Island geographic legibility enhances a place's historical comprehensibility, making it easier to date, imagine, essentialise, and mark arrival, conquest and colonisation."[9] This phrasing speaks to both message and medium: the historical processes that drove map-making in this period, and the practice of inscription in textual and visual form. Prince Edward Island is a vital place to explore dynamics of environmental definition. The editors of *Time and a Place: An Environmental History of Prince Edward Island* echo Grydehøj's concept of legibility: "Viewed from a distance, it seems encouragingly self-contained, an unambiguously bounded, curiously homogeneous space, a place where nature has largely been tamed and where time has meshed its people with their environment."[10] But why do we see it this way? How have we made it this way? These are the stories that environmental historians seek to unravel. And this is why the eighteenth century deserves more attention.

At a time when territorial boundaries were fluid and contested, the more recent fiction of island insularity fades into larger shapes of ecological zone (coastline/littoral), waterbody (Gulf of St. Lawrence), or geopolitical context (Mi'kma'ki, French Acadie, British Atlantic). While there is a strong tradition of historical geography in Atlantic Canada, and there have been some important explorations into coastal environmental history on the Atlantic Seaboard, attention to the eighteenth century—when coastlines were so crucial to the course of human events—calls out for more. The geographical literature, with its emphasis on land (or sea) *use,* also gives us space as humanists to consider the effects of certain practices of imagination, understanding, and representation.[11] What are the conventions that render unfamiliar places intelligible, and why are these accepted? Why does the gaze from above, for example, graft so easily into the imperial imagination? Why does a terrestrial audience believe it can understand maritime space?

Historical maps are widely popular as wall-hanging because they present time and place in ways that are accessible enough to be understood, different enough to be intriguing, visually alluring to the point of art, and imprecise (by our estimation) enough to be entertaining. But we owe them more. Reading maps requires both close scrutiny and a critical distance to recognize how they work as texts, in representing the natural world and in circulation, and as intellectual commodities. But where did they come from? Who and what were they for? Most of the maps shown in this chapter were published and printed for sale as part of the burgeoning literary and cartographic marketplace of the eighteenth century and were aligned to the imperial identity of their intended audience. Thanks to expanding digital collections, such as Island Imagined at the University of Prince Edward Island, we can access more sources than ever before; we can linger over details, breaking apart the completed map and its *legible geography* to query presence, absence, and meaning.[12] That said, digital collections raise new questions, not just about maps themselves, but also about how we write about them. Indeed, this essay was written during the COVID-19 pandemic, when researchers could access archival materials *only* online. What does this mean for our research practice? What does it mean, for example, to assemble a story from images already designated by an unknown archivist as worthy of digitization? Perhaps only the acknowledgement that all record-keeping and history-making is selective, and digital records are no different. On the other hand, if digital access allows us to sample from archives thousands of miles apart, is this integrative (and a responsible conservation of carbon), or idiosyncratic? How do we make use of these collections in careful and judicious ways?

## ISLE SAINT-JEAN AND THE ARCHIPELAGO OF ACADIE

At the start of the eighteenth century, Isle Saint-Jean was generally featured as part of an Acadian archipelago of French colonies in the Gulf of St. Lawrence, along with Acadie, Isle Royale (later Cape Breton), *les isles de la Madeleine*, and parts of Terre-Neuve (Newfoundland). This was partly because of the small size of the island but, more importantly, because of its relational place in a regional economy. Strategically, the gulf was the entrance to New France and the continental interior, but these islands and shorelines were more prized as bases for the transatlantic fishery. Maps carefully delineate offshore banks, such as the Banc des Orphelins west of Isle Saint-Jean, and inshore fisheries; indeed, the undated *Carte du Fleuve et du Golfe de St. Laurens* marks the entire north shore of the island as such a zone (figure 6.2).[13] These maps were intended not for fishing captains and fleets but, instead, for outward-looking claims to areas of maritime wealth.

There is also greater detail to sandbars, shallows and shoals, and other hazards, particularly at the far ends of the island. Sketching banks under the water as if islands rest above them, crowding the gulf, is disorienting to us today, accustomed as we are to sharper delineations between blue and not-blue. But the visual effect is to draw the eye seaward, away from land, creating greater-than-terrestrial footprints: akin to satellite images that show the continental shelf lying beneath and

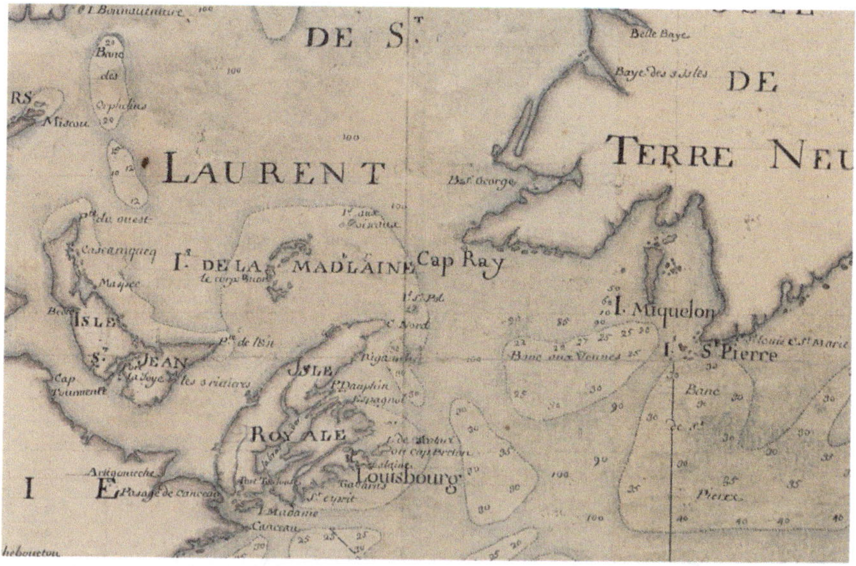


Figure 6.2 *Carte du Fleuve et du Golfe de St. Laurens*, detail. This probably dates after 1713, as it shows the Dauphin Gate at Louisbourg with a pennant. The French began constructing a fortress at Louisbourg after surrendering mainland Acadie to the British in the Treaty of Utrecht in 1713. Courtesy of Bibliothèque Nationale de France (BNF).

behind solid land. Just as that shelf becomes more and more valuable with offshore drilling, it was the more valuable real estate of the fishing age.

The places that first appear on early French maps mark older ports that likely had been known to Basque and French fishers for over a century. *Isle St-Jean, isles de Miscou, isles de la Madelaine, coste d'Acadie, isle Royale* (1713) notes a few unnamed ports and channels for *chaloupes* along the north shore, along with Havre St. Pierre, which was the most populous settlement on the island (figure 6.3). Several names indicate a Mi'kmaq presence: (Cascamquesques/Keskamskek, Macpec/Maqpa'q, Caccocpiche/Katewpijk), as these bays and estuaries represented crucial sites of seasonal harvests, whether of birds, sea mammals, fish, or shellfish.[14] But the map does not seek or serve to acknowledge Mi'kmaq people or territory; it is aligning the vernacular knowledge and shoreline footholds of fishermen into a more formal cartographic arrangement and a larger imperial network. In assembly and purpose, the epistemological register is firmly European.

Climate change cannot help but frame our reading of these maps. They signpost places to look for older cultural sites, where the landscape has invited and sustained similar use over time. As archaeologist Peter Pope said of northeastern Newfoundland's "French shore," "A good place to land a boat is a good place to

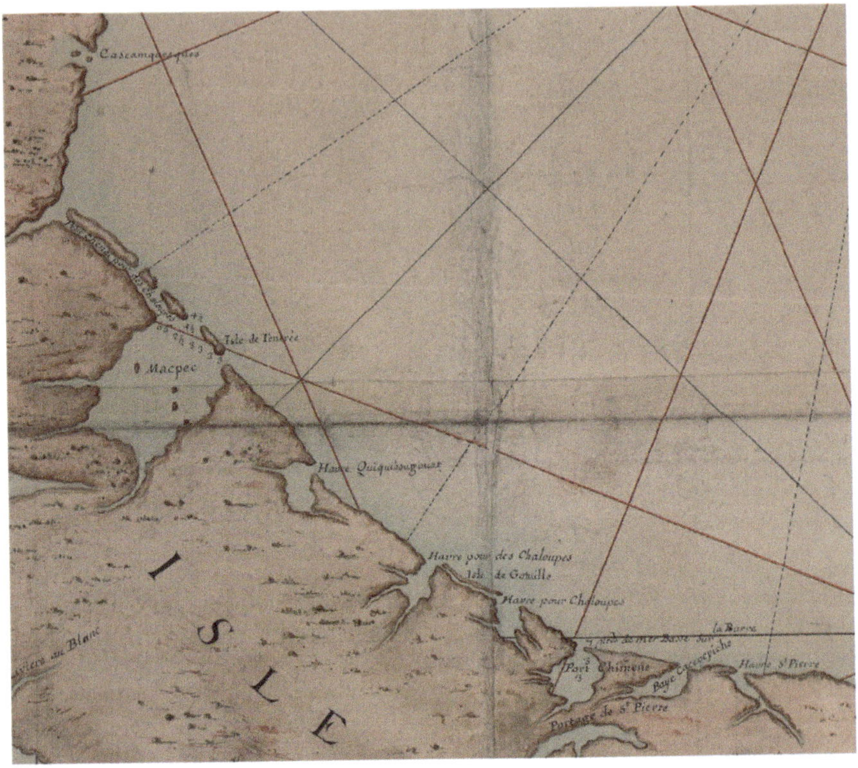


Figure 6.3 *Isle St-Jean, isles de Miscou, isles de la Madelaine, coste d'Acadie, isle Royale*, 1713, detail. BNF.

land a boat, and landing stages (fishers' wharves) have therefore been rebuilt in the same locations for centuries."[15] This is, of course, a palimpsest. At the same time, there is an urgency here, because the parchment is no longer holding together: these are the sites crumbling into a rising sea.

Storm surge and rising sea levels also give us a different view of the scale of interior waters like rivers. Seventeenth- and eighteenth-century maps exaggerated the size and detail of water routes because of their utility and familiarity as the primary means of travel—some, for example, making *la rivierère Nord-Est* nearly a full and straight channel between Northumberland Strait and the gulf. This is an important reminder for us, land- and asphalt-weighted, that earlier movement and measurement was immersed differently in the topography. But these wide waterways seem eerily prophetic when today, engorged by intense storms and sea level rise, they threaten to fracture the island into a micro-archipelago. It is as if the older maps are becoming more accurate with time.

Figure 6.4 *Carte de l'Isle St. Jean et d'une partie de l'Isle Royalle et de l'Accadie*, 1690, detail. This map is likely misdated, as it shows the fort at Port La Joye, established in 1720. BNF.

Acadian farming communities initially remained concentrated on the tidal marshes around the Bay of Fundy, but farm lots fronting freshwater on the island appear in the *Carte de l'Isle St. Jean et d'une partie de l'Isle Royalle et de l'Accadie* (figure 6.4). Port La Joye had been established as the major government presence on the island in 1720, as French authorities encouraged Acadians to migrate from the mainland of Acadie, now held by the British as Nova Scotia, to the islands of Saint-Jean and Isle Royale. Acadian agriculture relied on draining and diking coastal marshes for pasturage and cultivation, a practice that deeply shaped a terroir that would persist even after British occupation. Though arguably a kinder and gentler form of land-making than clearing forested uplands, as the British did, diking and *aboiteaux* still represented an active intervention in the landscape.[16]

Acadian farming also belonged to a seasonal rhythm of what we call occupational pluralism, where small-scale farming complemented subsistence fishing.[17] On the north shore this map shows Katewpijk, now translated literally as Havre aux Anguilles (eels), a mainstay of Mi'kmaw and northern European diets. Here the brick-red hatching visually exaggerates the stretches of shoreline around the Acadian communities, including Port La Joye, St. Pierre, and Trois-Rivières, in contrast with the interior to the west. As Sieur de la Roque, a census-taker and observer, wrote in 1752,

> There has for a long time been a mistaken belief, founded on a lack of experience of the conditions, that the settlers who follow the fisheries, neglect the cultivation of the soil. The harbours of Saint-Pierre and of L'Acadie are a certain proof in evidence to the contrary. Witness the extensive clearings which the settlers have made in those places, and I venture to affirm that the fishery is an incontestable means of promoting the culture of the soil, because it enables settlers to employ domestics, and to raise cattle and livestock.[18]

While affirming the prevailing environmental ethic of colonial settlement—an industrious "culture of the soil" that would be relentlessly idealized and promoted throughout North America—de la Roque nonetheless confirms the value of complementary seasonal work and the ecological productivity of coastal landscapes.

## AN ISLAND CONTESTED

In the 1740s and 1750s, the gulf became a major theatre of the seemingly endless conflict between the European powers. With mainland Acadie lost to the British in 1713, and with both powers (and their colonies) anticipating further conflict, it was in France's interest to bolster the cartographic profile of its remaining Atlantic colonies. Jacques-Nicholas Bellin, the royal hydrographer and most prominent cartographic authority of the French empire, did this in a number of ways in *Acadie, Isle Saint Jean and part of Isle Royale with the Baye Francoise* (1740).[19] The focus here is not cultivation or harvest, but possession and strategy. There is significantly less ecological detail—no coloring or hatching, a cruder rendering of rivers—but more crowding in *habitations* and pathways between them, most prominently the *portage de Cobeguit* linking Tatamegouche on the Northumberland Strait to the Bay of Fundy (figure 6.5).

This emphasizes the capacity for circulation between small communities—in, for example, supplying food from Fundy or Saint-Jean to the fortress at Louisbourg on Isle Royale. The map also labels "cabanes savage" at Malpeque and across the strait, siting the positions of a historic ally. Such markers are relatively rare, and, where they appear, they are a statement not of territorial acknowledgement but of European concerns. Today, however, they remind us of the long history of intrusions into Mi'kma'ki, as Nova Scotia grapples with issues of environmental injustice in precisely these communities.[20]

The French presence in the gulf was a source of perennial anxiety among American colonists. Fortress Louisbourg shadowed the Atlantic shore, guarding French movement to the Great Lakes and the trans-Appalachian interior, where Americans desperately wanted to expand. A colonial force led by Massachusetts and supported by the British navy successfully besieged Louisbourg in 1745, only to see it returned to France in a 1748 peace. The lack of familiarity with the area

Figure 6.5 *Acadie, Isle Saint Jean and part of Isle Royale with the Baye Francoise,* 1740, detail. Courtesy of Norman B. Leventhal Map & Education Center, Boston Public Library.

is apparent in *Island of St John . . . in the bay of St Lawrance, subject to King George by the capitulation of Lewisburg 17th June 1746,* in which Prince Edward Island looks less like an island than like a segment of lower intestine (figure 6.6). Its dating is as approximate as its depiction. The most distinctive feature is the exaggeration of the southern harbor, past the small garrison at "Port La Joy," and the sense of a back-of-the-napkin sketch.

Such surveys are marked by their economy of information, measured by and for approaches by sea. The *PLAN of the HARBOUR and Rivers of PORT LE JOYE in the Island of St. Iohn* offers more detail in sandbars and channel depths, but only a handful of houses, no farms or cultivated land, and no signs of life at the garrison—a rather convenient pretext for British occupation (figure 6.7).

In contrast, *Le Port de La Joye dans l'Isle S. Jean* (1750), a map of the exact same location, shows farm lots thick near the water linked by roads as well as clusters of houses, wharfs, and named points (figure 6.8).

*Le grand dérangement* saw the expulsion of thousands of Acadians, first from the mainland and then the islands of Saint-Jean and Royale. Though some families managed to remain, Isle Saint-Jean was occupied by British forces from 1758.

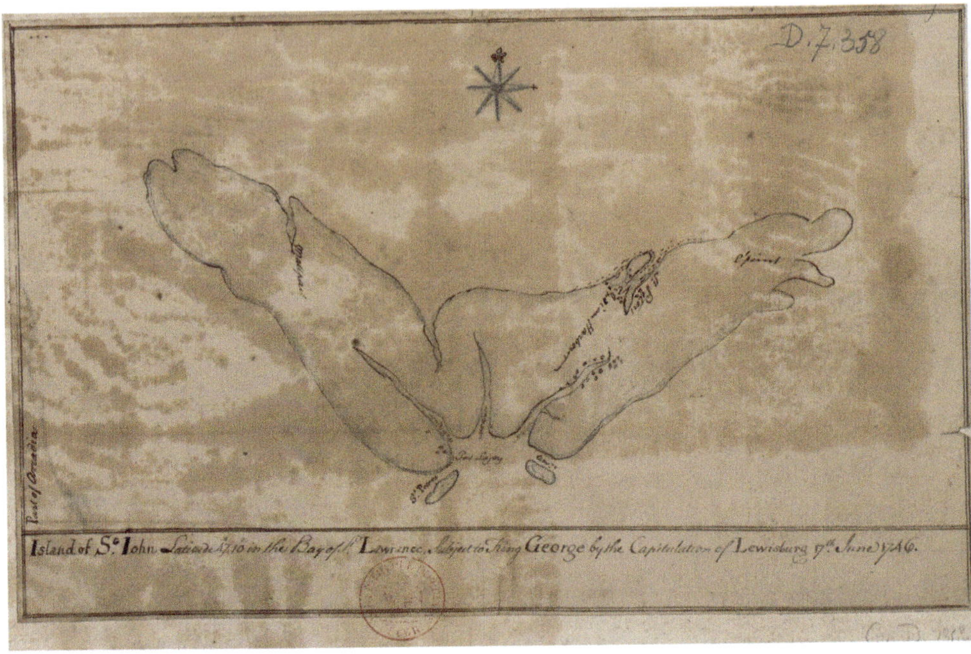


Figure 6.6 *Island of St John . . . in the bay of St Lawrance, subject to King George by the capitulation of Lewisburg 17th June 1746*. BNF.

Figure 6.7 *PLAN of the HARBOUR and Rivers of PORT LE JOYE in the Island of St. Iohn*, 1750–65. Norman B. Leventhal Map & Education Center, Boston Public Library.

Figure 6.8 *Le Port de La Joye dans l'Isle S. Jean*, 1750. BNF.

W. I. Hebert's *Plan of Ile Saint-Jean / the Island of Saint John* (1760) is a fascinating glimpse of the island at a pivotal moment of displacement and remaking (figure 6.9). The map is a hybrid of French names in the original and translation (e.g., "Village of St. Peter") as well as new British toponyms. Visually, Hebert flattens any topographical variance into a gentle green, with a system of roads between major settlements, extensive cultivation on both the northern and southern shores, mouths of freshwater, and relatively little coastal detail.

The soft watercolor palette and the pastoral details gloss over the traumatic violence of the preceding decade and present instead a garden awaiting proper

Figure 6.9 W. I. Hebert, *Plan of Ile Saint-Jean / the Island of Saint John*, 1760. Courtesy of Island Imagined/UPEI and Public Archives and Records Office, Prince Edward Island.

stewardship: part of a British tendency at this point to overestimate the island's agricultural capacity in crafting an image infinitely more appealing on both sides of the British Atlantic.[21] As Robert Stewart, then planning his bid for a property on the island, wrote to George Washington in 1764,

> This Island was deem'd the most pleasant fertile and best Cultivated in French America, it's Coasts abounds with immense quantitys of Fish, has two very fine Harbours vizt St Peters and Port Joy, is extremely healthy and a great deal of Land clear'd and laid out into fine Farms, the French having upwards of fifteen hundred Families who were Settled there for many years we propose that the environs of these Harbours on which the principal Towns must stand, likewise all the improv'd Lands shall be equally divided amongst us—Some who are well acquainted with this Island seem quite extravagant in it's praises . . . certain it is this Island is deem'd very valueable.[22]

## AN ISLAND APPRAISED

France surrendered Isle Saint-Jean along with most of the rest of its Atlantic territory in 1763 at the end of the Seven Years' War. Well before the treaty ink was dry, Samuel Holland had begun work on what would become the most famous map of the island and, arguably, the ultimate expression of cartographic possession.[23] Holland had been directed by the Board of Trade and Plantations to divide the island's interior from stem to stern in anticipation of large-scale British settlement. Holland laid out sixty-seven lots, which were awarded to well-placed British officials, and placed himself in Lot 28, as per *Plan of Prince Edward Island showing the names of Proprietors* (1767). The absentee landowners were then charged with settling tenant farmers on their island holdings. Holland's survey exemplified mapping in the service of colonization: an exquisite example of spatial organization for the purposes of legitimating property and imperial legibility (figure 6.10).

And yet, as scholars like Stephen Hornsby and Max Edelson have shown, the survey was very much a coastal exercise. With Holland headquartered at Port La Joye (now Observation Cove), four crews rowed around the island taking soundings and noting the location of sand beaches, bluffs, and wetlands.[24] Some of their work later appeared in the best-known cartographic project of eighteenth-century Europe, the *Atlantic Neptune* (figure 6.11). Unlike the Board of Trade, the *Atlantic Neptune*'s primary concern was navigation and "the safety, abridgement, confidence, and frequency of Voyages" of His Majesty's navy.[25] It included St. John's Island as part of the gulf, with two detailed profiles of the major ports at "Cardigan Bay, Island of St John" (Trois Rivières) and the "South East Coast of St. John's" (Port La Joye). The *Atlantic Neptune* speaks in a language of approach under sail: navigable channels, tidal direction, and inshore hazards. Keel depth and ship draw mattered more now than it had with fishing chaloupes pulled up on beaches. Gazing up from a

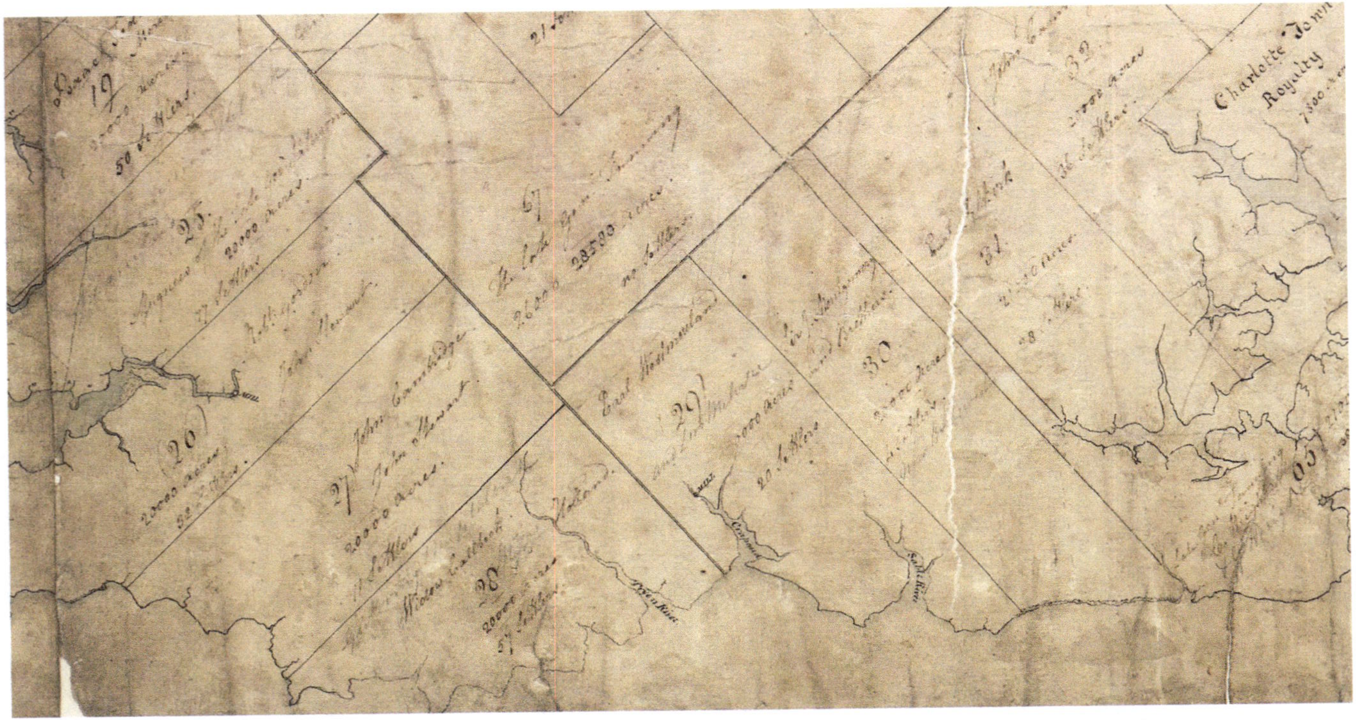


Figure 6.10 Samuel Holland, *Plan of Prince Edward Island showing the names of Proprietors*, 1767, detail. Island Imagined/UPEI and Public Archives and Records Office, Prince Edward Island.

Figure 6.11 J.F.W. Des Barres, "South East Coast of the Island of St. John's (Port La Joye)," *The Atlantic Neptune*, London, 1780, detail.

survey rowboat offered greater detail in shoreline topography but little information about the land beyond. Faint imprints of Acadian fields and tidal marshes were still visible but were diminished compared to the officious naming of bays and harbors.

Holland's map of the island was published in 1775, before the *Atlantic Neptune*, as *A plan of the island of St. John with the divisions of the counties, parishes, & the lots as granted by government, likewise the soundings round the coast and harbours* (figure 6.12). This was obviously designed to be a more public-facing version than the 1767 survey of and for landholders, and the differences are worth noting.

Despite the emphasis on lot lines and parish organization, this map maintained the long-standing, *seaward* value of the island. An inset located its position in the gulf; its title announced the soundings of inshore waters; and most obviously, the cartouche presented the proven wealth of the sea, in fish and nets, in equal measure as the aspiration of cleared woods, ploughed fields, and a seed frame promising future growth. True enough, the Anglo-Celts farmscape that became so symbolic of the island in the nineteenth century remained (like its Acadian predecessors) deeply entwined in seasonal fishing, as farmers capitalized on "hayfields that ran down to the sea."[26] This map even kept a few French names for major bays and harbors (perhaps a tacit acknowledgment that Holland's effusive rechristening of every point for his patrons had not yet registered in public lexicon).

Holland had also surveyed plats for three county seats, one of which would be the capital of the colony. At a glance, *A plan of Charlottetown, the capital of the Island delineated by order of Walter Patterson* by Thomas Wright (1771) follows the linear arrangement of the "model town" planted up and down the British American seaboard (figure 6.13).[27] Straight and spacious streets form regular blocks with a commons beyond for pasture. But water frames the plan, in ways both intentional and unexpected.

As with the other townsites, Holland chose the Charlottetown site for its harbor, which offered both known, sheltered anchorage and access to interior farm country via three navigable rivers. Freshwater was equally necessary for any settlement, and Charlottetown was bordered by Charlottetown Creek and West Creek. Both are now buried, along with several other ponds and runs in the city. I am always intrigued, unsettled, and strangely heartened to see how much water exists in colonial cities. What would happen if we unearthed these streams? And what can Wright's tidal line tell us about the original shoreline when rising sea levels and storm surge are flowing over two centuries of infill?[28]

A second generation of lot survey in the 1780s undertook to relocate and compensate Loyalists evacuated from the thirteen colonies. Like the New England Planters who had occupied Acadian farmland on the Bay of Fundy, these new arrivals were gifted lands already cleared and cultivated.[29] The *Plan of Lot 58* (1784) noted the location of an "Old French Saw Mill" on the north Pinette River and

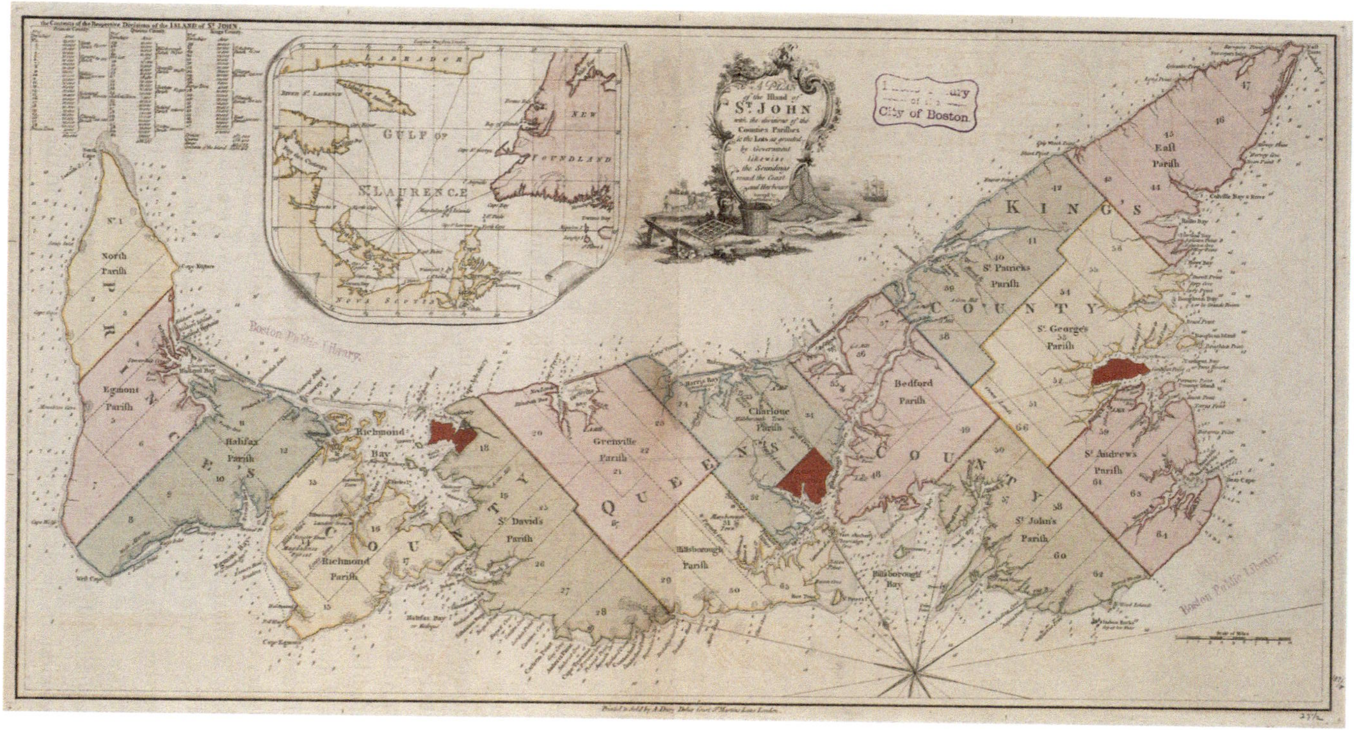


Figure 6.12 Samuel Holland, *A plan of the island of St. John with the divisions of the counties, parishes, & the lots as granted by government, likewise the soundings round the coast and harbours*, 1775. Norman B. Leventhal Map & Education Center, Boston Public Library.

Figure 6.13 Thomas Wright, *A plan of Charlottetown, the capital of the Island delineated by order of Walter Patterson*, 1771. Island Imagined/UPEI and Public Archives and Records Office, Prince Edward Island.

the footprint of an Acadian village fronting Pinette Bay. The plan's primary concern, of course, was naming property holders and the allotted acreage, but there are subtle clues to water, fresh and salt.

Surveyors noted freshwater runs and springs, and marked the heads of the tide in estuaries (figure 6.14).[30] There was no land reserved for the Mi'kmaq, who were increasingly isolated on Lennox Island and the barrier islands.[31]

By the middle of the eighteenth century, watercolors were supplanting maps as the dominant form of British landscape art. While "plain-style" maps increasingly signified reputable scientific cartography, watercolor painting—frequently by British military officers trained in topographic sketching—lent itself to public preferences for a Romantic aesthetic. The Picturesque preferred landscapes that were (or could be seen to be) pastoral, peaceable, and harmonious, at a time when rural politics in the United Kingdom was frequently not.[32] George Heriot offered the Island of Saint John as an ideal(ized) illustration of this in *Greenwich Park, Prince Edward Island* (1795).[33] A serpentine river draws our eye through rolling hills and lush deciduous green. A few head of cattle lie in repose in the foreground, a standard feature of the genre; a well-maintained stone wall and a bit of wooden fence suggest care and attention, but there is no other intrusion of human activity

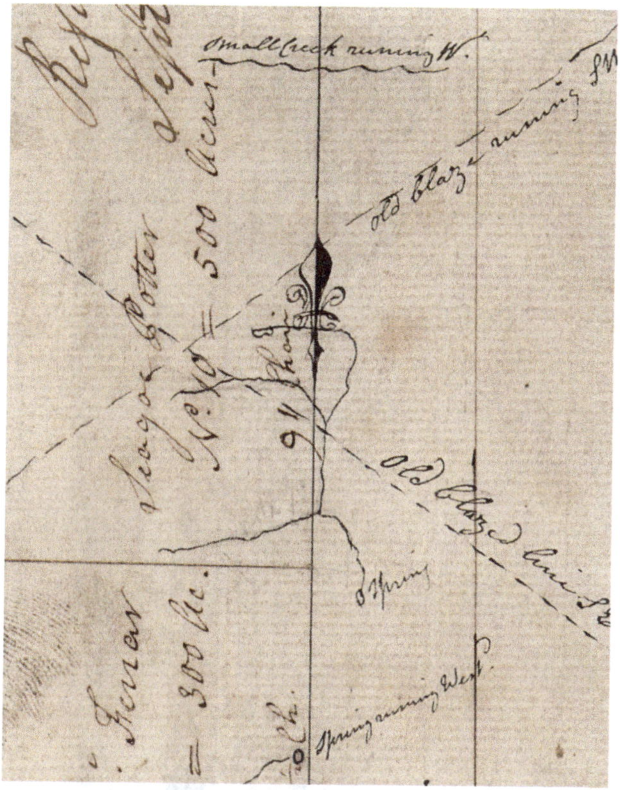


Figure 6.14 *Plan of Lot 30*, 1784, detail. Island Imagined/ UPEI and Public Archives and Records Office, Prince Edward Island.

(figure 6.15). By this point British settlers had settled on the shores of St. Peter's Bay in earnest; this is a rich but restful landscape.

John Stewart, on the other hand, was less satisfied by the accomplishments to date. Landowners like Stewart envisioned the island as what scholars now call *landesque capital*, or an environment transformed through human intervention for the purposes of intensifying use and, therefore, productivity.[34] Nominally a history of the island, Stewart's *An Account of Prince Edward Island in the gulf of St. Lawrence* (1806) was truly a call for "intelligent cultivation," to devote attention to the soil that—so unlike neighboring Nova Scotia—"presents everywhere a vast variety of fine situations for building and improvements" such "that there need not be a waste acre in the Island."[35] His book included the most recent map of the Island (figure 6.16), renamed as *Prince Edward Island: divided into Counties*

Figure 6.15 George Heriot, *Greenwich Park, Prince Edward Island*, 1795. Courtesy of RISD Museum, Providence, RI.

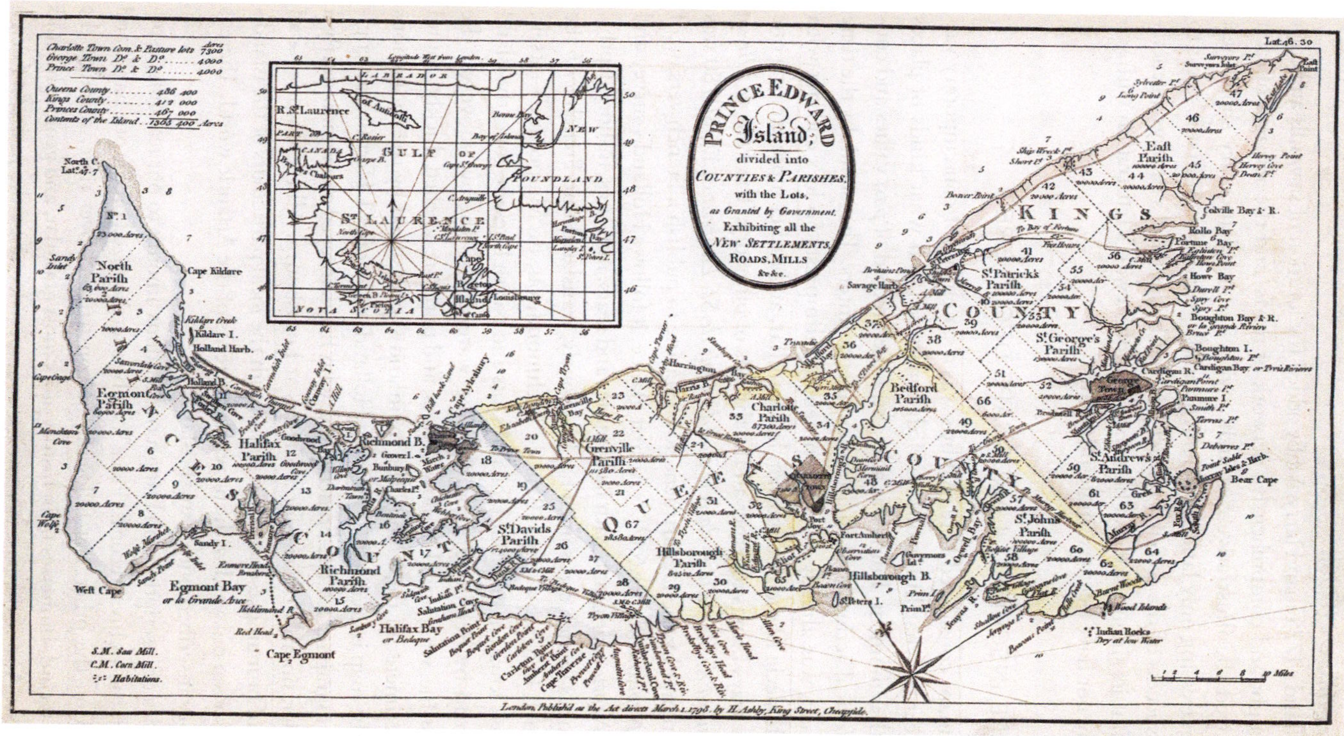


Figure 6.16 Harry Ashby, *Prince Edward Island: divided into Counties & Parishes, with the Lots, as Granted by Government. Exhibiting all the New Settlements, Roads, Mills, &c*, 1798. Courtesy of Prince Edward Island Museum and Heritage Foundation, James W. & Barbara MacNutt Collection of Historical and Antique Maps, HF.96.1.19.

*& Parishes, with the Lots, as Granted by Government. Exhibiting all the New Settlements, Roads, Mills, &c* (1798).

To Holland's 1775 map Harry Ashby added the location of sawmills and corn mills, the straight and comprehensive network of roads, and the acreages of each lot. He removed the illustrative cartouche, with its complementary shorelines, with the title stressing only landed settlements and agriculture. At first it seems to confirm the success of the colonial project laid out by Holland's surveys a generation before. But its singularity also betrays a hint of Stewart's frustration with settlers who may not have been as single-minded.

## LANDSCAPE INSCRIBED, LANDSCAPE DECIPHERED

Like all the maps considered here, Ashby's was a statement of contemporary values and concerns: what kinds of knowledge mattered and what kinds of places mattered at that moment. But it was also a stratified record of *past* values and concerns, of historical actions and environmental encounters. The island, by the time of Ashby's map, was already an environmental palimpsest. So let's end not with Ashby's fin de siècle island but another, less familiar, and rather more exciting one that in many ways brings together all the threads of our story. The *Plan de l'Île de St. Jean, au nord de l'Acadie et dans le sud du Golf de St. Laurent* of 1778/1780 was part of an attempt by the Départment de la Marine to publish a French version of the *Atlantic Neptune* (figure 6.17).[36] At first glance, it follows Holland's survey and the *Atlantic Neptune* original, with its counties and seats, mill sites, inshore depths, and place-names (which are, somewhat hilariously, translated into French *with* the original French beneath—e.g., "Baie de Richmond, ou, de Malpecque"—itself, of course, a Mi'kmaq name). But Holland's lots and parish lines are gone, the island divided instead by heights of land showing watersheds. The eye moves to the detailed coastline, where tidal shallows, rivers, and ports are labelled thickly around the whole perimeter of the island. The mainland shore is held by the Indigenous Mi'kmaq. The map is credited, along with Holland, to Jacques L'Hermite, an engineer and cartographer for the French navy, who had surveyed the region nearly a century earlier, notably the site for a fortress at Louisbourg. All these stories—all these marks of human presence—twine through the local particulars of the island's shoreline to sweep outward with the wider currents of the Atlantic world. There is a lot of history inscribed onto a small island.

This idea of inscription can bring a tremendous amount to our teaching, especially through the archival record. "Place" can be overtheorized on the one hand, and romanticized on the other; but following its construction over time, through the cartographic record, immerses us in history at its most granular, and geography at its most evocative. And if the past is a foreign country—let alone the past *of* a foreign country—we need as much immersion as the very best study abroad.[37]

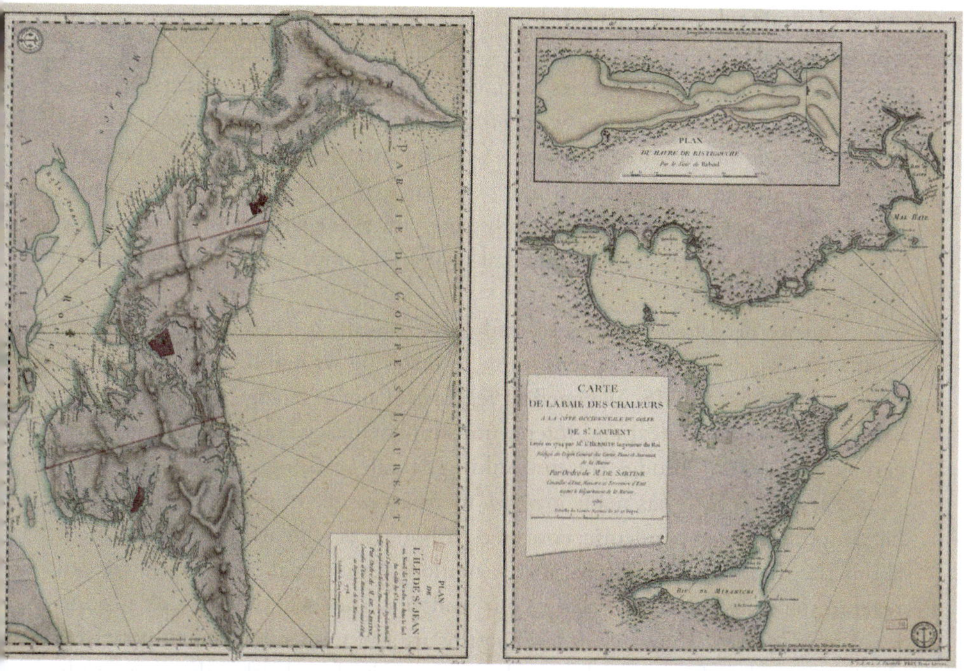


Figure 6.17 Départment de la Marine, *Plan de l'Île de St. Jean, au nord de l'Acadie et dans le sud du Golf de St. Laurent*, 1778/80. BNF.

I turned into teaching the eighteenth century when I moved from Nova Scotia to central Pennsylvania, largely a result of an American school not knowing where else to fit a Canadianist ("Maybe a course on the French and Indian War?" the chair asked). I am regularly staggered by the insularity of American education, and so, rather perversely, delight in teaching *around* and beyond Washington crossing the Delaware. I have written elsewhere about the challenges of teaching to Americans, so I will simply say here that early America cannot be understood from within its present borders.[38] A fortress guarding Île Royale and the Gulf of St. Lawrence troubled those in Boston for decades. The Richelieu River served as a conduit to the citadel at Quebec and its guardian winters. Loyalists landed by the tens of thousands in Nova Scotia, making Shelburne one of the largest cities on the continent. It seems parochial, irresponsible, and counterintuitive for a self-proclaimed superpower to have such a limited spatial imagination toward its history.

A cartographic reading—which places the seaboard along, well, the sea—expands the frame not just of the historical narrative, but the ecological one. Environmental history (let alone environmental humanities) is still a rare commodity in most college curricula, at any level, where "environmental studies" is anchored in the natural and social sciences. Likewise, and unsurprisingly, our approach to environmental issues remains defined by measures of the quantitative and nonhuman (degrees of warming, species loss). Maps and other historical texts, in contrast, repopulate the landscape with *human* actors, whose agendas and actions shaped the contours of those very crises we bequeath to our students. At the same time, they offer a record of the human experience of environmental opportunity, limit, and change. This seems especially poignant as we witness in real time the loss of coastlines that were so important to spatial knowledge and movement in the age of sail.

But we need to learn how to read these records, in their composition and their meaning. Such is the special value of teaching with and through archival maps, in contrast to GIS. We need to teach students to engage with maps critically, emphasizing context and intent, not simply efficacy and use; as I suggested earlier, we need to dislodge the unthinking default of Google. Somewhat ironically, the digital turn has made *archival* maps accessible in unprecedented ways, defying restraints of pandemic, cost, and carbon. If as scholars we are ambivalent about the digital avatar relative to the manuscript original, as teachers we can revel in the collections open now to our students. These are the maps we need to learn how to read. They are simultaneously art and text: students can pore over the iconography of a cartouche, or the gauge of harbor depths; they can study the annotations of promised riches, navigational hazards, or the deliberate erasure of older names and inhabitants. But maps are also artifacts, of ambition, epistemology, technology, and commerce. And they offer endless insights into humanity's relationship with the natural world. Accordingly, they require a special kind of literacy—or perhaps more precisely, a set of braided literacies. Is there anything more emblematic of the liberal arts?

The palimpsest anchors big and abstract questions in particular places and shows the changes of centuries and generations where we stand. Environmental history registers the weight of human presence over time, how many ways we have relied on, adapted to, and altered the world around us—and just how many generations of footprints need to be accounted for. It gives us a way to read climate change in historical context and in a longer narrative of causality that deepens our engagement beyond "crisis" and connects us across time and space. At the same time, it makes clear an ethical obligation, in how embedded we are, how much we owe to something greater than ourselves. Perhaps, in that, it may cultivate a little humility and care.

## NOTES

1. When I wrote this, in September 2020, five named storms were spinning in the Atlantic north of the equator. Dorian is particularly germane to our story—about the symbolic power of maps and the nature of knowledge—because of "Sharpiegate," when the president of the United States redrew a NOAA map of the storm's projected path.

2. For some sense of these changes, see M. M. McCulloch, D. Forbes, and R. D. Shaw, *Coastal Impacts of Climate Change and Sea-Level Rise on Prince Edward Island* (Dartmouth, NS: Geological Survey of Canada/Bedford Institute of Oceanography, 2002); Norm Catto, Kate MacQuarrie, and Marianne Hermann, "Geomorphic Response to Late Holocene Climate Variation and Anthropogenic Pressure, Northeastern Prince Edward Island, Canada," *Quaternary International* 87, no. 1 (2002): 101–117. D. L. Forbes et al., "Storms and Shoreline Retreat in the Southern Gulf of St. Lawrence," *Marine Geology* 210 (2004): 169–204; Natural Resources Canada, *Canada's Marine Coasts in a Changing Climate* (Government of Canada, 2016); Liette Vasseur, Mary Thornbush, and Steve Plante, "Climatic and Environmental Changes Affecting Communities in Atlantic Canada," *Sustainability* 9, no. 8 (2017): 1293; Abbas Farhat, Aitazaz A. Farooque, and Hassan Afzaal, "Homogeneity in Patterns of Climate Extremes Between Two Cities—A Potential for Flood Planning in Relation to Climate Change," *Water* 12, no. 3 (2020): 782.
3. As examples of cartographic biographies, see Richard Harris and Martin Luymes, "The Growth of Toronto, 1861–1941: A Cartographic Essay," *Urban History Review / Revue D'histoire Urbaine* 18, no. 3 (1990): 244–255; Robert Summerby-Murray, "Four Hundred Years of Mapping in the Upper Bay of Fundy: Changing Coastal Environments and Economies, 1550–1950," *Journal of the Royal Nova Scotia Historical Society* 15 (2012): 1–23; John Greelee, "Eight Islands on Four Maps: The Cartographic Renegotiation of Hawai'i, 1876–1959," *Cartographica* 50, no. 3 (2015): 119–140. On the environmental humanities, Denis Cosgrove, "Geography within the Humanities," *Envisioning Landscapes, Making Worlds: Geography and the Humanities*, ed. Stephen Daniels, Dydia DeLyser, J. Nicholas Entrikin, and Doug Richardson (New York: Routledge, 2011), xxii–xxv.
4. Patrick J. Augustine, "The Significance of Place in Textual and Graphical Representation: The Mi'kmaq on Lennox Island, Prince Edward Island, and the Penobscot on Indian Island, Maine," Island Studies master's thesis, University of Prince Edward Island, 2009.
5. I thank the ongoing conversation of C-JAM (Jennifer Bonnell, Andy Robichaud, and Melanie Kiechle) for their thoughts on this piece, and for this observation.
6. Jacques-Nicholas Bellin, "Remarks by M. Bellin in Relation to His Maps Drawn for P. Charlevoix's History of New France," In *The Gentleman's Magazine*, February 1746, 72. I thank Jeffers Lennox for this.
7. Matthew Edney, *Cartography: The Ideal and Its History* (Chicago: University of Chicago Press, 2019), 77. There has been exciting scholarship on mapping and Indigenous, imperial, and early national identities and territories, but these have originated from the dynamics of imperial or intellectual history, rather than environmental history. See, for example, The History of Cartography Project from the University of Chicago Press, https://geography.wisc.edu/histcart/; Stephen Hornsby, *British Atlantic, American Frontier: Spaces of Power in the Early Modern British America* (Hanover, NH: University Press of New England, 2005), and *Surveyors of Empire: Samuel Holland, J.F.W. Des Barres, and the Making of the Atlantic Neptune* (Montreal: McGill-Queen's University Press, 2011); S. Max Edelson, *The New Map of Empire: How Britain Imagined America before Independence* (Cambridge, MA: Harvard University Press, 2017), his website http://mapscholar.org/empire/ (accessed August 6, 2021), and https://earlycanadianhistory.ca/2018/11/14/colonizing-st-john-island-a-history-in-maps/; Alexander Johnson, *The First Mapping of America: The General Survey of British North America* (Bristol, UK: I.B. Tauris, 2017); Jeffers Lennox, *Homelands and Empires: Indigenous Spaces, Imperial Fictions, and Competition for Territory in Northeastern North America, 1690–1763* (Toronto: University of Toronto Press, 2017); Chet Van Duzer and Lauren Beck, *Canada before Confederation: Maps at the Exhibition* (Wilmington, DE: Vernon Press, 2017). Marilyn McKay, *Picturing the Land: Narrating Territories in Canadian Landscape Art, 1500–1950* (Montreal: McGill-Queen's University Press, 2011), includes a chapter on the art-map before 1800. On the early national period in the United States, see Martin Brückner, ed., *Early American Cartographies* (Chapel Hill: University of North Carolina Press, 2011), and *The Social Life of Maps in America, 1750–1860* (Chapel Hill: University of North Carolina Press, 2017).

8. I have always liked the term littoral, not least because it is a French term that has found its way into English use. There is ongoing debate over how to define the Anthropocene; I tend to follow the argument that dates to the latter eighteenth century, with industrialization and the adoption of fossil fuels on a transnational scale. For illustrations of how maps have been used to document environmental change in the northeast (with varying degrees of technological and quantitative emphasis), see, for example, K. D. Bromberg and M. D. Bertness, "Reconstructing New England Salt Marsh Losses Using Historical Maps," *Estuaries* 28, no. 6 (2005): 823–832; Mark Monmonier, *Coast Lines: How Mapmakers Frame the World and Chart Environmental Change* (Chicago: University of Chicago Press, 2008); Stephen T. Mague, "In Search of the Shawmut Peninsula: Using Modern Cartographic Analysis to Discover the 'Original' Boston Shoreline," in *Remaking Boston: An Environmental History of the City and Its Surroundings*, ed. Anthony Penna and Conrad Wright (Pittsburgh: University of Pittsburgh Press, 2009), 75–102.
9. Adam Grydehøj, Yaso Nadarajah, and Ulunnguaq Markussen, "Islands of Indigeneity: Cultural Distinction, Indigenous Territory and Island Spatiality," *Area* 52, no. 1 (2020): 14–22, a special issue dedicated to islands; also Grydehøj, "Islands as Legible Geographies: Perceiving the Islandness of Kalaallit Nunaat (Greenland)," *Journal of Marine and Island Cultures* 7, no. 1 (2018): 1–11. There has been relatively little on maps *as* environmental history, but consider Denis Wood and John Fels, *The Natures of Maps: Cartographic Constructions of the Natural World* (Chicago: University of Chicago Press, 2008) and—relevant to this location—K. J. Rankin and Poul Holm, "Cartographical Perspectives on the Evolution of Fisheries in Newfoundland's Grand Banks Area and Adjacent North Atlantic Waters in the Sixteenth and Seventeenth Centuries," *Terrae Incognitae* 51, no. 3 (2019): 190–218.
10. Edward MacDonald, Joshua MacFadyen, and Irene Novaczek, "Introduction: Promise and Premise: An Environmental History for Prince Edward Island," *Time and a Place: An Environmental History of Prince Edward Island*, ed. Edward MacDonald, Joshua MacFadyen, and Irene Novaczek (Charlottetown: McGill-Queen's University Press and Island Studies Press, 2016), 3–15, 7. A thorough account of the political history is J. M. Bumsted's *Land, Settlement, and Politics on Eighteenth-Century Prince Edward Island* (Montreal: McGill-Queen's University Press, 1987), though Bumsted consistently refers to the island as wilderness, which suggests we reconsider the period from the perspective of environmental history.
11. On environmental history in the Canadian Atlantic, see *Land and Sea: Environmental History in Atlantic Canada*, ed. Claire Campbell and Robert Summerby-Murray (Fredericton: Acadiensis Press, 2013), and *The Greater Gulf: Essays on the Environmental History of the Gulf of St. Lawrence*, ed. Claire Campbell, Edward MacDonald, and Brian Payne (Montreal: McGill-Queen's University Press, 2020); down the coast, see Richard Judd, *Second Nature: An Environmental History of New England* (Amherst: University of Massachusetts Press, 2014), and Christopher Pastore, *Between Land and Sea: The Atlantic Coast and the Transformation of New England* (Cambridge, MA: Harvard University Press, 2014). Andrew Hill Clark's *Three Centuries and the Island: A Historical Geography of Settlement and Agriculture in Prince Edward Island, Canada* (Toronto: University of Toronto Press, 1959) has become as central an object to the field of historical geography as to the island's history. A good introduction to island studies is Graeme Wynn's "Museums, Laboratories, Showcases: Islands in Environmental History," in *Time and a Place: An Environmental History of Prince Edward Island*, ed. Edward MacDonald, Joshua MacFadyen, and Irene Novaczek (Charlottetown: McGill-Queen's University Press and Island Studies Press, 2016), 36–56. Island studies tends to lean toward the social sciences, but R. A. Walshe and C. E. Stancioff, "Small Island Perspectives on Climate Change," *Island Studies Journal* 13, no. 1 (2018): 13–24, in a special issue on climate change, acknowledge the value of considering *la longue durée*, especially in the legacies of colonization. Elaine Stratford et al., "Envisioning the Archipelago," *Island Studies Journal* 6, no. 2 (2011): 113–130, speaks to Canada's archipelagic

coastlines and proposes archipelagic thinking as a form of counter-mapping. There has been an emergence of coastal history as distinct from maritime: as with John R. Gillis, *The Human Shore: Seacoasts in History* (Chicago: University of Chicago Press, 2013); David Worthington, ed., *The New Coastal History: Cultural and Environmental Perspectives from Scotland and Beyond* (Cham: Springer International, 2017); Paul Sutter and Paul M. Pressley, eds., *Coastal Nature, Coastal Culture: Environmental Histories of the Georgia Coast* (Athens: University of Georgia Press, 2018); Linde Egberts and Meindert Schroor, eds., *Waddenland Outstanding: History, Landscape and Cultural Heritage of the Wadden Sea Region* (Amsterdam: Amsterdam University Press, 2018).

12. This is especially true for restoring Indigenous place names, which may have been altered and then stripped *from* the archival record. I relied on information from the L'nuey project of Prince Edward Island (http://lnuey.ca/resources/education-and-reconciliation/, accessed August 6, 2021), and Margaret Wickens Pearce, and Brittany Luby and Stephen Hay's "Coming Home to Indigenous Place Names in Canada," *Canadian Historical Review* 100, no. 1 (2019): 101–103.
13. Similarly, the *Carte de l'isle de Saint Jean* (Bibliothèque nationale de France, http://catalogue.bnf.fr/ark:/12148/cb43640089b, undated: labeled as seventeenth century but likely dates to after 1720 as it shows a "Maison du gouvernement") highlights the *relative* value of the north shore, "Le port de Malbec est parfaittement [*sic*] beau et commode pour la correspondance de les isles de Magdelaine," and praises the harbor at Port La Joye as "le plus beau qui sait connu au monde." I want to acknowledge Jack Bouchard's reading of maps of this region in terms of early modern understanding of *terra* and maritime zones; his monograph, *Terra Nova: Food, Water and Work in an Early Atlantic World*, is forthcoming with Yale University Press.
14. David Keenlyside and Helen Kristmanson, "The Palaeo-Environment and the Peopling of Prince Edward Island: An Archaeological Perspective," in *Time and a Place: An Environmental History of Prince Edward Island*, ed. Edward MacDonald, Joshua MacFadyen, and Irene Novaczek (Charlottetown: McGill-Queen's University Press and Island Studies Press, 2016), 59–81; and Rosemary Curley, "Wildlife Matters: A Historical Overview of Public Consciousness of Habitat Loss on Prince Edward Island," in *Time and a Place: An Environmental History of Prince Edward Island*, ed. Edward MacDonald, Joshua MacFadyen, and Irene Novaczek (Charlottetown: McGill-Queen's University Press and Island Studies Press, 2016),109–139, 111. Allan Dwyer discusses the Beothuk use of inner coastal zones in "Liminality and Change in an Atlantic Borderland: Notre Dame Bay, Newfoundland in the Eighteenth Century," in *Environmental History in Atlantic Canada*, ed. Claire Campbell and Robert Summerby-Murray (Fredericton: Acadiensis Press, 2013), 27–44. For more on Indigenous coastal life, see also Andrew Lipman, *The Saltwater Frontier: Indians and the Contest for the American Coast* (New Haven, CT: Yale University Press, 2015).
15. Peter E. Pope, "Historical Archaeology and the Maritime Cultural Landscape of the Atlantic Fishery," in *Method and Meaning in Canadian Environmental History*, ed. Alan MacEachern and William Turkel (Toronto: Nelson, 2009), 36–54, 47.
16. Matthew Hatvany, "'Wedded to the Marshes': Salt Marshes and Socio-Economic Differentiation in Early Prince Edward Island," *Acadiensis* 30, no. 2 (2001): 40–55; Bumsted, *Land, Settlement, and Politics,* 8; Graeme Wynn, "Reflections on the Environmental History of Atlantic Canada," *Land and Sea: Environmental History in Atlantic Canada*, ed. Claire Campbell and Robert Summerby-Murray (Fredericton: Acadiensis Press, 2013), 235–256. There is more literature on Acadian dykeland farming in Nova Scotia, by such scholars as Sherman Bleakney, Jonathan Fowler, and Ronald Rudin.
17. Edward MacDonald and Boyde Beck, "Lines in the Water: Time and Place in a Fishery," *Time and a Place,* 218–221. Tracadie also borrows from the Mi'kmaw Tlaqatik.
18. "Tour of Inspection Made by the Sieur de La Roque, Census 1752" (translation), in *Report of the Public Archives of Canada, 1905, Part 2* (Ottawa: S. E. Dawson, 1906), 151–152. I am indebted to Daniel Samson for this source and his collegial support.

19. There was, further, an ongoing dispute over what constituted the borders of Acadia and so what, exactly, had been surrendered. Maps such as this or *Carte de l'Accadie et Pais Voisins pour servir a l'Histoire Generale des Voyages: Par M.B. Ingr. de la Marine* (1757) speak to French efforts to contain "Accadie" (and therefore British territory) to the Atlantic shore. See Lennox, *Homelands and Empires.*
20. Specifically, Boat Harbour and Pictou First Nation. Ingrid R. G. Waldron, *There's Something in the Water: Environmental Racism in Indigenous and Black Communities* (Black Point, NS: Fernwood Publishers, 2018).
21. Johnson, *First Mapping of America*, 48–49.
22. Stewart counted himself in a "Compy of 25 Gentn most of them Officers and severals of them Men of Fortune" petitioning for a grant on the island. Robert Stewart to George Washington, January 14, 1764. Stewart received part of Lot 18 in 1767; see Bumsted, *Land, Settlement, and Politics,* 50.
23. See Hornsby, *Surveyors of Empire.* Holland was commissioned in 1762 to undertake a general survey of the "Northern District" (north of Virginia) of the newly British North America, and began with the gulf and the old Acadian archipelago—Saint John and the Magdalen Islands, then Cape Breton and the Gaspe. As the archive "Island Imagined" says, "The Island would literally never be the same." See Earle Lockerby and Douglas Sobey, *Samuel Holland: His Work and Legacy on Prince Edward Island* (Prince Edward Island: Island Studies Press, 2015).
24. Edelson's MapScholar project has digitized the fieldwork notes and sketches by assistant surveyors Henry Mowatt and Thomas Wright, which noted in precise depths and inshore tidal shallows, detail beaches, heights of bluffs, and coastal wetlands.
25. Maps published from 1775, but first titled in 1777. J.W.F. Des Barres, *The Atlantic Neptune* (London, 1777 and 1800).
26. From L. M. Montgomery's *Anne of Ingleside* (ebook) (London: Virago, 2017), chap. 18. See Claire E. Campbell, "'A Window Looking Seaward': Finding Environmental History in the Writing of L. M. Montgomery," in *The Greater Gulf: Essays on the Environmental History of the Gulf of St. Lawrence*, ed. Claire Campbell, Edward MacDonald, and Brian Payne (Montreal: McGill-Queen's University Press, 2020), 283–318. A public coastline—within five hundred feet of the high water mark—was reserved for fishing stages and use, resulting in a built, active coastline now difficult to imagine. I thank Edward MacDonald for the metaphorical reading on the seed frame.
27. These were Princetown on Malpeque Bay, Charlottetown on Hillsborough Bay, and Georgetown on Cardigan Bay. On the model town, see Hornsby, *British Atlantic, American Frontier.* Wright was surveyor-general of the colony for decades.
28. G. G. Thresher, *Charlottetown Harbour*, ca. 1830, Prince Edward Island Archives.
29. See, for example, Debra McNabb, "The Role of Land in the Development of Horton Township, 1760–1775," in *They Planted Well: New England Planters in Maritime Canada*, ed. Margaret Conrad (Fredericton: Acadiensis Press, 1988), 151–160.
30. Numerous freshwater springs rising in Lot 30, for example, would later be recognized as sources for the upper West River; the head of the tide is marked on the neighboring Dog (Clyde) River in Lot 31. See Plan of Lot 30 (No.1) showing Refugee Shares drawn for in Council, 18 September, 1784; Plan of Lot 31 (No. 2) shows refugee Share drawn for in Council, 12 February, 1784.
31. Augustine, "The Significance of Place," 2009. As with other areas of Mi'kma'ki, the island was never ceded. Today there are two recognized First Nations on the island: the larger Lennox Island First Nation and the Abegweit First Nation.
32. McKay, *Picturing the Land*, 32 and 48–49. See also John E. Crowley, *Imperial Landscapes: Britain's Global Visual Culture, 1745–1820* (New Haven, CT: Yale University Press, 2011). Pietro Piana and Charles Watkins, "Questioning the View: Historical Geography and Topographical Art," *Geography Compass* 14, no. 4 (2020): 111–134, discuss how art can reveal changes in weather, glaciers, and even geomorphology.

33. Held by the RISD Museum, Providence, Rhode Island, via Watercolour World, https://www.watercolourworld.org; Gerald E. Finley, "HERIOT, GEORGE," in *Dictionary of Canadian Biography*, vol. 7, University of Toronto/Université Laval, 2003, http://www.biographi.ca/en/bio/heriot_george_7E.html (accessed October 1, 2020). In his *Travels through the Canadas* (1805), the colonial administrator described the island less effusively than earlier admirers and grant-holders, but as "flourishing," fertile and well-watered, with safe harbors (18).
34. E. Clark and H. M. Tsai, "Islands: Ecologically Unequal Exchange and Landesque Capital," in *Ecology and Power: Struggles over Land and Material Resources in the Past, Present and Future*, ed. A. Hornborg, Brett Clark, and Kenneth Hermele (New York: Routledge, 2013), 22–67, 57–58.
35. John Stewart, *An account of Prince Edward Island in the gulph of St. Lawrence, North America* (1806), 24–26. An island petition proposed the renaming in honour of the Duke of Kent, commander in chief of British forces in North America, who never visited the place.
36. M. S. Pedley, *The Commerce of Cartography: Making and Marketing Maps in Eighteenth-Century France and England* (Chicago: University of Chicago Press, 2005), 146–155.
37. L. P. Hartley, *The Go-Between* (New York: Stein and Day, 1953), 1. This line was popularized in the literature about heritage and history by David Lowenthal, *The Past Is a Foreign Country* (Cambridge, UK: Cambridge University Press, 1985).
38. Claire Campbell, "National Interest and Continental Geography: Teaching about Canada through an Environmental Lens," *The Construction of Canadian Identity from Abroad*, ed. Christopher Kirkey and Richard Nimijean (Palgrave Macmillan, 2022), 47–67.

7

# OF WATER, WINDS, AND STORMS

## The Elemental Regimes of the Buccaneer Journal

JASON M. PAYTON

THE HISTORY OF PIRACY in the long eighteenth century has been traditionally understood in political terms.[1] This focus has generated significant insights within historical and literary studies about the complex relationship between piracy and empire and about the role that pirate narratives played in the development of New World literary forms.[2] And yet, piracy studies has been relatively inattentive to the role of the environment within the archive.[3] I redress this problem by exploring the representation of environmental assemblages within the buccaneer journal tradition.[4] In particular, I focus on seas, winds, and storms to limn the contours of what I call the "elemental regimes" of buccaneer narratives. The elemental focus of this essay highlights a structural ambivalence within these narratives between the desire to represent the sea as the setting for human stories of mobility and mutability and the recognition that oceanic environments can emerge as actants capable of denying the primacy and viability of human endeavor.[5] This focus brings the ecopoetics of the buccaneer journal into view and reframes the pirate narratives of the long eighteenth century as environmental texts.[6]

Centering the environment in this way helps perform the work Tiffany Lethabo King describes as "shoaling" by disrupting the "movement and flow" of settler colonialism.[7] Despite their seeming opposition to imperialism, buccaneer authors draw frequently on the same taxonomic logics of environmental mastery and racial subjugation that animate the literatures of empire. By foregrounding both the function of these logics and the moments when they break down, I challenge the anthropocentric orientation of what Felicity Nussbaum has called "critical global studies" by underscoring the central role the environment plays in the critique of emergent modernity undertaken by global eighteenth-century studies.[8] The payoff of an environmentally attuned reading of the buccaneer journals is twofold. First, such a reading contributes to the work of "historicizing the oceans" called for by scholars across the humanities and sciences by foregrounding the role of oceanic environments in pirate narratives.[9] Second, it places piracy studies in conversation with decolonial critique. As Sylvia Wynter has argued, the Caribbean was an especially important site for the exercise of colonial power and colonialist ideas of racial difference. I use the buccaneer journal to highlight the use of colonialist tropes by Caribbean pirates, while also highlighting the decolonial possibilities entailed by the narratives of environmental resistance embedded within the form.[10]

I begin by considering the agency of water in the first English translation of the original buccaneer journal, Alexander Exquemelin's *Bucaniers of America* (1684).[11] The buccaneer journal has been deeply engaged with questions about the relationship between human and environmental agencies from its inception. I argue that oceanic currents in particular display what Jane Bennett has called a "conative intention" by exhibiting the features of trajectory, efficacy, and causality normally associated with conceptions of human agency.[12] I then place William Dampier's *Voyages and Descriptions* (1700) within the genealogy of buccaneer journals and read the "Discourse of Winds" contained therein as a text that simultaneously uses the language of science to achieve epistemic mastery over nature and recognizes the capacity of nature to resist such mastery. Wind-water assemblages emerge as a potent force in the "Discourse" that is capable of laying waste to colonial settlements and the ships of travelers, traders, and pirates. These assemblages introduce an aesthetic of chaos into the narrative and articulate a global network of oceanic and riverine spaces that resist and repel human incursion. The chaos embodied by the winds, storms, and floods that riddle Dampier's text asserts what I call a principle of "environmental sovereignty" that precedes and supersedes human claims to sovereignty over terraqueous spaces.[13] I conclude with reflections on the pedagogical implications of these arguments, which encourage students to reimagine the history of piracy as environmental history and to see the figure of the pirate as situated within an ecological context that complicates popular understandings of piratical agency.

## THE HYDROGRAPHICAL REGIME OF *BUCANIERS OF AMERICA*

Originally published in Dutch in 1678 as *De Americaensche Zee-Roovers*, Exquemelin's firsthand account of the exploits of infamous Caribbean pirates such as François L'Olonais and Henry Morgan became an instant international sensation. In less than a decade, translations had appeared in German, Spanish, English, and French, demonstrating a broad cultural interest in the history of seventeenth-century Caribbean piracy. The reasons for its popularity are not hard to fathom. Readers were generally interested in what Exquemelin had to say about the politics of empire and the possibility of life beyond the control of the state; statesmen were interested in what he had to say about the state of Spain's colonial economy and defenses; and naturalists were interested in the details he recorded about Caribbean flora and fauna. The popularity Exquemelin's text enjoyed created an international literary marketplace for similar tales, of which there were several.[14] Its English translation in 1684 as *Bucaniers of America* sparked significant interest in related narratives. Richard Frohock has shown that these narratives "contributed significantly to England's imaginative, literary rendering of the Americas in the late seventeenth and early eighteenth centuries" by providing "a venue for public dialogue about sea voyagers and their position within empire."[15] This dialogue represented the buccaneers variously as inimical to civil society, as vehicles for the critique of imperial violence, and even as tools for empire-building.[16] My interest here is in exploring the relationship between the geopolitics of *Bucaniers* and its ecopoetics. I argue that *Bucaniers* establishes a hydrographical regime that recognizes the agency of the sea as the material condition of possibility for piracy and that recognizes the precariousness of all human endeavor in the face of its indomitable force.

The opening chapters frame *Bucaniers* as a story of political liberation by chronicling Exquemelin's escape from indentured servitude to join the buccaneers at Tortuga. This framework primes readers to see the text as a commentary on imperial labor politics and the possibilities of resistance, but the story proper begins at sea. Exquemelin begins by noting that he departed for the West Indies in May of 1666 from Havre de Grace on the *St. John* and anchored at the Cape of Barfleur while awaiting the arrival of company ships from Dieppe. The company ships were bound for Senegal as well as various Caribbean islands, and they were joined by a convoy of other vessels bound for Europe and North America. These vessels formed a "Fleet of thirty sail" that assumed "a convenient posture of defense" in anticipation of a conflict with English frigates cruising near the Isle of Ornay.[17] Exquemelin's fleet got under way with help from "a favorable gale of wind" an was cloaked from predators by "some mists arising" that "totally impeded the English Frigats, from discovering our Fleet at Sea."[18] Wind and fog conspire to allow their departure from Barfleur without incident, but Exquemelin raises the specter of a counterfactual history of his departure in which company ships suffer depredation by pirates for want of cover, as did a Flemish ship from Ostend.[19]

The possibility that the *St. John* could suffer the same fate as the ship from Ostend grounds the entire narrative of *Bucaniers* in a recognition of the company ships' dependence on favorable environmental conditions. Exquemelin meditates further on this state of dependency in his description of the ships' passage into the open sea. Once past the threat of rovers, Exquemelin notes that the company ships were forced to pass by the "*Ras of Fonteneau*" to avoid exposure to English pirates near the Isles of Scilly. He describes the "*Ras*" (Race) as a "current very strong and rapid, which rowling over many Rocks, disgorgeth itself into the Sea" and notes that its passage is "very dangerous" because the location and extent of its reefs is "not thoroughly known."[20]

Currents, coastlines, reefs, winds, and fogs enter the narrative as co-determinants of its main action by serving as gatekeepers for the passage of the *St. John* and other company ships into the open sea. The benevolence of the elements toward the company ships raises the counterfactual possibilities of disastrous results for other ships passing the Race at other times. While the opening of the narrative foregrounds human agency by representing the sea as a conduit for travel and trade, the threats of depredation by coastal pirates and shipwreck under less favorable circumstances reveal the agentic force of the environment. The environmental assemblages present in this passage write themselves into *Bucaniers* as what Lowell Duckert calls "onto-stories" that emphasize the ontological primacy and independence of the environment.[21] The configuration and duration of these assemblages create the necessary conditions for the company's errands but remain entirely beyond human control. The trajectory of the Race in particular introduces a counternarrative of modernity's relation to the sea that denies fantasies of mastery over the sea by the mapping of winds and currents in favor of what Steve Mentz calls a "shipwreck modernity" in which the era of global mercantile expansion is comprehended within the horizon of the pervasive threat of destruction by the elements.[22]

This brief passage on the good fortune of Exquemelin's convoy during its departure from Europe establishes the contours of a hydrographical regime that emphasizes the dangers of life at sea. The fate of the ships in the convoy is determined by the will of nature. The counterfactual possibilities raised in this passage suggest that the sea cannot be mastered by force of will, nor can its precarious movements be predicted by observation and experience. This way of representing the sea was prominent in early modern travel literature and cartography, for while European powers had embraced the political necessity of sea power by the mid-fifteenth century, cultural understandings of the sea as the traversable space of merchant capitalism took longer to coalesce and emerged as dominant only in the eighteenth century.[23] Sixteenth- and early-seventeenth-century representations of the sea often depicted it as a "wild zone filled with untamable elements, animals, and people."[24]

Fear of the sea is also visible in literary texts from the period. The shipwreck scene in Shakespeare's *The Tempest* (c. 1610), for example, opens with a boatswain pleading to Gonzalo to "command these Elements" if it is within the scope of his "authoritie," which, of course, it is not.[25] As Daniel Brayton observes, *The Tempest* evinces a "particular ideology of territorial mastery" that sought to link terrestrial and oceanic environments as spheres malleable by the force of human will.[26] But Shakespeare's ocean remains characterized by what Rob Watson calls a "fundamental indeterminacy" and Mentz characterizes as a "terrifyingly lost paradise" that "remains inimical to human life," despite Renaissance attempts at mastering nature through science and technology.[27] Though the recognition of the alterity and hostility of the sea gave way to a more benign view of oceanic space as amenable to global mercantile expansion, the recognition of environmental agency in *Bucaniers* shows that this process was nonlinear and uneven.[28] While Exquemelin uses the journal form as a way of inscribing the narrative within what Margaret Cohen calls the "near present" by relating actions and observations in a factual, plain style, the sea exceeds the managerial poetics of the journal form and suggests an ecopoetics of environmental resistance in its place.[29] The hydrographic regime of *Bucaniers* thus falls within an early modern literary genealogy that bears greater resemblance to the early-seventeenth-century ambivalence about the mastery of the sea expressed by Shakespeare than to the rationalized view of oceanic space that characterized eighteenth-century maritime literatures.

The micronarrative of ecological resistance to which I have drawn attention here sits in tension with Exquemelin's explicit interest in identifying the merchantable commodities of Tortuga.[30] His interest in commodifying nature encodes Tortuga, and the Caribbean by extension, within the "hydraulic web" that Henri Lefebvre identifies as the spatial structure of Western modernity.[31] Seas, rivers, ports, and coastal towns connect terrestrial and water spaces within this web, which allows for the co-emergence of the modern state and merchant capitalism. The construction of this web required physical and epistemic violence, as territories were mapped, claimed, and appropriated for the enrichment of imperial states. In the economic domain, a "quantifying rationality" cloaked the violence of accumulation and encouraged the global expansion of networks of travel and trade.[32] There is a noteworthy antagonism, however, between the quantifying rationality used in *Bucaniers* to commodify Caribbean environments and the representation of the sea itself as a precarious and dangerous force. This sense that the sea both connects the buccaneers to nodal points of exchange within the "web" of imperial expansion *and* resists the epistemic mastery on which this expansion relies is a central theme in *Bucaniers*, and it emerges as a central theme in Dampier's *Voyages and Descriptions* as well.

## DAMPIER AND THE POETICS OF ECOLOGICAL RESISTANCE

The international success of Exquemelin's narrative produced a spate of English buccaneer journals, including Dampier's *New Voyage Round the World* (1697).[33] Dampier's *New Voyage* enjoyed commercial success and helped create a market for his subsequent *Voyages and Descriptions* (1699) and *A Voyage to New Holland* (1703). Like Exquemelin's text, these narratives addressed a range of topics, including the history of the buccaneers, their current exploits, and the natural history of the Caribbean. And like Exquemelin, Dampier frequently grapples with questions of narrative authority by crafting an authorial persona that draws on the experience of life among the buccaneers while disavowing any direct connection to the violence they perpetrated.[34]

The rhetorical connections between the *New Voyage* and *Voyages and Descriptions* show that Dampier's self-fashioning as a legitimate subject of empire depends on his ability to represent nature itself as amenable to imperial expansion, whether in his accounts of the "natures" of indigenous populations or of the geographies and natural products of particular regions. The same quantifying rationality on display in *Bucaniers* works in *Voyages* to represent nature thusly. Of interest here is the way this rationality works in the "Discourse of Winds," which makes up the third section of *Voyages and Descriptions.* Though not as frequently studied as the preceding sections detailing his time in the East Indies and among the logwood cutters of Campeche, the "Discourse" made important contributions to recent marine studies of interest to the society done by Robert Hooke, Isaac Newton, Edmund Halley, and Robert Boyle.[35] Though designed "primarily for the practical benefit of voyagers," the "Discourse" made major advancements in the scientific understanding of the relationships among tides, currents, and winds. It was also accompanied by the first global maps of the trade winds and prevailing tides and currents.[36] These advancements entailed an ecopoetics of environmental availability centered on the belief that the global motions of the sea were divinely ordained to operate according to rational principles and the concomitant belief that these principles could be discerned and leveraged by men of particular genius such as Dampier.

This ecopoetics emerges in the early chapters of the "Discourse," in which Dampier discusses the relationship between the trade winds and what he calls land-breezes and sea-breezes. He observes that the true trade winds blow both day and night and enable deepsea travel. They are complemented by sea-breezes, which blow in on the shore in the morning but are dormant during the evening, and land-breezes, which blow out from the shore in the evening but are dormant during the morning. These countervailing breezes perform their "Offices of the Day" "by the order of Divine Impulse," breathing the breath of life into archipelagos and rendering them accessible to mariners.[37] Dampier's use of liturgical language makes global wind systems intelligible within a theological framework that depends on Dampier's unique "Genius and Experience" for explication.[38] He cites as a case in point the confusion suffered by the sloopmen of Jamaica, whom he says "are commonly put to their Trumps, when they come there in their Voyages: For if they meet no Land-Wind they are obliged to beat about by turning windward against the Sea-Breeze in the Day time."[39] The frustration they experience leads them to "curse these Points of Land" and renders them "foolishly apt to believe that some *Daemon* haunts there."[40] Dampier further notes that "some Captains of Privateers" caught beating about the eastern end of Jamaica "have stood close to the Point, and fired their Guns to kill the old *Daemon* that they say inhabits there to disturb poor *Seamen*."[41]

Dampier's gesture toward a view of the sea as animated by supernatural forces speaks to the early modern recognition of the "divine mystery" of the sea, even as the sea became comprehensible during the period as the aqueous substrate of the global through cartographic images of the *orbis terrarum*.[42] Dampier straddles the epistemic divide between enchanted and disenchanted views of nature in the "Discourse" by rebuking Jamaica sloopmen for their ignorance while also forwarding a providentialist interpretation of the arrangement of breezes and winds—a point that Adam Sweeting likewise addresses in chapter 3. These sloopmen are for Dampier examples of "how ignorant Men are that cannot see the Reason of it" when attempting to divine where and why the winds blow about the island.[43] To avoid leaving readers "in the dark," Dampier includes additional examples of complex geographies that produce mystifying wind-effects, including the north end of the Yucatán Peninsula and the Peruvian coastline near Cabo Blanco.[44] Theological and scientific discourses demystify deep sea and coastal travel by exorcising the "Daemon" that haunts perplexed mariners, revealing the principles of order that govern the natural world.

This exorcism produces a narrative of environmental availability whereby the entire globe becomes traversable for travelers possessing the expertise necessary to sail with and against the trade winds. The Atlantic and Caribbean are of particular interest to Dampier due to the presence of land- and sea-breezes and "other helps, such as Providence has supplied these seas with."[45] These "helps" enable travel along the African and South American coastlines, and they also enable the deep-sea voyages against the trades made by Guinea and West Indies ships, illustrating what Dampier calls the "Benefit of an open Sea."[46] Though not all seas are as "open," because of variable local conditions, Dampier's commentary on the openness of the Atlantic and Caribbean has global significance: it signals an epistemic commitment to representing winds, breezes, tides, and currents as providentially configured to facilitate mercantile expansion, with particular focus on the transatlantic slave trade, and it reframes earlier seventeenth-century debates about the openness of the sea in ecological rather than juridical terms. As a result, the sea appears as the handmaiden of empire, poised to produce racial violence as a "natural" outcome of environmental assemblages.[47]

Racial violence is cloaked in the "Discourse" by a principle of abstraction that sublimates the physical violence of slavery into a discussion of the shipping routes and navigational techniques. Dampier discusses, for example, Portuguese voyages to Brazil, which require expert knowledge of currents and calms when crossing the line and navigating the Brazilian coast, as well as the voyages of English Guinea ships traveling from the Bight of Biafra to the West Indies, which likewise require a deft use of the differential flows governing westward and eastward trips across the Atlantic.[48] References to "our" Guinea and West Indies ships litter the early chapters of the "Discourse" and appear precisely at the moments when Dampier is most intensely focused on dispelling superstitious understandings of the sea in favor of a better understanding of the material causes of marine phenomena. His preoccupation with wedding narratives of environmental availability to racial violence makes the salve trade a tacit interest of the "Discourse," but earlier texts such as the *New Voyage* are much more explicit in their articulation of the centrality of slavery to the production of White subjectivity in the burgeoning world of transoceanic mercantile and imperial expansion.

Though Dampier is invested, in both the *New Voyage* and *Voyages and Descriptions*, in rhetorically distancing himself from the violence of the buccaneers with whom he travels, the *New Voyage* contains a curious passage wherein Dampier muses about the possibility of a transoceanic, archipelagic quasi-state comprising multinational pirates whose sustenance would depend on piracy, marauding, and slavery. This passage occurs within the larger context of a buccaneer raid on Guayaquil led by Captain Edward Davis in 1684. Dampier muses on the possibilities of establishing a buccaneer presence in the Caribbean and South Sea linked by footholds on the South American coast and on the Isthmus of Panama. He states, "There was never a greater opportunity put into the Hands of Men to enrich themselves than we had, to have gone with these Negroes, and settled our selves at Santa Maria, on the Isthmus of Darien, and employed them in getting Gold out of the Mines there."[49] The Darien settlement would have been created by the stolen labor of the "1000 Negroes" Dampier claims to have ready to "work for us" and would have connected the South Seas pirates with English and French Caribbean pirates to form a stronghold that crossed the isthmus and would eventually enable a takeover the Spanish mines in Central and South America.[50] The ephemeral nature of this "opportunity" leads Dampier to couch this passage in the language of reverie by noting that "these may seam to the Reader but Golden Dreams" and promising to "leave them therefore."[51]

Dampier's explicit interest in using slavery as a means of self-fashioning sheds important light on the passages in the "Discourse" that emphasize the "providential" arrangement of winds and currents in the Atlantic world. Placing the "Discourse" in the context of Dampier's earlier writings about race shows that Dampier had an abiding interest in perpetuating slavery, whether he was acting as a buccaneer, as in the *New Voyage*, or as an imperial apologist, as in *Voyages and Descriptions*. Such interest reveals the colonialist logic of both texts, which use slavery to produce White modern subjectivity, or what Wynter calls "Man1," and it reveals the centrality of African diasporic subjects in or proximate to the Caribbean to this process.[52]

Despite the overt interest in the "Discourse" in producing an orderly view of oceanic space as conducive to racialized narratives of political and economic expansionism, there are countercurrents in the "Discourse" that allow for a decolonial reading centered on the power of the environment to deny colonialist impulses. The very chapters devoted to establishing the orderly function of winds and currents are littered with references to storms, which disrupt the project of mercantile expansion and colonial settlement. The destructive power of tornadoes and monsoons are of particular interest to Dampier. He first mentions tornadoes when discussing the crossing of the equinoctial line. Dampier notes that the autumn months are typically the best, whereas the summer months bring "nothing but Calms and Tornadoes."[53] Though the tornado's wind could potentially deliver a ship stranded in the calms, "few Commanders will endeavor to take the Advantage of the Winds that come from them" because these storms are "very violent and fierce, so that a Ship with her sails loose, would be in danger to be overset by them."[54] Equatorial tornadoes threaten travel and trade across the hemispheric divide and become as constitutive an element in the "Discourse" as the sea-breezes and land-breezes that make island navigation possible. Tornadoes trouble trade in the East Indian Sea, the South Sea, and the Atlantic.[55] Taken together, these sites make visible a global geography of ecological resistance to human incursion. The contours of this geography become visible when we read the "Discourse" with an eye toward those moments when the brute materiality of the sea "shoals" the processes of colonial expansion.

One such moment occurs when Dampier describes Caribbean hurricanes. He offers the relation of one Captain Gadbury, who endured a four-hour storm while in the Mosquito Cove at Antigua. This storm drove one ship ashore and tossed it so the "Head of her Mast sticking into the Sand."[56] Another was "carried up a great way into the Woods," while yet another was "strangely hurl'd on two Rocks that stood close by one another."[57] The hovel Gadbury and his men sheltered in was left with "little beside the Walls standing," having lost its ridge and thatch. Dampier notes, "It was not the Ships only that felt the fury of this Storm, but the whole Island suffered by it"—houses were blown down, trees uprooted, and the foliage of the island stripped so bare that "all look'd like Winter."[58] Ships visiting the island after the storm found it nearly unrecognizable because of the extent of the devastation, which is punctuated for Dampier by the image of a shoreline strewn with the corpses of porpoises, sharks, and "an abundance of Sea fowls."[59]

Another such moment occurs when Dampier describes the East Indies monsoons. These storms have caused English East India merchants to suffer "very considerable losses," as "the stress of the Winds blows right in upon the shore, and often hurries the ships from their Anchors, and tosses them in a moment on the sandy Bay."[60] He stresses the losses suffered at Fort St. George, "a Place doubtless designed by the English from its Original to be the Center of the Trade" in the East Indies.[61] The frequency of the monsoons leaves ships without a "secure

Place to Ride in," acting as the "greatest Inconvenience" to English designs there and subordinating the entire English trade east of Cape Comorin to their vicissitudes.[62] The Dutch, too, once "had a place of Consequence, called *Pallicat*" on the coast of Coromandel that they were forced to abandon because of the "raging Winds" of the monsoons.[63] Similarly, mariners at Portuguese Goa had difficulty traveling safely during the monsoon season until the passing of the "Eliphanta," which was the Portuguese name for "one very terrible Storm" that signals the end of the stormy season.[64]

These storms form a global network of catastrophic weather events whose destructive power challenges claims to epistemic and physical mastery of seas, islands, and coastlines. Wind-water assemblages assert environmental sovereignty over domains claimed by imperial functionaries such as Dampier. If early modern conceptions of political sovereignty required states to make territorial claims on terrestrial spaces and the waterways that connected them and to extend the rule and reach of law across those spaces, the wind-water assemblages that permeate the "Discourse" render such claims subject to natural forces that transgress the power of states, first by crossing the physical boundaries of states and empires, and second, by repelling attempts to possess and appropriate nature.[65] The reach of these assemblages extends beyond the deep sea and coastal waters into riverine spaces that penetrate the terrestrial sphere. The monsoon seasons that threaten the East Indies with floods are part of a larger global phenomenon. Floods also beset Egypt via the Nile and cause a "vast havock" in the Americas via the Campeche, the Rio Grande, and the Ilo.[66] Contrary to the taxonomic and cartographic logic that guides *Voyages* and Dampier's naturalist writing more broadly, the watery spaces in the text defy the principles of order and predictability that organize eighteenth-century conceptions of global space. Though, as Philip A. Steinberg notes, the late seventeenth century saw the emergence of a view of the sea as "an empty space to be crossed by atomistic ships" and "the surface across which transpired much of the channeled circulation that characterized merchant capitalism," the havoc wrought by global storms and floods in Dampier's text suggests that this rationalized view of ocean-space developed along uneven temporal and geographic axes.[67] Centering the materiality of water and watery assemblages in our reading of *Voyages* allows an aqueous aesthetic of chaos to emerge as a crucial, if critically submerged, component of the text.

## PEDAGOGICAL APPLICATIONS

I teach buccaneer narratives at the undergraduate and graduate level in special topics courses on the literatures of piracy, the history of the novel, and oceanic studies. Buccaneer narratives signify differently in each of these contexts, though they never fail to elicit extra enthusiasm thanks to the appeal of the buccaneer as a transgressive figure. Buccaneer narratives—and piracy narratives more broadly—help us understand how imperialism and colonialism work. They also allow us to explore the possibilities of escaping the hegemony of empire. As important as these topics are, I argue that we must explore the representations of the environment in piracy narratives so we can better understand the historical contexts for our own age of ecological crisis. Buccaneer narratives shed light on the tensions between the desire to master the environment and the recognition that the power of sea in particular poses an existential threat to global trade and to colonization.

In our own moment, environmental questions are ineluctable and bear urgently on the future of our species and our biosphere. One of the challenges of teaching piracy narratives is that passages related to the environment are often seen as the *least* interesting parts of the story. In *Bucaniers*, the detailed narration of the exploits of famous buccaneers and their harrowing assaults on Spanish ships and settlements draws the reader's attention to questions about the instrumental use of violence and the moral character of the pirate. In Dampier's writing, the violence of buccaneer life is intentionally elided in favor of a "mixt Relation of Places and Actions" that foregrounds his observational and synthetic powers as a navigator and naturalist.[68] In both cases, the narratives center human action, which means that students have to be taught to "shoal" their readings of the overt interests of buccaneer authors and learn to read for the ecopoetics entailed by their texts.

This move is counterintuitive, given the current orientation of piracy studies, and it demands a theoretical framework that makes the need and the benefit of an environmentally focused reading strategy clear. One such framework can be found in Timothy Morton's theory of enmeshment, through which Morton emphasizes the "radical intimacy" humans share with "other beings, sentient and otherwise" within the biosphere.[69] Another may be found in Steven Mentz's theorization of "blue ecology," which recognizes the "sea's overwhelming presence in the natural environment" and reminds us that our bodies and our planet are filled with water.[70] The vistas offered by Morton and Mentz create a theoretical climate in which environmental questions take on contemporary and historical importance. Teachers can use these and other related works in ecocriticism to help students move from thinking about the environment as scene and setting and toward seeing it as a co-determinant of human action and experience. My emphasis here on "elemental regimes" can be used to consider how particular texts construct or construe their environmental contexts and to consider how environmental contexts shape particular narrative forms.

The elemental focus I take in this essay has broad utility for teachers in a wide range of fields. The archives of British, American, and transoceanic literatures from the early modern period through the nineteenth century are replete with narratives amenable to an ecopoetical reading. In my own disciplinary context of early American studies, the storms and flotsam that pop up in Columbus's logbook invite a reading of the colonial archive as troubled by the "unsettling" experience of environmental incursion into the anthropocentric epic of imperial expansion.[71] Likewise, texts as distant historically as Poe's "Descent into the Maelstrom" invite a consideration of the influence of archipelagos and the oceanic flows they create on the literary imagination of the American Renaissance. Attending to the various material assemblages that reach into and out of the archive is both elementary and essential to the forging of an informed and ethical relation to the environment.

## NOTES

1. See Marcus Rediker, *Villains of All Nations: Atlantic Pirates in the Golden Age* (Boston: Beacon Press, 2004); Peter Linebaugh and Marcus Rediker, *The Many-Headed Hydra: Sailors, Slaves, Commoners, and the Hidden History of the Revolutionary Atlantic* (Boston: Beacon Press, 2000).
2. See Robert C. Ritchie, *Captain Kidd and the War against the Pirates* (Cambridge, MA: Harvard University Press, 1986); Paul Baepler, ed., *White Slaves, African Masters: An Anthology of American Barbary Captivity Narratives* (Chicago: University of Chicago Press, 1999); Richard Frohock, *Buccaneers and Privateers: The Story of the English Sea Rover, 1675–1725* (Newark: University of Delaware Press, 2012); Mark G. Hanna, *Pirate Nests and the Rise of the British Empire, 1570–1740* (Chapel Hill: University of North Carolina Press, 2015); Gretchen J. Woertendyke, *Hemispheric Regionalism: Romance and the Geography of Genre* (New York: Oxford University Press, 2016).
3. Notable exceptions to this rule include Steve Mentz, "Toward a Blue Cultural Studies: The Sea, Maritime Culture, and Early Modern English Literature," *Literature Compass* 6, no. 5 (2009): 997–1013; Mentz, *Shipwreck Modernity: Ecologies of Globalization, 1550–1719* (Minneapolis: University of Minnesota Press, 2015); and Lowell Duckert, *For All Waters: Finding Ourselves in Early Modern Wetscapes* (Minneapolis: University of Minnesota Press, 2017), especially chap. 1.
4. I draw here on the theorization of material assemblages within new materialist philosophy. See Jane Bennett, *Vibrant Matter: A Political Ecology of Things* (Durham, NC: Duke University Press, 2010); Kyla Wazana Tompkins, "On the Limits and Promise of New Materialist Philosophy," *Lateral: Journal of the Cultural Studies Association* 5, no. 1 (2016): 1–6. C.f. Gilles Deleuze and Félix Guattari, *A Thousand Plateaus: Capitalism and Schizophrenia* (Minneapolis: University of Minnesota Press, 1987).
5. I draw and expand here on Duckert's theorization of "hydrographies" as the "onto-stories" that "non/human bodies of water" write into historical and contemporary archives by foregrounding the co-constitutive agency of water and wind within the literatures of piracy. On the emergence of air and wind as prominent elemental forces within the British literary imagination, see Jayne Elizabeth Lewis, *Air's Appearance: Literary Atmosphere in British Fiction, 1660–1794* (Chicago: University of Chicago Press, 2012).
6. I draw here on the theorization of "ecopoetics" articulated by Hume and Osborne as an investigation of "enduring assumptions about what nature has been, might be, or will be and about which objects, bodies, people, and experiences count as natural." See Angela Hume and Gillian Osborne, *Ecopoetics: Essays in the Field* (Iowa City: University of Iowa Press, 2018), 2.
7. Tiffany Lethabo King, *The Black Shoals: Offshore Formations of Black and Native Studies* (Durham, NC: Duke University Press, 2019), 1, 31.
8. Nussbaum, "Introduction," *The Global Eighteenth Century*, ed. Felicity Nussbaum (Baltimore, MD: Johns Hopkins University Press, 2003), 1–20, 1, 9.

9. See Mentz, "Toward a Blue Cultural Studies," 998; Bernhard Klein and Gesa Mackenthun, eds., *Sea Changes: Historicizing the Ocean* (New York: Routledge, 2004); Helen M. Rozwadowski, *Fathoming the Ocean: The Discovery and Exploration of the Deep Sea* (Cambridge, MA: Belknap Press, 2005); Jeffrey W. Bolster, "Opportunities in Marine Environmental History," *Environmental History* 11, no. 3 (2006): 567–597; Duckert, *For All Waters*.
10. C.f. Sylvia Wynter, "Unsettling the Coloniality of Being/Power/Truth/Freedom: Towards the Human, after Man, Its Overrepresentation—an Argument." *CR: The New Centennial Review* 3, no. 3 (2003): 257–337; King, *Black Shoals*, especially chap. 2 and "Epilogue."
11. All citations here come from the first English edition, published by William Crooke. Crooke published a second edition in 1684 and saw his editions rivaled in the marketplace by the 1684 publication of the *History of the Bucaniers* by Thomas Malthus. On the characteristics and contexts of these early English editions, see Frohock, *Buccaneers*, especially chap. 3.
12. Bennett, *Vibrant Matter*, 23.
13. I use this phrase to highlight the capacities of nature to resist the claims to "ecological sovereignty," or human sovereignty over the natural world, that Mick Smith argues are central to the formation of territorial states. I also draw inspiration from Murray Boochkin's emphasis on the "sovereignty of nature" as a "higher power" than human sovereignty and from the theorization of the "Symbiocene" articulated by Glenn A. Albrecht. See Mick Smith, *Against Ecological Sovereignty: Ethics, Biopolitics, and Saving the Natural World* (Minneapolis: University of Minnesota Press, 2011); Murray Boochkin, *Post-Scarcity and Anarchism* (Palo Alto, CA: Ramparts Press, 1971), 59, qtd. in Albrecht, 14; Glenn A. Albrecht, "Exiting the Anthropocene and Entering the Symbiocene," *Minding Nature* 9, no. 2 (Spring 2016): 12–16, https://humansandnature.org/minding-nature/.
14. Other widely read buccaneer narratives include Bartholomew Sharp's *South Sea Waggoner* (1684), Raveneau de Lussan's *Journal of a Voyage Made into the South Sea* (1689), William Dampier's *New Voyage Round the World* (1697), Woodes Rogers's *A Cruising Voyage Round the World* (1712), George Shevelocke's *A Voyage Round the World by Way of the Great South Sea* (1726), and William Betagh's *A Voyage Round the World* (1728). See Frohock, *Buccaneers*.
15. Frohock, *Buccaneers*, 2.
16. Frohock, *Buccaneers*, 4–5.
17. Exquemelin, *Bucaniers of America* (London: Printed for William Crooke, 1684), 2, *Early English Books Online*, E3894.
18. Exquemelin, *Bucaniers*, 2.
19. Exquemelin, *Bucaniers*, 2.
20. Exquemelin, *Bucaniers*, 3.
21. Duckert, *For All Waters*, xii.
22. Mentz, *Shipwreck Modernity*, esp. ix–xxiii.
23. Philip Steinberg, *The Social Construction of the Ocean* (Cambridge, UK: Cambridge University Press, 2001), 75.
24. Steinberg, *Social Construction of the Ocean*, 99.
25. Steinberg, *Social Construction of the Ocean*, 101.
26. Daniel Brayton, *Shakespeare's Ocean: An Ecocritical Exploration* (Charlottesville: University of Virginia Press, 2012), 51.
27. Qtd. in Brayton, *Shakespeare's Ocean*, 55; Steve Mentz, *At the Bottom of Shakespeare's Ocean* (New York: Continuum, 2009), 5.
28. Steinberg, *Social Construction of the Ocean*, 105.
29. On the managerial poetics of the journal form, particularly in the context of Dampier's writing, see Margaret Cohen, *The Novel and the Sea* (Princeton, NJ: Princeton University Press, 2010), 69–71.
30. See *Bucaniers*, chaps. 2–5.
31. Henri Lefebvre, *The Production of Space* (Malden, MA: Blackwell, 1991), 266.

32. Lefebvre, *Production of Space*, 277–280.
33. On the proliferation of English buccaneer narratives, see Frohock, *Buccaneers*.
34. See Anna Neill, "Buccaneer Ethnography: Nature, Culture, and Nation in The Journals of William Dampier," *Eighteenth-Century Studies* 33, no. 2 (2000): 165–180, 168; Frohock, *Buccaneers*, 105–108.
35. See Diana Preston and Michael Preston, *A Pirate of Exquisite Mind: Explorer, Naturalist, and Buccaneer: The Life of William Dampier* (London: Corgi Books, 2005), 330–331.
36. Preston and Preston, *Pirate of Exquisite Mind*, 334–337.
37. Citations are taken from the second edition (1700). See William Dampier, *Voyages and Descriptions* (London: Printed for James Knapton, 1700), vol. 2, part 3, 29, *Early English Books Online*, D166.
38. Dampier, *Voyages*, dedicatory epistle.
39. Dampier, *Voyages*, vol. 2, part 3, 31.
40. Dampier, *Voyages*, vol. 2, part 3, 31.
41. Dampier, *Voyages*, vol. 2, part 3, 32.
42. Mentz, *At the Bottom of Shakespeare's Ocean*, 4.
43. Dampier, *Voyages*, vol. 2, part 3, 32.
44. Dampier, *Voyages*, vol. 2, part 3, 32–33.
45. Dampier, *Voyages*, vol. 2, part 3, 24.
46. Dampier, *Voyages*, vol. 2, part 3, 24–25.
47. On the early modern debates about the openness of the sea, see Steinberg, *Social Construction of the Ocean*, esp. chap. 3; Lauren Benton, *A Search for Sovereignty: Law and Geography in European Empires, 1400–1900* (Cambridge, UK: Cambridge University Press, 2011), esp. chap. 3.
48. See Dampier, *Voyages*, vol. 2, part 3, chap. 1.
49. Dampier, *A New Voyage Round the World* (London: Printed for James Knapton, 1697), 158, *Early English Books Online*, D162.
50. Dampier, *A New Voyage*, 158.
51. Dampier, *A New Voyage*, 159.
52. Wynter, "Unsettling," 264.
53. Dampier, *Voyages*, vol. 2, part 3, 6.
54. Dampier, *Voyages*, vol. 2, part 3, 6.
55. Dampier, *Voyages*, vol. 2, part 3, 6–8.
56. Dampier, *Voyages*, vol. 2, part 3, 69.
57. Dampier, *Voyages*, vol. 2, part 3, 70.
58. Dampier, *Voyages*, vol. 2, part 3, 70.
59. Dampier, *Voyages*, vol. 2, part 3, 70–71.
60. Dampier, *Voyages*, vol. 2, part 3, 73.
61. Dampier, *Voyages*, vol. 2, part 3, 73.
62. Dampier, *Voyages*, vol. 2, part 3, 73.
63. Dampier, *Voyages*, vol. 2, part 3, 74.
64. Dampier, *Voyages*, vol. 2, part 3, 74.
65. On the role of mariners—including pirates—in extending sovereign claims across oceanic spaces, see Benton, *Search for Sovereignty*, chap. 3.
66. Dampier, *Voyages*, vol. 2, part 1, 34–35.
67. Steinberg, *Social Construction of the Ocean*, 105, 109.
68. Dampier, *A New Voyage*, "Preface."
69. Timothy Morton, *The Ecological Thought* (Cambridge, MA: Harvard University Press, 2010), 8.
70. Mentz, *Shipwreck Modernity*, xxix.
71. I borrow the language of "unsettlement" from Anna Brickhouse's *The Unsettlement of America: Translation, Interpretation, and the Story of Don Luis de Velasco, 1560–1945* (Oxford: Oxford University Press, 2015).

# Part Four

## EIGHTEENTH CENTURY + INDIGENEITY AND DECOLONIALITY

# 8

# "SUPPORTING SINKING NATIONS"

## Indigenous Women and Their Disasters in John Dennis's *Liberty Asserted*

MATTHEW DUQUÈS

JOHN DENNIS'S TRAGEDY *Liberty Asserted* (1704) opens with an unsettling first scene. Two Indigenous women evoke ecological disasters as they argue about the Indigenous and European nations currently warring on Mohawk land (southeastern Canada). Sakia, a Huron woman, wife to a French soldier, and a long-standing captive of the Mohawk, calls out the English and the Mohawk through her wish that "just Heaven" swiftly revenge her by causing both nations to succumb to the "sharpest Plagues!"[1] Sakia's interlocutor, a Mohawk woman named Okima, describes the Huron and the French as "gay, / But rank and filthy Weeds from Canada; / Which lately crept usurping thro' the Corn, / T'oppress the genuine and the noble Seed."[2] On account of such expressed anger, critics have rightly connected the tragedy's subsequent heavy emphasis on positive liberty—the freedom to self-govern—with the suppression of its characters' disturbing, private feelings.[3] As these opening lines suggest, though, the focus on managing emotions hinges on the glaring display of Indigenous women's rage toward the nations they despise. Their volatility—invoking widespread harm through the microbial forces of disease and crop blight—frames the play's main premise: if Huron, Mohawk, and French characters are mindful of a narrowly defined public weal, like Anglo men supposedly are, they can overcome the personal traumas caused by the warring nations as well as by the harsh American elements. Indigenous women's vengeful, disaster-promoting remarks set the stage, literally and figuratively, for the privileging of a moderately rousing, masculinist Anglo-Protestant patriotism that, Dennis's tragedy stresses, distinguishes good monarchs and their freshly empowered peoples from corrupt leaders and what the play's only English character provocatively calls "sinking Nations."[4]

With its initial projection of environmental perils onto Indigenous women, *Liberty Asserted* launched a modern Western subgenre that continues to be prevalent in twenty-first-century sci-fi and cli-fi. Kyle P. Whyte terms this subgenre an ancestral dystopia. He explains that ancestral dystopias prominently feature Indigenous characters, imparting the reputed wisdom of their ancestors, who empower white characters to fix a perilous future that is upon us. In this way, ancestral utopias make white characters alone seem destined to save the world from the disasters their colonizing ancestors have sparked.[5] Dennis's tragedy similarly exploits indigeneity—specifically, Indigenous women's work as ecological mediators—in order to advance the idea of a world-to-come rescued by English men. In doing so, *Liberty Asserted* popularized a growing form of settler colonial patriarchalism, which naturalized policies and practices of dispossession. This patriarchalism holds that only properly educated English men were emotive and rational enough to appreciate the environmental and social problems allegedly attested to by Indigenous people, and, therefore, only these men were equipped to lead a charge in amending damaged ecologies in places foreign to them (an imperial rhetoric that Christopher Black identifies in chapter 10 of this volume). Meant to set the English apart from their putatively more aggressive colonial forebears (the Spanish and the French), this ideology ironically aided new forms of harmful Anglo-American territorial conquest, predicated as it was on a facile understanding of the confluence of human and more-than-human causes and effects of global disasters. Understanding how Indigenous women's rhetoric in *Liberty Asserted* advances this ideology helps question the profound sway such thinking has had, and continues to have, on modern culture. Such questioning is warranted as it can, in turn, foster redress for historic crimes committed on Indigenous homelands and inspire collaborative strategies for resilience based on respect for people, nations, and natural forces.

## BACKDROP

During much of the seventeenth century, unusually long winters, pandemics, drought, shipwrecks, warfare, crop blight, forest fires, and starvation too often hindered England's early incursion efforts in the Americas and, at the same time, adversely affected England at home.[6] By the beginning of the Enlightenment, though, England began to have more New World successes than failures, in part because environmental conditions improved ever so slowly during this period. A pronounced warming across much of the Global North occurred: an initial waning of the Little Ice Age, which climate scientists and historians increasingly see as closely tied to colonization, specifically to the aggressive extension of American settlement, dispossession, enslavement, shipping and trade routes, and resource mining and exporting.[7]

*Liberty Asserted* was performed around the outset of this warming epoch. It debuted in London at Lincoln's Inn Fields Theater's newly renovated stage and had a very successful run throughout the spring, likely because its subject matter was so topical. The plot revolves around the Iroquois, or Haudenosaunee, mourning war practice of prisoner adoption and assimilation, which was complicating ongoing European-Indigenous conflicts in North America. Mourning wars served to replace human losses from battle and from harsh environments and grow the five nations that made up the Iroquois Confederacy, while also separating non-Iroquoian conflict survivors from their families, kinship groups, and homelands.[8] From the English perspective, the result of ongoing mourning wars was twofold: it provided a renewed sense that they had chosen the appropriate Indigenous allies, and it fueled an early Enlightenment interest in the social and political questions of identification raised by their Indigenous allies' aggressive means of incorporation. Early in the tragedy, the significance of these developments become apparent. We learn that twelve years ago, when the Mohawk (one of the five Iroquoian nations) and the English were fighting the French and the Huron, the former won a battle that enabled them to take Sakia and her son Ulamar from their Huron homeland and family as well as from their attachments to their French allies. This deracination included, most poignantly, a removal from the Frenchman Miramont, Sakia's beloved husband and Ulamar's father. The tragedy's rising action stems from Sakia and Ulamar's differing reactions to their abduction and to the increasing power and provenance of their Iroquoian captors. Sakia continues to blame the Mohawk for her inability to return home. Ulamar, who has come of age without full knowledge of his past, has fully assimilated to the Mohawk Nation and thrives as a warrior fighting with the English against the Huron and the French. Tensions build in *Liberty Asserted* as Sakia tries to keep her son from unwittingly murdering his French father Miramont, and Ulamar questions his mother's motives and competes for the "Iroquoian princess," Irene, with his compatriot, the English general Beaufort. Iroquoian and French delegations threaten existing alliances when they enter the play. Modeling a warmer blend of stoicism and wit, Beaufort reunites the separated Huron-French family, in large part by stepping aside from the heated contest to let Ulamar and Irene marry (importantly keeping the English man from intermarriage). Humanity in Canada now acts precisely in accordance with the supposed natural order of things qua unfettered respect for English positive liberty, the tragedy ultimately trumpets.

While *Liberty Asserted* is hardly a historically accurate dramatization of early-eighteenth-century Indigenous-European relations, as this summary makes clear, it is not a baseless, imperialist fantasy either.[9] In addition to its inclusion of common Iroquoian prisoner adoption practices, it is grounded by current events. The plot is influenced by late-eighteenth and early-eighteenth-century conflicts in Mohawk territory, which pitted the Iroquois and the English against the Huron and the French, conflicts that were, by 1704, part of Queen Anne's War, the North American theater of the War of the Spanish Succession. The Iroquois were growing during these years of conflict while the Huron saw significant declines.[10] The English, who had lagged behind the French for decades, were becoming nearly as powerful in the colonial northeast as their European adversaries. *Liberty Asserted* builds on these trends. It conveys recent military developments, while predicting a plausible outcome: an English-Iroquois victory in Queen Anne's War resulting in England's theoretical suzerainty over the Iroquois and England's unilateral rights to trade with Indigenous nations in the continental interior.

The tragedy's relative accuracy can be explained by its investment in first-hand accounts of French explorers who had experienced the sights and sounds of Mohawk territory, unlike the playwright. Dennis, for instance, drew in creative ways on place-specific terms and ideas found in contemporary French colonial literature.[11] He relied heavily on Baron de Lahontan's popular travelogue *Nouveau Voyages et Norte Amerique* (published in English in 1703). "Angie," the name for where the action occurs and for the main Indigenous nation represented therein is, according to Lahontan, the French designation for the Mohawk homeland and the Mohawk themselves. Dennis notes that he chose to alter the word "Agnie" to "Angie," for "the sake of better sound," improving the more guttural French transliteration.[12] Frontenac, one of two French characters in the play and the name of an important military leader, appears in Lahontan's travelogue. The names Sakia and Okima come from an Algonquin lexicon appended to Lahontan's text; Sakia means "love" and Okima means "leader." Ulamar likewise comes from this same source. It means "red powder." Irene, Ulamar's Iroquoian bride and the initial love interest of the English Beaufort, means "peace."

These name choices index the ecological implication of this tragedy. Dennis draws on the ostensible power of indigeneity, the wisdom of First Peoples. In keeping with the features of Whyte's ancestral dystopia, he manipulates that wisdom for non-Indigenous ends, altering the French transliteration of the Algonquin word for Mohawk territory, enacting linguistically what his English compatriots were carrying out territorially: physically changing the landscape and its meaning to fit their desired colonialist end while also expurgating presumptive human and more-than-human obstacles to trade and settlement. Dennis suggestively anthropomorphizes his setting in the preface as the "back of New York and New England." The phrase evokes a supine entity ripe for conquest *and* a source of physical support for the colonies, a mixture reflecting the moderate environmental state imagined for the region and the matching moderate disposition of its English male protagonist. Dennis ultimately brings together the young couple, Ulamar and Irene, into lockstep with this Anglo-centric ecological framework, under Beaufort's guidance, a central point symbolized by the couple's names ("red earth" and "peace"). Ulamar, the Huron-French-turned-Iroquoian-English Protestant warrior becomes an Indian Adam, whose name means "red" and whose affinity to the Hebrew word Adamah, or red-ground or red-clay, signifies a new union between humankind and the earth. Wedding Ulamar to Irene rounds off the idea that a more peaceful treatment of American land and inhabitants than under past colonial compacts between French and Native people is now possible under English control. Because the drama touts a conception of English-positive liberty under which all compacts grow healthfully, this narrative about conflicting Euro and Native American figures who rise up out of their circumstances popularized a colonialist mentality that serve to attenuate Indigenous peoples' ties to their physical homelands while strengthening English rights to those homelands. How the promotion of such reformed ecological bonds happens in *Liberty Asserted* depends significantly, though, on what its two main Indigenous women characters declare when the curtains first open.

## LOVE AND LEADER

Sakia and Okima catalyze the English-led regenerative ecology Dennis's tragedy builds toward by conveying the shocking idea that "Angie" has been characterized by surmountable, forcefully obscured kinship and other climate and culture-induced traumas associated with displacement. It is, at least initially, a kind of dystopia, a story suggesting to its inhabitants that an even worse world than the one they find themselves in might be imminent. The two women animate this conception through their heated exchange in the first and, importantly, the longest scene in *Liberty Asserted*. Sakia's first, in medias res, lines of the play, "speak on, thy looks seem big with something that's important," highlight her ancestral authority, but also, importantly, the limits of her speech and understanding (note the imprecise diction), associating her character with a traumatic experience, which has heightened her sensitivity to external phenomena yet also compromised her rhetoric.[13] In response to Okima's news that a recent battle is over, and the Mohawk and English forces are about to return victorious, Sakia uses her lineage to explain what causes her to have such feelings:

> SAK.
> Hurons we once were call'd, and once were thought
> To be descended from no Vulgar Stock,
> But now, alas! are sunk to wretched Slaves
> OKIM.
> To Slaves, *Sakia!*
> Your Son at least has other Sentiments.
> SAK.
> Howe'er alas, my Son may be deceiv'd,
> I am a Slave, a miserable Slave;
> Who far remov'd from my sweet Native Soil,
> Far from the dearer Partner of my Heart,

Have for twelve tedious Years been now confin'd
To drag the galling Yoke of loathsome Life,
In this accursed Place:
A Burthen which I never could support,
But that I had some glimpse of Hope remaining;
That cruel Fortune might at length relent,
And might at length restore me with my *Ulamar*,
To the dear Partner of my faithful Heart:
Those Hopes have lately been reviv'd and animated
By the prophetick Visions of the Night,
And Expectation of this fatal Action;
But this Relation has o'erthrown them all:
My Son's mistaken Valour has undone us,
And thou, O *Miramont*, art lost for ever![14]

This exchange underscores Sakia's suffering. But it also does more than this. It encourages the audience's identification with Sakia because of her suffering in a way that enables them to transcend it and, by definition, her. Sakia creates strong feeling for her situation through her note about her dramatic decline in status from "no vulgar Stock" to being a "slave," which she emphasizes as entailing a lost homeland and then a separation from her husband, "the dearer partner of her heart." The contrast in the descriptive words for where Sakia was and where she is now ("sweet" versus "accursed") highlights the importance of this lost homeland as does a shift in nouns from the more elemental "soil" to the more general "Place" and the meaning of Sakia's husband's name, Miramont, or mountain view. This powerful, affective ecological language is couched in familiar early-eighteenth-century British abolitionist rhetoric designed to show Sakia's change from a resistant slave-like figure to a grateful slave-like figure.[15] We witness this at the beginning of the exchange, when Okima interjects to remind Sakia that there are other valid feelings about her displacement from her homeland and her loss of status and her husband: "your son at least has other sentiments." And we see reminder at the end, when we learn that Sakia's "prophetick Visions of the Night"—a phrase associated with witchcraft and Catholic priestcraft since the English Civil War—have been dispelled by Okima's "relation," a term associated with accuracy and rational explanation. The exchange then is not directly questioning Sakia's pain. It serves to flag her suffering, even to the point of emphasizing that her son, who has "other sentiments," may indeed be "deceived" or "mistaken" while encouraging the audience to question Sakia's attachment to reality, to the soil, to what is to come.

The scene continues to advance English audiences' compassion for Sakia in a way that elevates them above her as well as above an American binary of "sweet soil" and an "accursed place":

OKIM.
Forget your Hurons and become an Angian [Mohawk]
SAKIA.
O may the whole accursed Race by Fate
Be rooted ev'n from Human Memory!
Perish their very Names too with their Persons,
Excepting thine, for thou art wondrous good.
OKIM.
*Sakia*, you forget,
To curse your Benefactors thus is impious.
SAK.
To curse our mortal Enemies is just.
OKIM.
The Angians are your Friends and your Defenders.
SAK.
My Country's mortal Enemies are mine.
OKIM.
*Iroquian Angie* is your Country now.
SAK.
My Prison never can be call'd my Country.
OKIM.
The Angians gave you Liberty at first;
The Hour that made you Captives, saw you free.[16]

A critical difference between the two positions emerges as Okima and Sakia go back and forth with matching ancestral concepts. Okima makes the commands, taking the leadership role from Sakia using rhetoric. These commands seem grounded despite the fact that she asks her to "forget" her people and to overlook a contradiction in terms (a captivity that means liberty). In contrast, Sakia's prayer for vengeance, her hope, in this instance, for the removal of all human and more than human reminders of her captivity, is a less grounded, more clearly Lockean-inspired projection: the dream of a blinkered tabula rasa, a pre-sensate reality already defined implicitly by moderately progressive Anglo-Protestant values qua natural values.

Because of its otherworldly resonance, what is essentially a curse evokes the early modern associations with sin, superstition, witchcraft, and priestcraft. It adheres to the notion of disaster (*dis-astre*, bad star, as in born under a bad star), a point of moral significance that Okima underscores in order to firm up her case. Sakia's reference to outliers ("excepting thine") and her heavy use of the first person further weaken her case. They make an audience feel for what she calls her "Yoke of Slav'ry," even as they become invested in seeing themselves as uniquely positioned to do something constructive with her passions. *Liberty Asserted* primes audiences, particularly English male audience members, to be the sui generis ones equipped to make Sakia's age-old psychological, physical, and material conditions (her "detested clime") more palatable.

Okima tempers Sakia's anger by suggesting that Ulamar has been fostered by the people she curses. Both he and the Iroquois are now victors:

> OKIMA.
> Think that our Conquest to your Son we owe,
> Then curse us if you can.
> SAK.
> Therefore the barbarous Iroquois I curse.
> OKIM.
> Upon your Son's account you ought to bless us,
> Think on our Bounties heap'd upon his Head.
> SAK.
> Think on the Trophies by his Valour won,
> With which tenfold your Bounty he repays.
> OKIM.
> 'Tis to our Favour that he owes his Glory,
> Which, tho' he scarce has reach'd his twentieth Year,
> Surpasses that of all our ancient Warriours.[17]

Here Okima shifts the narrative from Sakia's pining after her homeland and husband and toward the sociological positives of her confinement: her son's successful military accomplishments. This move confirms Okima's position that God is on the side of the current winners: "You curse the Victors, Heav'n has curst the vanquish'd," lending the winners in this contest an environment that is conducive to healthful growth.[18] Here, Sakia wishes that "just Heaven by the sharpest Plague revenge," and Okima expresses her frustration with Sakia's unwillingness to assimilate when she speaks of the French and Huron as "gay, / But rank and filthy Weeds from Canada; / Which lately crept usurping thro' the Corn, / T'oppress the genuine and the noble Seed."[19] Both lines evoke a familiar (to the English) seventeenth-century image of an America whose inhabitants are in grave ecological danger from pervasive death caused by disease, starvation, and of course a lack of proper faith—which Adam Sweeting also locates in chapter 3. Sakia's words align future perils with uncontrollable early modern forces while Okima's words align current perils with more modern proto-Romantic human forces that can be altered by other human forces who can create a healthier, more "natural" world defined by a less-fettered pursuit of liberty, fecundity, and bravery, the likes of which have not been seen by "Westerners" since classical antiquity.

In its own way, the dialogue in this crucial opening scene establishes the idea that this contested sliver of North America was, at this moment, not just a conceptual middle ground but a material one, suited to properly arouse English men and their othered proteges.[20] "Angie" is, the scene establishes, an alluring halfway between a limitless Eden-like garden defined by untouched and unknown growth and beauty and a hellish tundra defined by indiscriminate death of human and plant life. It is a middle ground allowing for the possibility of measured, humanity-controlled growth and removal because its moderate environmental elements are conducive to fostering the kinds of Anglo-male modeled people who are disinclined to aggressively confine and expand, kill and starve matter. Such a middle-ground hypothesis was based on a neo-Aristotelian Western theory of environs. This theory held that dry heat yielded wit and science, whereas wet cold environments fostered corporeal fortitude and an aggressive disposition. And, for this reason, people inhabiting "the temperate regions who blended the wisdom of the south with the fortitude of the north" were considered physically and morally superior, as Londa Schiebinger explains, "predictably . . . account[ing] for the dominance of the hotter male over the colder woman."[21] Drawing on this principle, eighteenth-century English male writers advanced the position that the remote American outposts that they had found were difficult yet also fecund environments, an alluring position aided by the new modern climate in the Global North, which likely witnessed the earliest and most dramatic changes in temperature on colonial American ground around the outset of the Enlightenment.

In late-seventeenth-century promotional writings, English colonists presented desirable areas of the American backcountry as could-be Aristotelian temperate regions. They did so often by projecting the deadly effects of ecological disasters onto Indigenous people whose own ecological views and traditionally matrilineal modes of governance were elided as they complicated theories of a found, temperate region. As Amy Den Ouden notes, in his efforts to construe New World environments in moderate terms, English colonist Daniel Gookin described Indigenous women's theories of natural phenomena, including their explanations for outbreaks, accounts of the origins of deadly bogs, the import of disturbing weather patterns, animals, and spirits. Den Ouden contends that such descriptions "evoke [that knowledge] only so that it may be invalidated and silenced."[22] Silencing, she underscores, "enmeshed with the material and military processes of dispossession."[23] It served to cut ties between sacred, environmentally unstable, places and Indigenous women, who traditionally had such knowledge because they were responsible for both passing on this essential information to children and using it to make political decisions. Local accounts of natural phenomena, Den Ouden states, "embody the histories and identities of their people and bind them to their homeland."[24] Invalidating such accounts denies the domestic/educational transmission of environmental knowledge from one generation to the next and discredits the use of such knowledge as a civic and social tool.

A metropolitan dramatist rather than a colonial ethnographer, Dennis presented North America and its First Peoples in a different manner from his English contemporaries abroad, but with the same aim in mind. He conveyed Sakia's passions in response to her trauma through a wish for plague, and he conveyed Okima's assessment of Sakia's trauma through her view of the Huron and French as acting much like crop-blight-inducing weeds. Unlike the examples Den Ouden provides, these comments harbor views of natural phenomena; they are not Indigenous women's claims about natural phenomena. Dennis does not describe and then discredit Indigenous women's environmental knowledge. He uses their loaded disaster comments to flag and then temper a Huron woman's spirited resistance to her captivity and displacement by contrasting it with a Mohawk woman's ethnocentrism, exemplified by her characterization of outsiders as invasive plant-like others. Sakia furnishes the strong, aggressive disposition associated with cold and wet northern climes and with women. And while both characters display wit, it is Okima who displays the better part of it, indexed by a sense of humanist protoscience in her rhetoric. Together, though, through extreme negative examples, tantamount to expressions of genocidal violence, typified by their references to natural disasters, they introduce the idea that "Angie" could become a temperate zone. It is, therefore, an ideal environment for English men, like Beaufort, and English-tutored Indigenous men like Ulamar to realize their fullest potential, or become seminal, godlike leaders therein. Dennis's combination of Indigenous women's voices—progressively silenced in the succeeding scenes by the competition between men for women and by military victory—positions metropolitan English male subjects as morally and physically superior prospective settlers and supporters of a moderate, or less aggressive, less indifferent process of colonization. A documented seventeenth-century history of unstable environs in northeastern North America and what would soon be conceived as Britain were causing the English and their allies who subscribed to Aristotelian views of the temperate zones to think they were becoming either too bellicose or too meek. Dennis, perhaps sensing that the weather was warming in the Global North, and that the conditions of living were becoming more hospitable across the Atlantic, penned *Liberty Asserted* to represent an affirming American middle ground that would, by virtue of its incorporation as territory, raise up England and, just behind them and, therefore, with them, French and Indigenous subjects as well.

To explain the political bent of *Liberty Asserted* in more conventional Western literary terms, Sakia functions as the shocking, more abstract (pre-Enlightenment), Miltonic ancestral figure associated with divinity—hence her name, which means "love." Her role in the tragedy is principally to energize and transform more literal European characters (Frontenac, Beaufort, Miramont) whose plausibility is indexed by the fact that they refer to actual historical figures and common names.[25] Okima is the more reasonable conduit ancestral figure; her rhetoric, antagonistically humanist rather than antagonistically theist, creates a plausible national field for Sakia's powerful emotions, hence Okima's name meaning "leader." Okima's name choice, as Bridget Orr notes, may be yet another sign (along with the focus in the first scene on the power of Indigenous women) that Dennis tried to accurately represent the fact that Iroquois are a matrilineal people, a fact that he could have found in a host of Anglo and French colonial texts, a fact, moreover, that an ardent supporter of England's reigning Queen Anne might have found intriguing. Dennis's self-serving accuracy, this first scene shows, though, is part of his effort to manipulate Indigenous women characters for a desired sociopolitical reformation of his national culture: the play's female "leader" is a confidant for "love," a lady-in-waiting who evokes the passions and councils, then recedes into the background of the production once she has drawn out and made the divining yet potentially destructive power of "love" instrumental to *Liberty Asserted*'s English plan. Such a dramatic process was meant to fire up a London audience with a moderate national passion, but it also limited their emotional and fiscal investment in reforming ongoing colonizing efforts. The rising action demanded that audiences oppose the agonies of captivity and enslavement and the human and more-than-human forms of mass death associated with both—priming them to see indifference to such pervasive tragedies as unchristian. The rising action, however, also inclined audiences to tolerate such practices if only because it afforded the opportunity to see themselves as putting the displaced and traumatized on the spiritual and material path to redemption.

The ecological implications of Sakia and Okima's haunting ancestral message become still clearer if we consider their initial exchange as reflecting not just patterns of representation found in early Enlightenment colonial ethnography, but also patterns found in early Enlightenment diplomacy. Early eighteenth-century diplomatic exchanges between the French, English, and the Iroquois, as we might expect, entailed much posturing and positioning. Full of complex, heated emotion on account of the huge stakes, these exchanges unfolded like good theater: highly formal performances, characterized by the heavy use of tropes and props used to avoid translation issues. For example, in New York in 1689, a French diplomat named La Barre and an Onondagan elder named Garangula each delivered a speech (printed in London two decades later and reissued extensively during the French and Indian War) in which La Barre claimed the Iroquois were weakened by disease and famine.[26] In turn, Garangula called out the French diplomat for wishing "that the lakes had overflown the banks that had surrounded our castles."[27] Garangula pointed out that it was not the Iroquois who were "lost in the ground" near the "Great Tree of Peace," but rather the French contingent who were "submerged" and "clouded" because they were led by a man who "speaks as if he were dreaming," a man "raving in a camp of sick men."[28] While speeches like these by men were archived, Indigenous women often led or advised in these negotiations, playing the same roles, though they were ultimately omitted from print records. In *Liberty Asserted*, Dennis does not present Sakia and Okima as playing the important diplomatic roles that Indigenous women often played. But their introductory exchange resembles La Barre and Garangula's in their use of natural disasters to index the weakness of opposing nations and the power and provenance of their own nations. As with La Barre and Garangula's speeches, so too with Sakia and Okima's: the loser represents the diseased and deadly as well as the floundering and rootless, the latter exemplified by Sakia's threat to look for her husband whom she presumes is in France, to "trespass the ocean that separates these climes."[29] Meanwhile, the winner promised not only sociopolitical reform but environmental reform accompanying their nation's divine favor and victory: Okima conveys Anglo-followed-by-Mohawk-led physical and material rejuvenation in Mohawk territory—the uninhibited growth of freshly sprouting "noble seed," as Okima proposed and Ulamar exemplified through his marriage to Irene.

Dennis underscores this distinction in his tragedy. Sentimental marriages and political alliances, with English Beaufort's help, wrap up the play. Sakia remains in "Angie" reunited with her French husband. Ulamar and Irene marry, and peace is achieved between Indigenous and European nations. Even the French can be reformed, the drama concludes. Dennis's unabashed Anglophilia suggests that these feelings and actions and struggling environments will be fixed by the English, a point best expressed in the concluding line of Sakia's son Ulamar: "O Great *Britannia*," Ulamar wonders, "thro the world renown'd / For propping falling liberty / supporting sinking nations."[30] Such an invocation has ecological implications. It weds the topographical and moral status of non-English nations in order to emphasize their sociopolitical instability and, in contrast, to underscore the immaterial, oral power of the English to overcome environmental obstacles. This colonialist humanist position is reinforced by the distinction between this final image of propped up "sinking nations" and several intriguing references in *Liberty Asserted*'s prologue to "streams of despair"—an emphasis on perilous waters that enlivened the spirit of the English even as it was also potentially applicable to the Five Iroquois Nations who viewed their original homeland (Iroquoia) as a spirited island.[31] This position of English superiority becomes paramount in the first scene through references to plague and crop blight from Sakia and Okima. The final invocation to "Great Britania" confirms the diplomatic solution toward which Sakia's and Okima's morbid, divisive comments have oriented the tragedy's characters. It is a parting reminder that such a diplomatic solution entails a fundamental ecological adjustment to affect political and economic changes. Plague and crop blight induce not just the mass death of people, but the transformation of everything more than human to which humans give meaning and with which they are ineluctably entangled. New relationships between organisms and their surroundings must take root, or all seems lost. Dennis's tragedy promotes such new relationships at the expense of the Huron and the Iroquois, rejecting their provenance and their ecological theories and practices. *Liberty Asserted* achieves this effect by cultivating a desire for a warming Mohawk homeland and its peoples, both of which are now ostensibly less prone toward catastrophic animal and plant mortality and, therefore, more grounded, reliable, and responsive to humanity's selective whims. Enflamed by Sakia's passionate fortitude and organized by Okima's domineering, scientific wit—together the bearings of collective ancestral dystopic baggage—we have the next viable temperate zone, and only English men can save it.

## HOW TO USE THIS CHAPTER

Reading *Liberty Asserted* critically can teach us how to imagine collaborative counter-colonialist approaches to the ecological challenges now associated with climate change. Students can first come to learn where Dennis's late Restoration drama fits in a long history of Anglo and white U.S. reformist responses to environmental devastation associated with settler colonization. This history includes, in recent memory, Keep America Beautiful's 1970 "crying Indian" advertisement, Disney's *Pocahontas* (1995), and sensational media on the Dakota pipeline protest and the Indigenous Peoples March in Washington. Evocative illustrations of U.S. environmental problems, they discover, have long been, and continue to be, tied to the emotional appeals and insights of Indigenous people and framed in a way to privilege non-Indigenous people who take stock of the ancient wisdom (ostensibly) inherent in Indigenous tears, songs, and speeches. Such powerful messages for environmental reform tend to present Indigenous people as impoverished, unfortunate vestiges of a more natural era, at best scorned and/or receptive love interests, and, in the case of Indigenous women, anything but complex collaborators and future equals. This pattern directs behaviors. It makes those of us who are not Indigenous presumptuous about how to clean up everyone's acts and how to respond to environs that ought to belong to all of us, or none of us. Recognizing the effects of this history can inspire revised cultural and practice-based local knowledge about environs: Indigenous and non-Indigenous sustainable commons traditions set collectively on a level longhouse table as we work together to rectify our current climate disasters.[32]

Scholars of the eighteenth century can use this chapter to inspire such revised thinking, thereby advancing the activist aims of eighteenth-century environmental humanities. To do this, one must use compelling storytelling to ground Dennis's play in its metropolitan context. Try to re-imagine London at the beginning of the eighteenth century as a freshly crowded, globally connected city where literary debates about the stage were highly politicized. Dennis, a prominent London literary critic and dramatist, sought to set the rules for English drama and to impress the court and his audiences with his tragedies, all for partisan nationalistic ends. To help better understand what Dennis's activities meant at the end of the Restoration period in London, one could have advanced undergraduate or graduate students create an annotated version of *Liberty Asserted*. No annotated version exists. This editing activity will immerse you and your students in this newly modern, urban place by way of Dennis's pioneering literary criticism, such as "The Usefulnes of the Stage" (1700), which left its deep mark on his contemporaries and on the later Romantics. Such contextualizing editorial activities ought to include reading some of Dennis's influences for this particular tragedy, including the writings of Longinus, Aristotle, John Locke, Mary Astell, Richard Baxter, Aphra Behn, John Dryden, Baron de Lahontan, and the Jesuit authors of the multi-tome early American classic, *Jesuit Relations*.

There are also juicy biographical tidbits about Dennis worth noting. Dennis's spat in print with Alexander Pope, who would later eviscerate him in verse for being a literary martinet, is well known and worth getting a handle on. But there are two great tall tales still associated with Dennis. The first tale holds that, following the successful staging of Dennis's third anti-Jacobin play, he was walking along the London port and saw a ship in the offing with a French flag and presumed that the ship was coming into port specifically to arrest him for his play's political sentiments. The second story holds that Dennis invented a machine specifically for the stage to create the sublime sound of thunder. When he heard tell that a competing theater company had stolen his machine, he supposedly uttered (and coined) the idiomatic phrase, "they have stolen my thunder!"[33]

Such dramatic context helps us begin to think about the scenes of *Liberty Asserted* ecologically. We might learn, for instance, that, through its use of the latest "technologies," the stage was seen in London as a powerful international political agent commensurate with the ship, the cannon, and the sword for the fight for a global Protestantism.[34] Arguably, this was because it was a more materially oriented arena through which to represent the world than the increasingly fashionable medium of print. The stage could bring matter to life in more visceral, embodied ways than the more abstract combo of the pen and paper. For this reason, drama contributed to debates about the vitality of objects (vitalism—the eighteenth-century precursor to new materialisms), debates often fueled by the assumption that lively things could work wonders either against or in service of nationalist modern humanist impulses. How, we might ask, did Dennis wield the formidable role of the English stage, propelling his topical production to be ideologically and materially complicit or resistant to the environmental depredation accompanying colonization? In an effort to answer this question, imagine Dennis directing Sakia and Okima to wear the Turkish styles that the famous four "Indian

Kings" (three Mohawk, one Mohican) were asked to wear when they visited London a half a decade after *Liberty Asserted* was first staged. Imagine him encouraging the two prominent English actresses to deliver his lines emotively in keeping with the neo-Aristotelean strictures of tragedy. Imagine how such deliveries might have positioned Englishmen as sensitive enlightened models suited to overcome the ecological concerns raised by the disasters that had crippled North America and England since the advent of colonization: the so-called starving time in Virginia, a smallpox outbreak in New England, London's bout with the plague, the seven years famine in Scotland. Think, moreover, about how such dramatic deliveries presented Indigenous women as inferior, even necessarily expendable, fodder for ecological reform. Finally, consider how such scenes might be rewritten to show a less lopsidedly Anglo-centric transcendent outcome.

To round out a revisionist, or a more accurate, perspective, one will want to draw on recent scholarship. Consider, for example, Susan Sleeper-Smith's *Indigenous Prosperity and American Conquest* (2018); chapter one of Caroline Wigginton's *In the Neighborhood* (2016); Daniel Richter's works, particularly *Trade, Land, Power* (2013) and *The Ordeal of the Longhouse* (1992); Lisa Brooks's *Common Pot* (2008), Drew Lopenzina's *Red Ink* (2012), and Jace Weaver's *Red Atlantic* (2014), to name a few illuminating "early" Native American monographs focused on the first quarter of the long eighteenth century. One might also turn to early American/modern Global North environmental cultural histories, such as Thomas Wickman's *Snowshoe Country* (2018), William Cronin's touchstone monograph, *Changes in the Land: Indians, Colonists, and the Ecology of New England* (1973), and Katherine Rigby's *Dancing with Disasters* (2015). To this list of secondary sources, one could also include recent work by and about the Indigenous peoples and places represented by Dennis, including Kevin Connolly's work on Onondaga thought; Audra Simpson's *Mohawk Interruptus* (2014); and Shirley N. Hager and Mawopiyane's *The Gatherings: Reimagining Indigenous-Settler Relations* (2021).

With this secondary work in Native American studies and Global North climate history in mind, one can further assess how *Liberty Asserted*—particularly the play's opening scene—may establish a firmer sense of local and transatlantic environments as de facto characters pushing the tragedy toward an English colonialist outcome. Consider how these forceful more than human characters shaped a drama founded on both the idea of colonial land and trade in masculinist capitalistic terms rather than matrilineal and communalistic terms and on the accompanying idea of Indigenous people as "savages." Imagine how modern Western extractive and touristic behaviors grew out of such ecologically inspired productions and the ethnographic conventions on which they were based. In culture and in practice, exploiting Indigenous women's powerful traditional roles fomented English male-dominated settler colonization, encouraging English administrators and financiers to treat America's land and its First Peoples as transferable and irrevocably reformable. These practices uprooted Indigenous women from their homelands and, in many cases, disabused them of their roles as local political leaders and ecological knowledge creators and conveyors. One might also sit with the provocative notion that, at this juncture in history, the London stage, Mohawk pow-wows, diplomatic speeches by Indigenous elders and colonial leaders were all more alike in their complex formal features than most might presume. Northeastern Indigenous people and the English saw environmental phenomena, including diseases, fires, drought, and famine as signs of divine favor or judgment, a shared perspective that lent their cultural expressions philosophical as well as formal resemblances. Moreover, the Atlantic world from which these cultural expressions originated was also more well connected than we might think. Influential ideas, texts, and people moved back and forth across the ocean. Northeastern Indigenous people participated actively in the shaping of this world through their deft work with a range of media and print forms and through their travels, as Brooks, Lopenzina, and Wigginton show. Similarities and connections between northeastern Indigenous people and European people were stifled by patriarchal English settler colonial regulation and abstraction, a process mirrored in Dennis's play. Returning *Liberty Asserted* to its context directs us to a vital red (Indigenous) Atlantic as well as an English and a French Atlantic that could have enabled shared, resilient roots in the eighteenth century and may well inspire us to more ethically seek out survival in the twenty-first century.

## NOTES

1. John Dennis, *Liberty Asserted: A Tragedy* (London: George Stahan and Bernard Lintott, 1704), 3.
2. Dennis, *Liberty Asserted,* 3.
3. Brett Wilson, *A Race of Female Patriots: Women and Public Spirit on the British Stage, 1688–1745* (Lewisburg, PA: Bucknell University Press, 2011); Julie Ellison, *Cato's Tears and the Making of Anglo American Emotion* (Chicago: University of Chicago Press, 2001).
4. Dennis, *Liberty Asserted,* 72.
5. Kyle Powys White, "Indigenous Science Fiction for the Anthropocene: Ancestral Dystopias and Fantasies of Climate Change Crises," *Environment and Planning E: Nature and Space* 1–2 (2018): 224–242.
6. See Kathleen Donegan, *Seasons of Misery: Catastrophe and Colonial Settlement in Early America* (Philadelphia: University of Pennsylvania Press, 2013).
7. J. M. Grove, *Little Ice Ages: Ancient and Modern* (London: Routledge, 2004); A. Koch, C. Brierley, M. Maslin, and S. Lewis, "Earth System Impacts of the European Arrival and Great Dying in the Americas after 1492," *Quaternary Science Reviews* 207, no. 1 (2019): 13–36.
8. Jon Parmenter, "After the Mourning Wars: The Iroqouis as Allies in Colonial North American Campaigns 1676–1760," *William and Mary Quarterly* 64, no. 1 (2007): 39–76.
9. With the exception of Orr, scholars have tended to read the play, more or less, as a state of nature allegory. For example, Richard Braverman, *Plots and Counterplots, Sexual Politics and the Body Politic in English Literature,* 1660–1730 (Cambridge, UK: Cambridge University Press, 2008).

10. See Jon Parmenter and Thomas Wickman, *Snowshoe Country: An Environmental and Cultural History of Winter in the Early American Northeast* (Cambridge, UK: Cambridge University Press, 2018). On French early modern discourse about the Huron and French as the pious good guys and the Iroquois and the English as the impious bad guys (an early modern discourse, which Dennis turns on its head), see Zachary Yuzwa, "The Fall of Troy in Old Huronia," in *Classical Reception in the Early Americas,* ed. Adam Goldwyn, Maya Feile Tomes, and Matthew Duquès (Leiden: Brill, 2021).
11. See Orr, *British Enlightenment Theatre: Dramatizing Difference* (Cambridge, UK: Cambridge University Press, 2019). Orr notes, "Sakia's stubborn adherence to her original loyalties figures the internal threat to the state when an immigrant retains alien affiliations" (123).
12. Dennis, *Liberty Asserted,* iv.
13. Dennis, *Liberty Asserted,* 1.
14. Dennis, *Liberty Asserted,* 3–4.
15. George Boulukos, *The Grateful Slave: The Emergence of Race in Eighteenth Century British and American Culture* (Cambridge, UK: Cambridge University Press, 2008).
16. Dennis, *Liberty Asserted,* 4–5.
17. Dennis, *Liberty Asserted,* 5–6.
18. Dennis, *Liberty Asserted,* 5.
19. Dennis, *Liberty Asserted,* 5.
20. I draw on Richard White's germinal term. Richard White, *The Middle Ground: Indians, Empires, and Republics in the Great Lakes Region, 1650–1815* (Cambridge, UK: Cambridge University Press, 1991).
21. Londa Schienbinger, *Nature's Body: Gender in the Making of Modern Science* (Boston: Beacon Press, 1993), 186.
22. Amy E. Den Ouden, *Beyond Conquest: Native Peoples and the Struggle for History in New England* (Lincoln: University of Nebraska Press, 2005), 48. More recently, Drew Lopenzina calls this colonialist rhetorical act "unwitnessing." Drew Lopenzina, *Red Ink, Native Americans Picking Up the Pen in the Colonial Period* (Albany: SUNY Press, 2009).
23. Den Ouden, *Beyond Conquest,* 48.
24. Den Ouden, *Beyond Conquest,* 87.
25. See Steven Knapp, *Personification and the Sublime: Milton to Coleridge* (Cambridge, MA: Harvard University Press, 1985); and Catherine Callagher, "The Rise of Fictionality," in *Novel,* vol. 1, ed. Franco Moretti (Princeton, NJ: Princeton University Press, 2006), 336–363.
26. "Garangula, Speech to New France's Governor La Barre," in *The Literatures of Colonial America,* ed. Ivy Schweitzer and Susan Castillo (New York:Wiley-Blackwell, 2001), 352.
27. "Garangula, Speech," 352.
28. "Garangula, Speech," 352.
29. Dennis, *Liberty Asserted,* 57.
30. Dennis, *Liberty Asserted,* 67.
31. Dennis, *Liberty Asserted,* ii.
32. Peter Linebaugh, *The Magna Carta Manifesto: Liberties and Commons for All* (Berkeley: University of California Press, 2008); Peter Reimen, "Oeconomy and Ecology in Early Modern England," *PMLA* 132, no. 5 (2017): 1117–1133.
33. See Knapp, *Personification and the Sublime,* for this anecdote.
34. On theatrical machinery and its politics see Orr, *Empire on the Stage, 1660–1714* (Cambridge, UK: Cambridge University Press, 2001).

# 9

# IMAGINING DECOLONIAL FUTURES IN WILLIAM GILBERT'S *THE HURRICANE*

Ami Yoon

After three hundred years of oppression, the Indigenous dead of the Caribbean call up a tropical cyclone against imperial colonizers in Antigua-born poet William Gilbert's *The Hurricane: A Theosophical and Western Eclogue* (1796).[1] The titular hurricane apocalyptically "blows on Europe with unrelenting fury" (2.143), while sweeping revolutionary action led by Europeans who have become transfused with the spirits of Indigenous and enslaved persons over time—in a process of material and epistemological creolization—reorganizes the world into a "FIELD OF FREEDOM."[2] Whirling across the Atlantic, the hurricane administers corrective, distributive justice, such that the poem charts out an eschatology for Western empires, a fit judgment for sins committed: as Gilbert in his preface writes, "[Europe] depopulated America, and now AMERICA MUST depopulate her."[3] The poem thus literalizes the bold anticipation of Gilbert's fellow West Indian and poet Aimé Césaire, who would also prognosticate that Europe's decadent civilization, at once "morally [and] spiritually indefensible" for the disasters wrought through colonialism, will "[find] itself turned into a receptacle into which there flow all the dirty waters of history."[4] *The Hurricane* in effect animates Césaire's image of a civilization that becomes submerged by its own refuse, as a history of the exploitation and degradation of nonwhite, non-Western lands and populations ends in a watery apocalypse of its own making, positing an anticolonial critique and insurgent augury.

With its imagination of a future world order not dominated by those who have grown wealthy through the stripping and enslaving of others, *The Hurricane* places at stake more than the poetic justice of a violent society's destruction, as it reaches toward an alternative future that awaits beyond colonial structures. Opening up a future fomented by non-Western cosmologies and legacies, Gilbert's poem offers the prospect of a world in which the relationship between the human and the natural will be one of a balanced mutualism instead of a unilateral impulse for domination and possession. In America, "EQUILIBRATION will be found,"

Gilbert proclaims in one of his notes to the poem, both "in nature and in Man."[5] In this, the poet invokes "equilibration," or elsewhere "EQUILIBRIUM," in a physical sense, faithful to its denotation of a state of equal balance among powers or component parts, as the word was generally employed in chemical and physiological terminology in his time.[6] Equality in the poem operates as a principle of material reality as well as a political and social principle, as all matters work toward an "EVEN" condition without undue skewed power: hence destructions engendered by colonialism see reactions that become embodied by the empire-levelling hurricane.[7]

This essay argues that the environmental investments of *The Hurricane* construct the prospect of a wilding project for the American hemisphere as the commencement of a new world order, one based on a principle of equality that extends to characterize human-nonhuman relations as well as human socialities. The term *wilding* I borrow from contemporary environmental discourse, in which the concept of wilding (or sometimes "rewilding") names the effort to enable ecological revitalization in degraded or ruined land and to encourage complex ecological interactions to take place, working toward conservation or restoration. At the same time, I suggest, *The Hurricane*'s radical argument and vision become complicated by the poem's aesthetic and formal debts to, respectively, Caribbean natural histories and travel narratives, and a poetic tradition with roots in Virgilian classical literature. Raising the question of what creative forms or aesthetics are needed for realizing an anticolonial imaginary, these literary debts preserve lasting ties to the same Western culture that the poem repudiates, and materialize the challenge of balancing literary legibility with a radical political imagination. In exploring such difficulties of *The Hurricane*, my aim is to tease out some ways of reading that attend to this potentially alienating poem's prospective orientation, which anchors the importance of environmental politics to the corner of eighteenth-century literature *The Hurricane* represents.

## WILDING PROSPECTS

For all the grandness of his project, Gilbert remains a minor figure of literary history among more famous Romantic contemporaries, and *The Hurricane* is seldom read. Gilbert is best remembered for connections to poets such as Robert Southey, Samuel Taylor Coleridge, and William Wordsworth, whom he met in Bristol in 1795 through the bookseller and publisher Joseph Cottle.[8] Scholarship on Gilbert frequently rehearses how Coleridge published a portion of *The Hurricane* as "Fragment, by a West-Indian" in his periodical *The Watchman* (1796), and how Wordsworth, in *The Excursion,* quotes from one of Gilbert's endnotes; *The Hurricane*'s aesthetic influence has also been recognized in Coleridge's *Rime of the Ancient Mariner* and "Kubla Khan."[9]

Saliently, however, the poem straddles several key cultural and political interests of the late-eighteenth-century Atlantic world. Gilbert's father Nathaniel Gilbert was a lawyer and slaveholder, a third-generation owner of a sugar plantation in Antigua, who preached Methodist sermons to his enslaved labor force without freeing them; but, in contrast to his father's stance on slavery's continued ethical admissibility, William Gilbert sided with abolitionists, decrying slave traders as "Epicures in Assassination and Oppression" in an essay printed in the March 1790 edition of the *Bristol Mercury*.[10] Gilbert's sympathies emerge again in *The Hurricane*, a 441-line blank verse eclogue in the Virgilian dramatic style of pastoral allegory, with extensive notes that clarify an elaborate "theosophical" structure founded on Hermetic geography, aspects of Swedenborgianism, and traces of African spiritualism.[11] Within his notes, he reinforces his abolitionism by announcing a personal "strong predilection for Africa" and his commitment to "equal Liberty, equal Justice and equal Honor, to all Mankind," "against Europe."[12]

The plot of *The Hurricane* poses less of a difficulty than does its symbolic framing, in which aspects of the physical world connect to metaphysical properties that are laboriously explicated in pages of endnotes that outsize the verse. Since Gilbert offsets the straightforward action with the sophistication of his commentary, criticism has tended to elucidate the poem's philosophical design and the notes, with Paul Cheshire's 2018 *William Gilbert and the Esoteric Imagination*—a contextual gloss and annotated edition of Gilbert's poem—as a newly indispensable text for understanding its esoterism.[13] Over the course of his education and a youth spent in England, Gilbert had immersed himself in astrology, occultism, and esoteric arcana, including Hermeticism, which conceptualizes the four continents (Africa, America, Asia, Europe) as each corresponding to the four elements, cardinal points, and zodiac signs.[14] Surrounded by such metaphysical scaffolding, the poet speaker of *The Hurricane* introduces the island of Antigua and narrates how the spirits of exterminated Amerindian natives, who have "Survived Immortal, Vengeful and Creative" (1.44) in the atmosphere or under the sea in tombs safeguarded by mermaids, unleash the winds of a hurricane by chanting "a maze of song" (1.93). That a natural disaster becomes the agent of Europe's derangement as a delayed consequence of the destructions it has wreaked sounds perhaps not so outlandish to us today, under the shadow of an Anthropocene consciousness in which the unprecedented effects of climate change have altered global conditions. The hurricane heads for Europe, with vengeful spirits riding along to prove themselves "Victors on Europe, witherers of her might" (1.74), and they are joined by a "sweet alliance" of "Avengers" (1.57–1.58) from the similarly surviving spirits of other colonial victims, such as enslaved Africans and imported Asian laborers ("The many-murdered Innocence of Ind / Or East or West" [1.56–1.57]). As the hurricane roars, the speaker takes shelter and watches, and during a calm interval rescues a girl named Elmira from a shipwreck. The poem ends after two cantos with the pair gazing upon the splendid isle of Antigua while Europe topples into disaster far away.

Given the foundations provided by extant scholarship in addressing the most outstanding complexities of *The Hurricane*, I will turn instead to the prospects and intimations yielded by Gilbert's poetic project, which have remained obscured by the challenges of interpreting its occultism. Gilbert's focus on the natural environment's centrality in colonial order and the poem's embedded hope for a recalibrated sense of human and nonhuman relationality give expression to an anticolonial stance, delivering alternative conceptualizations of the reciprocities between the material and the human worlds. Within the poem, the hurricane registers the fatal intimacies between colonial practice and the material environment, for it is the murdered natives, from their "coral sepulchre" (1.40) in the Atlantic's "subterranean vaults where ocean roars" (1.47), who "ply the terrible Antistrophe" (1.100) that directs a storm toward Europe. Having outlasted their initial bodies, *The Hurricane*'s "American spirits" turn out to have continuously influenced the development of global civilization all along, both "infus[ing]" and "perfusing" their tendencies for freedom and equality into the bodies of western colonizers.

*The Hurricane* thus takes seriously the matter of the spirit as not just fanciful apparatus around poetic action, but as real and as enduringly affecting as any sensorily discernible environmental agent, such as climate. For "while AMERICAN *bodies* were destroyed by European bodies," Gilbert writes in one of his explanatory notes, "AMERICAN SPIRITS were subduing the Europeans—and much more effectually and really, though latently. . . . Though the inhabitants of AMERICA were destroyed, yet, as the SPIRIT or GENIUS, or PRINCIPLE of AMERICA infused itself into Europeans, and still goes on rapidly perfusing itself there, it could never be said, in the abstract or in the highest sense, that AMERICAN *bodies* were destroyed."[15] Such transformative spiritual transfusion, conceptualizing the human body as permeable to the influences of its material surround, resonates with eighteenth-century environmental thinking about physiological change in the face of the environment, per climate theory and humoralism, which believed in the power of physical conditions to affect bodies and characters. But here, the material includes the seemingly immaterial, which passes fluidly across bodies and environments: the spirit.

Later in his notes, Gilbert reaffirms a materialist understanding of that which is named the spiritual, remarking, "Strange it would be, that Mathematicians give no reality to Spirit, while they lay the basis of form in *Idea*."[16] By holding out an ontological formulation of the human as porously coterminous with the nonhuman at its most fundamental level, the poem's materialist argument dissolves the particular distinction of the human—crucial as it was to the colonial episteme—by opening up its security to essential transformations: a "creolized ontology," in

Monique Allewaert's summation, which considers bodies and selves disaggregated, mutable, and composite.[17] As Gilbert's language of chemical interactions proposes, the creolized ontology brought about by American spirits involves more than ghostly possession, causing integral changes that leave people different than they were before. With their interpolations, the American spirits and their active materiality provoke both ontological and historical revision, and reverse the prevalent European narrative of degeneracy caused by the American environment, for they cause *re*generation.

Further, in their refusal to disappear and their persistence in the unseen environment, these submerged beings—not undead; still sentient despite their imbrication with the coral and oceanic currents—give an image of resistance that does not end with bodily death. Exceeding historical narratives of their extinction, the singing Indigenous presences and their intercultural allies render *The Hurricane* an imaginative counter-archive to Western history's silences and absences. Indigenous and Black histories and legacies come into conjunction here, seeping into one another within the poem and demonstrating "the edgelessness of genocide and slavery" in the Americas, where the violence of "Black fungibility and Native genocide . . . moves as one."[18] Incommensurable, distinct experiences of violence nonetheless unfold confluently and linger, compelling simultaneous consideration. Gilbert's American spirits, who refuse to be appeased by anything less than a transformed world, call on us precisely to understand that their plights are shared among Natives and Blacks, and that their condition of pastness in the Western historical narrative does not leave them irrelevant to the present, or beyond ethical accountability. Their insurgency in the poem marks the ongoing claim they make on the present, and their bid for a future. Altogether, the spirits raise within the poem a demand that Christina Sharpe—writing of the afterlife of "the unfinished project of emancipation"—articulates as the need to "attend to, care for, comfort, and defend, those already dead, those dying, and those living lives consigned to the possibility of always-imminent death" under a world system organized by colonial power.[19] As Sharpe also states, the violence and subjections produced by Atlantic capitalism and slavery are atemporal but constitutive, reverberating across temporal distances: they form "the total environment," the pervasive "weather [that] necessitates changeability and improvisation [as] it produces new ecologies."[20]

In the way that the still-vibrant, still-furious dead of the poem show themselves to be "waiting for a future when all the slave marks would be gone," their anticolonial fervor and revolutionary longing indeed push toward new ecologies, both social and environmental, inextricably related as they were in plantation economies.[21] In the notes to Canto I, Gilbert in fact credits the Indigenous spirits as the true instigators of the currents of revolution for both the American and French Revolutions. *The Hurricane*'s Indigenous spirits join the ranks, within a futural historical imaginary, of Abbé Raynal's Black Spartacus and instantiate the true origin of freedom dreams in the Americas, where European attention in the last decades of the eighteenth century was already trained, especially with the ongoing Haitian Revolution (1791–1804).[22] Moreover, Gilbert explains that the "PRINCIPLE of AMERICA" brought on in the wake of the hurricane's justice will be one of "EQUILIBRIUM," signified by the "REIGN OF SATURN," under which humans and nature will coexist in harmony.[23] Through this astrological symbolism, his invocation of the classical Golden Age encodes an image of a future that outlines the possibility of ecological concord, insofar as the Golden Age defines an era of "an everlasting spring" and peaceful virtuousness among humans who do not excessively labor at harnessing the land.[24] Thus America, writ large, holds out the promise for a "FIELD OF FREEDOM," or an alternative state in which the remaking of society encompasses more than the human within its project for freedom.[25]

I want to press on Gilbert's environmental imagination of America as a *field* of freedom and suggest that what *The Hurricane* tantalizingly affords is a prospect of America as a wilding project, after colonialism's end. Adopted and proponed in twenty-first-century conservationist movements and biodiversity policy, the wilding of a place focuses on founding afresh relationships and interdependencies among material contexts, processes, and different kinds of life by allowing time and space for growth without intensive human intervention or cultivation.[26] *The Hurricane* stops short of depicting how the world will go on after the hurricane has finished its job, but its presentiment of a reconstruction, through abolition and empire's dissolution, of the social, environmental, and economic orders, implies possibilities of ecological renewal along non-extractive, non-territorial logics. After the hurricane's destruction, what remains will be free to grow beyond the hierarchies and oppressions of settler colonialism.

Although "rewilding" has become the increasingly common term, with the concept of wilding gaining traction as a resource management strategy in fields such as restoration ecology, conservation biology, and agricultural science, the term without the prefix that emphasizes a return—as if to some ideal prior state—better encapsulates what goes on within Gilbert's poem. Terminologically, *wilding* calls up multiple valences. Even besides the environmental ideas it activates, the word has been occasionally taken up in U.S. urban slang to refer to violent crimes by minority youths against whites.[27] As the violence of *The Hurricane* falls on white Europeans in their colonial privilege, the sense of uprising behind the slang use of the term also captures one facet of the poem's action, and attests to the fraught, protean, but continuous configurations of revolts against systemic marginalization. The risk and promise at once figured in the notion of wilding, the indefiniteness of its outcomes but the urgency of undertaking revitalizing pursuits, all develop a movement from anticolonial dissent to decolonial reimagination.[28]

More pertinent to the topic of *The Hurricane*'s prospective orientation is, however, the potential of a recalibrating wildness as a counter to curated plantation monoculture in its landscape. Early descriptions of Antigua before the tempest establish the West Indian island as a lush wildness kept barely in check by the "daily sweat" of the enslaved (1.181), with its imagery and turns of phrase evocative of John Milton's wild paradise: "Indian Groves of aromatic breath," and "spicy Thickets" with "ample flowers / Redolent of every various sweet that glows / Beneath the beams of Heaven's Eternal Sun" (1.146–1.149). These descriptions align with the irrepressibly regenerative Eden of *Paradise Lost*, rather than with the manicured pastoral scene that was the ideal of colonial planters, despite efforts at domestication through deployments of an enslaved labor force.[29] Gilbert's lines closely revise Milton's on the subject of the landscape that Adam and Eve enjoy as their habitation: a "blissful field" of "flowering odours, cassia, nard, and balm," where "Nature [is] / Wantoned as in her prime," "pouring forth more sweet, / Wild above rule or art."[30] By tapping into imagery that evokes Milton's Eden and its extravagant natural world, Gilbert hints that the West Indian land retains its wild potential and that its release from plantation systems' grids of environmental governance may bring the hope of an even greater realization of earthly paradise. Although it is true that Milton's paradise also calls for diligent pruning and tending by Adam and Eve, such agricultural labors scarcely bear comparing with systematic cash crop schemes run by slavery, and nowhere does *The Hurricane* liken enslaved people's work to georgic harmony—a situation dissimilar to the one that Christopher Allan Black outlines in other West Indian georgics in chapter 10. Once the hurricane passes, Gilbert's poem invites an imagination of a world no longer bound by the operations of private property and slave labor, which might compete with Eden's own "open field"—paradise regained, so to speak.[31]

After a dialogue in Canto II between the speaker and a shipwreck survivor named Elmira, in which the former informs the latter of what has happened, the concluding lines portray the pair looking at the archipelago stretched out before them after the storm, playing again on a Miltonic vein. The passage recasts the image of humankind's original couple at the end of *Paradise Lost*, staring out at "The world . . . all before them, where to choose / Their place of rest":[32]

> Shortly awake ELMIRA joined me soon,
> Treading with cheerful step and unrestrained
> The stately portico. 'Twas all enchantment
> To her soul. The sun burst brilliant forth and
> Welcomed her: All the Isle, the conquered ocean,
> Lay before her: Smaller Isles attract her:
> Unknown Diversities of Landscape strike:
> The distant Hills cite curiosity. . . . (2.184–2.193)

Whereas Adam and Eve are exiting a beautiful, unbridled wildness when looking out on the world left to them, Gilbert's speaker and Elmira occupy the reverse position of entering into one, now that plantation regimes need not interfere with natural "Diversities of Landscape" (2.190) taking on different shapes. Judgment now past, *The Hurricane* ends with "brilliant" (2.187) glimmerings for a beginning after the end of a world.

Wilding projects do not aspire toward structures of private property, and with Western empires and their entire economic systems swept away, a literally open field remains for the hurricane's survivors to configure alternative lifeways: the American spirits, the "Negroes" of Antigua (2.61) who help rescue Elmira, the poet-speaker who longs for "Love's and NATURE'S offspring pure" (1.139), Elmira guileless of any "mean lust of gain" (2.170), and the American land. Gilbert names the rule of the new world as "the Reign of just EQUALITY," but the charting of that just and egalitarian world remains an undrawn responsibility.[33] With the hopefulness suffusing its sunlit closing scene, the poem asks readers to imagine what it leaves undrawn and undescribed: it is this task for imagining an otherwise that I take to be the most radical aspect of *The Hurricane.* With its temporal attunement to futurity, the poem seeks to recuperate paradisal imagery and the idea of wildness from their usual applications within colonial ideology, wherein they engineer complicities with plantation management by highlighting plantations as desirable civilization. The wilding project that *The Hurricane* intimates for the future means to contemplate freedom from colonial structural impositions on the environment as well as on humans, and to consent to "unrestrained" (2.185) beauty in lieu of control.

The motley composition of the survivors reaccentuate the potential, too, of social and cultural interchanges and mixtures. The coalitions and formations created by Indigenous and African spirits, and by Indigenous spirits and European bodies that change under the American influence, constitute the basis of Gilbert's argument for America's special character, and such "relations of co-specificity" illustrate Sylvia Wynter's analysis of the New World's "transversal character."[34] In their readiness to cooperate and transfuse for the overturning of colonial order, what the spirits of *The Hurricane* already intuit is Wynter's sense that all of the exchanges that brought the "New World" into existence have not precluded the opportunity for forming new relationships that would not have taken place otherwise.[35] As Césaire likewise asserts, despite the horrors the New World has seen: "I admit that it is a good thing to place different civilizations in contact with each other; that it is an excellent thing to blend different worlds; that whatever its own particular genius may be, a civilization that withdraws into itself atrophies; that for civilizations, exchange is oxygen."[36] The decolonial hopefulness inaugurated by *The Hurricane* arises from its early apprehension of Wynter and Césaire's insights. But crucially, the poem stops on the threshold of a prospect, and the novelty that *The Hurricane* summons through its central ideas encounter a tension with the text's formal properties.

## *THE HURRICANE* AND IMPERIALIST POETICS

While the cyclone brews, the speaker of *The Hurricane* takes refuge in a "safe Recess" (1.138) and waits for the storm to pass. On a metanarrative level, this "safe Recess" adverts reflexively to poetry itself, as Gilbert finds poetry to be a cultural refuge from which to articulate his critique of culture and history. But even as *The Hurricane* pitches itself as an artifact of opposition to European worldviews and practices, the poem relies on the very traditions it would drown for its literary intelligibility and poetic force. In other words, *The Hurricane* advances an anticolonial argument, but it falls short of formulating an anticolonial poetics. The poem's narrative, philosophical, and political thrusts enact a repudiation even as its aesthetic infrastructure preserves an authoritative link, dramatizing how a disjunction between iconoclastic content and conventional form possibly checks an audacious vision. Beneath its metaphysical schema and insurrectionary ideas, *The Hurricane* remains closely bound up with the formal strategies of visual representation endemic to Caribbean voyage narratives and natural histories, and to a classical poetic tradition that cannot distance itself from European culture.

Reading some of *The Hurricane*'s descriptive passages on the Caribbean islands alongside treatments of the same topic in earlier natural history texts sheds light on the conventionality of Gilbert's aesthetic delivery of his American setting. The first canto introduces Antigua with a panoramic first view of a tropical world that lays down a sense of its incommunicable impressiveness, an impressiveness that increases rather than satisfies aesthetic desire:

> Near where with Tropic heats bright Cancer glows,
> And sun beams glitter with perennial force,
> Girt with the azure wave an Island lies,
> Called by the Spaniards ANTIENT. Its breadth is
> Measured by the eye; which, still unsatisfied,
> Strikes far beyond the reach of land, Northward
> When turned. Its utmost length doubles its breadth. (1.1–1.7)

The eye roves and takes in the island voraciously, and "still unsatisfied" (1.5) when it spies more "Islands, faint seen among the adjacent seas," it absorbs their "social and romantic scene" with wonder: "They give the wing for amplest thought to range / On all the mighty wonders of the world! / Scenes undiscovered, uncreate to man" (1.5, 1.8–1.13). The prefixal repetition of descriptors ("unsatisfied," "undiscovered," "uncreate") at once emphasizes the inexpressible sensations aroused by Antigua and its nearby isles, and their indescribable quality. With such an effect, these descriptive passages adhere to what Jonathan Lamb has called the je ne sais quoi rhetoric of South Sea voyage narratives, in which a "vaunted failure to communicate" details and experiences "utterly new and unparalleled" on the part of the voyager is leveraged as a paradoxical selling point, with its acts of witnessing "reducible to the assertion 'Nothing you can imagine is equal to what I saw.'"[37] So it goes with John Gabriel Stedman's popular *Narrative, of a Five Years' Expedition, against the Revolted Negroes of Surinam*, published in the same year as Gilbert's poem, which represents Suriname as a "luxuriant flourishing spot" of "delicious sensations" "that nothing could equal," not unlike the rhetoric of the wondrous in *The Hurricane*.[38] The orthodoxy of Gilbert's presentation of Antigua may also be traced on the level of narratorial progression. In as far back as Richard Ligon's 1657 narrative, *A True and Exact History of the Island of Barbados*, we find a similar movement from an external, Western voyager's perspective of the West Indian archipelago to an interior perspective that revels in ineffable wonders. In lieu of mounting poetic originality, the optics derive entirely from the conventions of colonial accounts. In describing the island of St. Jago, Ligon, too, draws out the initial panorama of the island as it comes into sight, and follows it with claims of the island's superlative internal beauties. St. Jago appears first "full of high and steep Rocks . . . whilst we [Ligon's crew] lay before it" on a ship; but once on the island, Ligon finds it "so fresh, so full of various greens . . . [and] so full of varietie, of the most beautifull colours, as if nature had made choyce of that place to shew her Master piece."[39] The gratification engendered by the combination of a prolonged aesthetic attainment and subsequent sensual delights constructs a climaxing effect, while anaphora underscores the delectations.

It is no surprise, then, that Gilbert's Romantic compatriots expressed appreciation for such sections of *The Hurricane*, since they poetically reprise familiar, pleasant scenes and ideas from the texts of natural history and travel writing that proliferated in the eighteenth century, bringing the exotic tropics back to the metropole for readers.[40] In some descriptive moments, the poem's speaker solicits wondering and wandering minds in a manner that offers up the West Indies as an exotic package. Proceeding to zoom in on Antigua's surrounding islands, for example, the speaker enthuses about verdant fields where "summer pleasure spreads the cocoa shade" and "Invites the mind, that springs to Nature's charms, / Or loves to class what she diffusely throws" (1.159, 1.164–1.165). For whom does the poet extend this invitation?

Modulating some of the political potential in the emphasis on wildness that I have earlier discussed, Gilbert's conjuration of rural natural delights participates in conventions of the eclogue, his chosen form, descended from Virgil in the English poetic tradition and heavy on bucolic arrangements.[41] Like the carefree shepherd Tityrus reclining under a tree in Virgil's first eclogue, Gilbert's speaker sits in a recess and evades personal misfortune while watching the motions of history play out catastrophically for others, and *The Hurricane*'s futural leaning conforms to the Virgilian eclogue's anticipatory bent.[42] An underused poetic genre today, the eclogue saw consistent deployment over the course of eighteenth-century

English poetry, particularly following Alexander Pope's 1709 eclogues based on seasonal themes.[43] Early on, William Diaper's "sea-eclogues" (1712) and Moses Browne's twice-reprinted piscatory eclogues (1729, 1739, 1773), which take the typical bucolic setting to sea, extended an early modern subgenre interest.[44] Mid- and late-century eclogues incorporated varying degrees of idiosyncratic modifications in setting and subject matter, meaning that *The Hurricane* appears less of an outlier within its cultural moment than its American loyalties might seem to indicate. Some, such as William Collins' *Oriental Eclogues* (1757), Thomas Chatterton's three *African Eclogues* (1770), or John Scott's *Oriental Eclogues* (1782), adopted contexts and figures as exotic as Gilbert's West Indian ones—and American and African eclogues become increasingly visible in the 1770s and 1780s, protesting slavery. A set of *American Eclogues* in the December 1783 and January 1784 issues of the *Gentleman's Magazine*, by a Reverend Gregory, features enslaved persons in Pennsylvania decrying slavery's cruelties, and later in 1784, a submission to the same magazine from "H.M." of Liverpool offers "The Lovers, An African Eclogue," an exchange between the eponymous African lovers who long for an end to white tyranny.[45] So contextualized, it is evident that *The Hurricane* in its formal identity stands as only a variation of a type, though it champions greater changes than the abolishment of the slave trade. Before Gilbert, Edward Rushton in 1787 had already published *West-Indian Eclogues*, set in Jamaica, which was met with an unfavorable response. For its passages in which enslaved speakers plot revenge, the *Critical Review* expressed especial discomfort.[46]

Gilbert also plainly bases his poetic authority on another classical model. Besides opting into a Virgilian legacy with the eclogue as his platform, Gilbert defends his metrical irregularities in syllabic stress—one of the less conventional aspects of his poetic construction—in a preface by appealing to Horace as his sanction. Having explained his accentual distributions and line breaks, Gilbert declares, "If, after all, the ear is fastidiously offended . . . let it feed upon my Motto, attack Horace, and let me go free—*Carmina non prius / Audita*."[47] The gesture grafts Gilbert's poetic authority and freedom on Horatian precedent, and by placing poetic innovation under a classical dispensation, the foundation of cultural production remains tied to imperialist civilizations, from the Roman to the British.

In spite of its promulgation of a liberating "American spirit" and a materialist philosophy, *The Hurricane*'s formal conventionalities and debts undercut its thematic stress on alternatives to Western imaginaries and epistemologies. Conveyed in the bracingly English meter of blank verse, the poem embodies its strong embeddedness in the aesthetic and literary tradition of colonial culture, missing the occasion of confronting even the "forced poetics" to be found in the aporia between desire and form for a creole literature.[48] In the way it wavers between the conservatism of its literary construction and the progressivism of its anticolonialism, however, *The Hurricane* provides productive grounds for weighing the challenges involved in pushing the work of thought into praxis—for when the conditions of our thinking and expressing set limits and contradictions on either critique or imagination.

## IMAGINING DECOLONIAL FUTURES

*The Hurricane* urges a dialectic reading by allowing both its potential force and cultural embeddedness to inform a critical interpretation that seeks to generatively engage with multiple "possibilit[ies] of existence, analysis, and thought."[49] Exemplifying how possibilities opened up by revolutionary thinking may conflict with how such thinking is enacted, the poem proffers a study in the difficult work of imagining decolonial futures. Consequently, *The Hurricane* also can be opened up *for* study, in the way that Fred Moten and Stefano Harney invoke "study" as a mode of interaction between a text and a community of learning, as a rich site for pedagogical engagement.[50] Given how it weaves together topics such as non-Western epistemologies and ontologies, abolition, indigeneity, and the fundamental relevance of the material environment to colonialism, *The Hurricane* permits inquiries into what analytics, modes, or practices may help us move away from settler frameworks and futures. As Walter Mignolo and Catherine Walsh assert, although "each of us, endorsing and embracing decoloniality, is responsible for our own decolonial liberation," ultimately such a "task is not individual but communal."[51] *The Hurricane*'s complexity renders it compatible with the kind of collective undertaking that Mignolo and Walsh describe, to be unfolded in the classroom.

*The Hurricane* suits a variety of course framings. Courses thematically centered on transatlantic ecologies or natural disaster events form some examples. In courses geographically and historically organized, focusing on early Caribbean or early Atlantic literature, comparative studies with other West Indian neoclassical long poems that index a range of tensions between poetic schemes and political desires, such as James Grainger's *The Sugar-Cane* (1764) or John Singleton's *A General Description of the West-Indian Islands* (1767), may cue productive explorations.[52] The specific coordinates lent by similar texts can illuminate how New World colonialism affected literary production and poetic subjects in an age when liberal Enlightenment ideas were mainstream. Given the length of such texts, I have previously excerpted strategic portions for assignments, with in-class conversations funneling into inquiries on how aesthetics, race, capitalist extractions, and refusals thereof come intertwined. Neoclassical verse generally tends to be a stumbling block for students unused to the heavy rhythms of measured feet and the cultural layers of classical allusions, even when poems come without dense annotations by the poet, posing reading challenges for individual parsing. But collective analysis as a class may disclose new reading horizons and diversify students' sense of the politics encompassed by eighteenth-century literature, beyond canonical syllabus inclusions.

For particular lessons or class session topics, the concept and framework of abolition within Gilbert's poem supplies another pedagogical touchpoint. In light of the pointed references to the American and French Revolutions in the prose notes, and the historical context of the dense waves of enslaved people's revolts across the Americas during the 1790s, *The Hurricane* lends itself organically to discussions of incomplete struggles for abolition.[53] That Gilbert makes an open question of imagining what exactly a future lived in a field of freedom looks like, while nonetheless keeping his eye fastened on its prospect and expecting its ineluctability, corresponds with the forward-looking slant of continued abolition movements today. The *longue durée* of abolition movements and abolitionist thinking—from the abolition of slavery and property to that of prisons, capital punishment, or borders—renders such discussions intellectually imperative rather than presentist. My experience in broaching these conversations with undergraduate students in the classroom has shown that they are keen to make space for considering how the knowledge they are gaining spills into relevance for the America that they know, and quick to make the connections without prompting; planning for such discussions facilitates shared reflection instead of leaving them for individual students' contingent interest. With a text like *The Hurricane* at hand, conversations could evolve with related additional questions: How does a future beyond slavery and settler colonialism unfold ecologically? How does its prospect expand our categories of inclusivity, for humans and nonhumans alike?

On the note of inclusivity, the poem provides the opportunity for recognizing and exploring indigeneity as a key conceptual category for reading New World literatures. Despite acknowledging my non-expert disciplinary relation to Indigenous studies, I take the willingness to engage the thought of Indigenous positions and perspectives as a way of practicing an intentional commitment to decolonial inclusivity, since "native" and "nonnative" categories relate inextricably to territorializing and exclusionary logics of the kind that *The Hurricane* incites us to abandon. As sociologist Nandita Sharma points out, indigeneity as a form of thought and representation requires nuanced probing as "a form of subjectivity that emerged *because of* the devastation wrought in the aftermath of 1492," and the Indigenous spirits of *The Hurricane* emblematize how the significance of their indigeneity in responding to European domination treads between the bitter "colonial state category of subjugation" and the patient "category of resistance"—triumphant resistance, in their case.[54] Their willingness to join with other victims of expropriation in their pursuit of justice puts the focus on collective action, returning us to the radical potential of collectives gathered around prospects of a decolonial world. Through shared study and praxis that can occur through classroom communities, seemingly unwieldy texts that do not slide easily under standard literary rubrics dislodge the disproportionate centrality of the settler colonial imaginary as the preeminent epistemological and cultural heritage for us, and reveal that this dominant mode has always been contested.

## NOTES

1. For their invaluable feedback and insights that have enriched the development of this essay, I am indebted to Naomi Michalowicz, Diana Newby, Dustin Stewart, and the 2020–2021 members of the Brown Bag Seminar Series at the McNeil Center for Early American Studies.
2. William Gilbert, *The Hurricane: A Theosophical and Western Eclogue to Which Is Subjoined a Solitary Effusion in a Summer's Evening*, in *William Gilbert and Esoteric Romanticism: A Contextual Study and Annotated Edition of* The Hurricane, ed. Paul Cheshire (Liverpool: Liverpool University Press, 2018), 103–162 (150). All citations of *The Hurricane* are from Cheshire's edited text, and passages are cited parenthetically.
3. Gilbert, *The Hurricane*, 105.
4. Aimé Césaire, *Discourse on Colonialism*, trans. Joan Pinkham (New York: Monthly Review Press, 2000), 32, 64.
5. Gilbert, *The Hurricane*, 151.
6. *Oxford English Dictionary Online*, s.v. "equilibrium," http://www.oed.com/view/Entry/63762 (accessed March 15, 2021).
7. Gilbert, *The Hurricane*, 149.
8. Gilbert, *The Hurricane*, 42–45.
9. Cheshire, *William Gilbert and Esoteric Romanticism*, 1–2; Sue Thomas, "Catastrophic History, Cyclonic Wreckage and Repair in William Gilbert's *The Hurricane* and Diana McCaulay's *Huracan*," in *Tracking the Literature of Tropical Weather: Typhoons, Hurricanes, and Cyclones*, ed. Anne Collett, Russell McDougall, and Sue Thomas (New York: Palgrave, 2017), 228–229; John Livingston Lowes, *The Road to Xanadu: A Study in the Ways of the Imagination* (Boston: Houghton Mifflin, 1955 [1927]), 149; Paul Kaufman, "*The Hurricane* and the Romantic Poets," *English Miscellany* 21 (1970): 99–115.
10. Thomas, "Catastrophic History," 228; Cheshire, *William Gilbert and Esoteric Romanticism*, 19, 202.
11. Cheshire glosses Gilbert's idiosyncratic Hermetic geography, which shares similarities with the eighteenth-century natural philosopher and mystic Emanuel Swedenborg's theologico-philosophical scheme of the world, in chapter 6 of his book. Swedenborgianism spread widely and attracted some interest among other Romantic poets: see Robert Rix, *William Blake and the Cultures of Radical Christianity* (London: Routledge, 2007). For a discussion of theosophy, see Kaufman, "*The Hurricane* and the Romantic Poets." Gilbert does not explicitly name his African influences, but mermaids, who aid in preserving the spirits of murdered Indigenous and enslaved persons in the poem, were figures of protection and power in an African-Atlantic spiritual and folk tradition, with the ability to control water—see Ras Michael Brown, "'But the Mermaid Did Not Rise Up': The Death of a Simbi in the Carolina Lowcountry," *Southern Quarterly* 47, no. 4 (2010): 120–150.
12. Cheshire, *William Gilbert and Esoteric Romanticism*, 144, 149.
13. See Paul Cheshire, "The Hermetic Geography of William Gilbert," *Romanticism* 9, no. 1 (2003): 82–93; Deidre Coleman, *Romantic Colonization and British Anti-Slavery* (Cambridge, UK: Cambridge University Press, 2005), 101–105; Patrick Curry, *Prophecy and Power: Astrology in Early Modern England* (Cambridge, UK: Polity Press, 1989), 104; Joscelyn Godwin, *The Theosophical Enlightenment* (Albany: State University of New York Press, 1994), 116; Ellic Howe, *Astrology and the Third Reich* (Wellingborough, UK: Aquarian Press, 1984), 24; and Marsha Keith Schuchard, "Rediscovering William 'Hurricane' Gilbert: A Lost Voice of Revolution and Madness in the Worlds of Blake and the Romantics," William Gilbert

Home Page, last modified April 25, 2019, http://www.williamgilbert.com/GILBERT_Schuchard.htm.

14. Cheshire, *William Gilbert and Esoteric Romanticism*, 166–182; Cheshire, "Classical Elements: Darwin, Gilbert, Blake, and Coleridge," *Wordsworth Circle* 50, no. 2 (2019): 153–157.
15. Gilbert, *The Hurricane*, 134–135.
16. Gilbert, *The Hurricane*, 137.
17. Monique Allewaert, *Ariel's Ecology: Plantations, Personhood, and Colonialism in the American Tropics* (Minneapolis: University of Minnesota Press, 2013), 6.
18. Tiffany Lethabo King, *The Black Shoals: Offshore Formations of Black and Native Studies* (Durham, NC: Duke University Press, 2019), 202, x.
19. Christina Sharpe, *In the Wake: On Blackness and Being* (Durham, NC: Duke University Press, 2016), 5, 38.
20. Sharpe, *In the Wake*, 5, 105–106.
21. Saidiya Hartman, *Lose Your Mother: A Journey along the Transatlantic Slave Route* (New York: Farrar, Straus and Giroux, 2007), 204. My use of the word "vibrant" is with reference to Jane Bennett's *Vibrant Matter: A Political Ecology of Things* (Durham, NC: Duke University Press, 2009).
22. Besides the Haitian Revolution, the 1790s saw revolts and wars in Jamaica, St. Vincent, St. Lucia, Dominica, Dutch Demerara, Curaçao, Cuba, Grenada, and Venezuela. See Thomas, "Catastrophic History," 231.
23. Gilbert, *The Hurricane*, 149–150.
24. Cheshire, *William Gilbert and Esoteric Romanticism*, 187. Cheshire cites Ovid.
25. Gilbert, *The Hurricane*, 150.
26. Nathalie Pettorelli, Sarah M. Durant, and Johan T. du Toit, eds., *Rewilding* (Cambridge, UK: Cambridge University Press, 2019); Néstor Fernández, Laetitia M. Navarro, and Henrique M. Pereira, "Rewilding: A Call for Boosting Ecological Complexity in Conservation," *Conservation Letters* 10, no. 3 (2017): 276–278; Mihnea Tanasescu, "Field Notes on the Meaning of Rewilding," *Ethics, Policy and Environment* 20, no. 3 (2017): 333–349.
27. Monica Erling, "Wilding," in *Encyclopedia of Race and Crime*, ed. Helen Taylor Greene and Shaun L. Gabbidon (Los Angeles: SAGE Publications, 2009), http://dx.doi.org/10.4135/9781412971928.n370.
28. Queer potentiality inheres in wilding. The multiple aspects and potentials of wildness have been activated in queer theory by Jack Halberstam in *Wild Things: The Disorder of Desire* (Durham, NC: Duke University Press, 2020). Although José Esteban Muñoz does not theorize wildness specifically, in *Cruising Utopia: The Then and There of Queer Futurity* (New York: New York University Press, 2009), he writes of the queer utopian project in resonant terms, as "the rejection of a here and now and an insistence on potentiality or concrete possibility for another world" (1).
29. On imperial landscape ideals in the colonies, see Jill Casid, "Inhuming Empire: Islands as Colonial Nurseries and Graves," in *The Global Eighteenth Century*, ed. Felicity A. Nussbaum (Baltimore, MD: Johns Hopkins University Press, 2003), 279–308.
30. John Milton, *Paradise Lost*, Book 5, lines 292–297, in *The Complete Poetry and Essential Prose of John Milton*, ed. William Kerrigan, John Rumrich, and Stephen M. Fallon (New York: Modern Library, 2007). See also Book 4, lines 132–166, 235–275.
31. Milton, *Paradise Lost*, Book 4, line 245.
32. Milton, *Paradise Lost*, Book 12, lines 646–647.
33. Gilbert, *The Hurricane*, 149.
34. Nandita Sharma, "Strategic Anti-Essentialism: Decolonizing Decolonization," in *Sylvia Wynter: On Being Human as Praxis*, ed. Katherine McKittrick (Durham, NC: Duke University Press, 2015), 166.
35. Sylvia Wynter, "1492: A New World View," in *Race, Discourse, and the Origin of the Americas: A New World View*, ed. Vera Lawrence Hyatt and Rex Nettleford (Washington, DC: Smithsonian Institution Press, 1995), 5–57.

36. Césaire, *Discourse on Colonialism*, 33.
37. Jonathan Lamb, *Preserving the Self in the South Seas, 1680–1840* (Chicago: University of Chicago Press, 2001), 23–24.
38. J[ohn] G[abriel] Stedman, *Narrative, of a Five Years' Expedition, against the Revolted Negroes of Surinam, in Guiana, on the wild coast of South America; from the year 1772, to 1777* (London: J. Johnson, 1796), vol. 1, 14.
39. Richard Ligon, *A True and Exact History of the Island of Barbados* (London: Humphrey Moseley, 1657), 8–9.
40. Almost two hundred years after Gilbert brought out *The Hurricane*, another Antiguan, Jamaica Kincaid, would critique precisely the habitual aestheticization of the Caribbean by and for Western eyes, commenting wryly, "Antigua is beautiful. Antigua is too beautiful. Sometimes the beauty of it seems unreal. Sometimes the beauty of it seems as if it were stage sets for a play." Jamaica Kincaid, *A Small Place* (New York: Farrar, Straus and Giroux, 2000 [1988]), 77.
41. Richard F. Jones, "Eclogue Types in English Poetry of the Eighteenth Century," *Journal of English and Germanic Philology* 24, no. 1 (1925): 35.
42. Timothy Saunders, *Bucolic Ecology: Virgil's Eclogues and the Environmental Literary Tradition* (London: Bloomsbury, 2008), 1–2.
43. Saunders, *Bucolic Ecology*, 39.
44. Killian Quigley, "The Pastoral Submarine: William Diaper and Eclogue's Marine Frontier," *Eighteenth-Century Studies* 53, no. 1 (2019): 109–127; Nicholas D. Smith, "Jacopo Sannazaro's *Eclogae Piscatoriae* (1526) and the 'Pastoral Debate' in Eighteenth-Century England," *Studies in Philology* 99, no. 4 (2002): 432–450.
45. In *Gentleman's Magazine*, December 1783, January 1784, and March 1784; also cited by Jones, "Eclogue Types in English Poetry," 54–55.
46. Anonymous review, in *The Critical Review: or, Annals of Literature*, vol. 64 (London: A. Hamilton, 1787), 434–435.
47. Gilbert, *The Hurricane*, 107. Gilbert quotes from Horace's *Ode*, from which he also takes his epigraph to *The Hurricane*. Cheshire translates the quoted lines: "Songs never previously / Heard."
48. Édouard Glissant, "Poetics," in *Caribbean Discourse: Selected Essays*, trans. J. Michael Dash (Charlottesville: University Press of Virginia, 1989), 120–121.
49. Catherine E. Walsh, "The Decolonial *For*: Resurgences, Shifts, and Movements," in *On Decoloniality: Concepts, Analytics, Praxis*, ed. Walter D. Mignolo and Catherine E. Walsh (Durham, NC: Duke University Press, 2018), 17.
50. Stefano Harney and Fred Moten, *The Undercommons: Fugitive Planning and Black Study* (New York: Minor Compositions, 2013), 107–110.
51. Walter D. Mignolo and Catherine E. Walsh, "Introduction," in *On Decoloniality*, 10–11.
52. The first Barbados edition of *A General Description* contains annotations comparable to those in *The Hurricane* or *The Sugar-Cane*: see Kimberly Takahata, "'Follow me your guide': Poetic Empire in John Singleton's *A General Description of the West-Indian Islands*," *Studies in Eighteenth-Century Culture* 49 (2020): 45–64. On tensions encoded by *The Sugar-Cane*'s georgic form, see Cristobal Silva, "Georgic Fantasies: James Grainger and the Poetry of Colonial Dislocation," *ELH* 83, no. 1 (2016): 127–156.
53. Thomas, "Catastrophic History," 231.
54. Sharma, "Strategic Anti-Essentialism," 170–171 (italics in original), 172.

# Part Five

## EIGHTEENTH CENTURY + GREEN UTOPIAS

10

# SLAVERY AND PLANTATION STEWARDSHIP

## The Eighteenth-Century Caribbean Georgics of James Grainger and Philip Freneau

CHRISTOPHER ALLAN BLACK

As a classical form of ecological writing, the georgic analyzes the practical aspects of agriculture and rural affairs by celebrating the industrious planter. In the mid-eighteenth-century British West Indies, environmental concerns over cultivation and the fungibility of enslaved persons contributed to the repurposing of Virgil's georgics in order to instruct planters in stewardship. While the development of sugar plantations returned profitable yields of cash crops, enclosure and over-farming contributed to the commodification of enslaved laborers and soil degradation. Caribbean agriculturalists interested in promoting conservation believed that practicing Virgil's methods of sustainable farming would make plantations profitable while avoiding further degradation. From the 1760s through the 1780s, neo-georgic poets adapted Virgil's didactic poetry to expose the impact of cultivation on the West Indian agroecosystem and allegedly advocate for humane treatment of enslaved persons. Reflecting growing calls for environmental reform of plantation agriculture and amelioration, James Grainger's Caribbean georgic *The Sugar-Cane* (1764) and Philip Freneau's georgic satires "The Beauties of Santa Cruz" (1776) and "To Sir Toby" (1791) expose the injustices of slavery in the West Indies through their description of the individual's ecological relationship to the plantation. While Virgil's georgics celebrate the renewable abundance of the Edenic farm, Grainger and Freneau's agrarian labor poems prophesize that through over-farming the land and overworking (and further exploiting) enslaved persons, sugar plantations will become an unfertile wasteland incapable of producing staple crops. Grainger believed that, if planters practiced the theories of responsible soil conservation and crop rotation promoted by Virgil, the laboring conditions of enslaved persons would be improved and the land would produce an abundance of cash crops. However, Freneau critiques this approach through his own georgic poetry. Witnessing the widespread failure of Caribbean planters to conserve natural resources, land, and labor, Freneau argued that Grainger's attempts to instruct growers in georgic methods of sustainable agriculture and stewardship did nothing to preserve the ecology of the West Indies and led to widespread abuse of enslaved labor, ultimately rendering the Caribbean a dystopian wasteland.

In the West Indies during the 1760s, the need to promote sustainable farming and to conserve natural resources came into conflict with the desire of British planters to mass-produce cash crops through the commodification and exploitation of enslaved labor. For Grainger, georgic methods of agricultural conservation provided an alternative method of settler colonialism that would, in theory, protect farmland from further degradation and prevent the exhaustion of enslaved laborers. British planters imagined their efforts of enclosure and slavery as more humane than those of their Spanish counterparts. The Spanish investment in geological extraction (rather than strictly agricultural) transformed enslaved persons into a fungible commodity exploited through mining. In *A Billion Black Anthropocenes or None*, Kathryn Yusoff charts a similar geologic focus and asserts that a transmutation of matter occurred during the era of Spanish imperialism in the West Indies, which augured the exploitable commodification of Black and Brown bodies. For Yusoff, the biopolitical category of Black nonbeing "is established through slaves being exchanged for and as gold. Slavery was a geologic axiom of the inhuman in which nonbeing was made, reproduced, and circulated as flesh."[1] The enslavement and devaluation of Black peoples is directly related to the Western European desire to control and extract agricultural and mineral profit from the land. The material practice of extraction that occurs alongside the racial categorization of matter, Yusoff reasons, displaces Black, Brown, and Indigenous peoples through commodification. On a surface level, georgic methods of sustainable agriculture were promoted as a more ecologically sound form of natural resource cultivation by attempting to remediate the fungibility of Black labor across colonial extractive projects. However, this methodology, which purportedly upheld environmental sustainability and conservation of enslaved labor, only led to the further simultaneous degradation of soil and exploitation of enslaved persons.

Imperial systems of slavery justified themselves through agricultural production, while Grainger and Freneau's poetry emphasize that planters treat enslaved persons humanely: their methods for doing so are juxtaposed. Proper stewardship of enslaved persons would prevent field hands from being treated as an exhaustible source of machine labor, as the Spanish had done. Grainger hypothesized that overworking slave labor led to unsustainable plantations. Spain and France minimized labor costs by instituting West African slavery; the British, unwilling to be trumped, followed suit.[2] David S. Shields notes that, throughout the eighteenth century, British West Indians, including Grainger, lamented their dependence on slavery, declaring, "The impossibility of doing without slaves in the West Indies will always prevent this traffic from being dropped."[3] This common apologist stance makes clear that plantation slavery and the slave trade would not be abolished.

For Grainger, the desire of planters to treat their slaves as fungible commodities may initially result in an abundant crop. Over time, however, this commodification and overworking of the cane fields and Black labor results in degraded unproductive soil. Courtney Weiss Smith contends that Grainger, in applying Virgil, "seriously instructs the husbandman to adjust himself, his labor, and his choice of seed to suit the soil's inherent tendencies."[4] Equating the care of sugarcane with the humane treatment of African enslaved persons, Grainger argues that for prospective planters to be successful growers they must not only maintain an ecological balance practicing stewardship of the cane, but also take care of the laborers who harvest. The georgic poet promotes stewardship over exploitation by encouraging the planter to treat enslaved persons humanely and plant his crop where the cane may receive the proper amount of wind and rain. Grainger advises the planter:

> Let not thy Blacks irregularly hoe:
> But aided by the line consult the site
> Of thy desmesnes; and beautify the whole.[5]

This logic equalizes commodity producers and commodities in ways that induces forms of anthropomorphism and dehumanization simultaneously. For Grainger, the planters who benevolently care for their slaves will be rewarded with a healthful and abundant crop, whereas the growers who overwork enslaved persons will produce an inferior yield. The environmental logic here is thus intimately tethered to an alleged commitment to the humane treatment of enslaved bodies. While Grainger tried to convince planters that humane treatment of enslaved persons would lead to more efficient and productive methods of cultivation, Caribbean growers worked Black laborers as if they were inexhaustible machines. Attempts at georgic stewardship and amelioration ultimately failed because planters saw human and nonhuman organisms only as inert objects rather than as interrelated components of an ever-changing ecological system. Grainger believed that practicing stewardship and amelioration would cause planters to recognize the ecological benefits of conserving natural resources and humane treatment of enslaved labor. However, Caribbean growers employed georgic methods of sustainable agriculture to exhaust the soil, mass-produce cash crops, and further treat enslaved persons as fungible commodities, as the Spanish had done. Georgic methods of sustainable agriculture did nothing to improve the laboring conditions of Black field hands; rather industrial cultivation was employed to make enslaved persons work harder, which led to the ecological degradation that Philip Freneau would witness on the sugar plantations in St. Croix nearly twenty years later.

## AGRICULTURAL IMPROVEMENT AND GEORGIC STEWARDSHIP IN *THE SUGAR-CANE*

Georgic poetry, as evidenced by Grainger's use, provided planters with an idealized means of applying the "New Science" to agricultural improvement, which husbandmen hoped would promote stewardship. Anthony Low asserts that the georgic "is a mode that stresses the value of intensive and persistent labor against hardships and difficulties."[6] And the Virgilian model that inspires Grainger, Kevis Goodman observes, seeks to establish a culture of sustainable estate management and accompanying terracultural experiment.[7] This emphasis on ecological responsibility appears in the first six lines, when Grainger encourages the farmer to consider who actually performs agricultural labor. Grainger asks,

> What soil the Cane affects; what care demands;
> Beneath what signs to plant; what ills await;
> How the hot nectar best to christalize;
> And Afric's sable progeny to treat:
> A Muse, that long hath wander'd in the groves
> Of myrtle-indolence, attempts to sing.[8]

Using a structure resembling Virgil's, Grainger employs the first four lines to summarize the focus of each corresponding book. This organization has the effect of equating the labor that is required to cultivate the sugarcane to the processes of refining sugar as well as the management of slaves.

Grainger's reference to the treatment of African slaves in the fourth line resembles the "*cura bonum* (care of cattle) in line three of Virgil's *Georgics*."[9] Cristobol Silva asserts that Grainger makes an explicit connection between the care of slaves and livestock, yet the first six lines also demonstrate how stewardship of crops, animals, and resources is intricately linked to managing systems of human, dehumanized, and nonhuman labor. Although Grainger encourages Caribbean planters to extend the procedures for the care and sustenance of livestock to enslaved persons, in reality humane supervision does not apply to enslaved field hands. Joshua Bennett argues that the concept of the "*cura bonum*" is illogical, because livestock like enslaved laborers have no "chronological orientation outside [their] relationship to the slaver's clock."[10] Cattle, horses, and the enslaved are "constantly moving between occupying a space of self-determination and being configured as a living commodity."[11] Livestock and slaves are saleable commodities, both of which possess an interiority that is denied. As Frederick Douglass would argue in his speech "Agriculture and Black Progress," delivered in Nashville, Tennessee, in 1873, "Not only the slave, but the horse, the ox, and the mule generally shared the general feeling of indifference to rights naturally engendered by a state of slavery. . . . The master blamed the overseer; the overseer the slave, and the slave the horses, oxen, and mules; and violence and brutality fell upon animals as a consequence."[12]

Grainger urged planters to treat livestock and enslaved laborers as co-laborers and partners in the field. Yet, the planter treated cattle and enslaved persons as machine laborers and fungible commodities. Grainger's poem encourages the prospective planter to reflect on the labor and cultivation that the cane demands along with establishing a plan for the equal distribution of labor and estate management. However, alongside this encouragement, Grainger promotes a model of conservation stewardship in which, theoretically, enslaved peoples became viewed as valued laborers and less of an exploitable commodity. This is the mental gymnastics of slavery apologists. When planters applied the concept of the *cura bonum* to human enslaved labor, stewardship becomes a means to extract as much labor from human capital as possible, invariably leading to exploitation.

Grainger's poetry continually equalizes the care of enslaved persons and the care of botanic life; the logic of the poem assumes that in cultivating and humanely treating one, the treatment of the other will follow suit. Shaun Irlam asserts that Grainger zoologizes "slave labor as some additional species of exotic fauna alongside the extensive botanical and zoological information he accumulates."[13] For Grainger, as Silva observes, "there is a parallel between slaves and the natural world, . . . ultimately marking Africans as passive commodities that, like sugar, circulate in the Atlantic world."[14] Grainger believed that if planters could come to view enslaved persons as a valued resource, similar to the cash crops they produced, they would naturally come to treat their laborers with dignity. Grainger hopes that his instruction in plantation husbandry will contravene the instrumentalization and commodification of enslaved labor in the slave economy. Yet, the stewardship of crops and slaves ultimately benefits the planter's profit margin rather than improving ecological laboring conditions. There are limits to this utopianism.

Despite Grainger's ecotheory, the planter's desire for abundant crops causes him to exploit the environment through the overworking of enslaved persons. At the close of Book One, Grainger introduces Montano, a Caribbean planter, who is unfamiliar with hard planting. As Richard Frohock describes him, Montano, "exiled from his native land, cheerfully takes up the hoe and cultivates a garden of West Indian vegetables, such as cassava, for his small family's sustenance."[15] Montano initially attempts to produce cane without enslaved persons. However, he comes to believe that slave labor is the only way to efficiently mass-produce. The yeomen planter soon reinvents himself as a successful freeholder. After a generation, the planter ultimately achieves "wealth beyond avarice" through industrial cultivation of sugar.[16] Unlike the planters who have overworked enslaved people and over-farmed the land, practicing stewardship has rewarded Montano with a thriving estate. Montano's sugar plantation is cultivated by a large number of "happy slaves" who respect their master for his benevolent treatment. The later life of the planter is "distinguished by benevolence and generosity freely bestowed on the needy."[17] Near the end of his life, Montano advises his eldest son on how to obtain wealth:

> Be pious, be industrious, be humane;
> From proud oppression guard the laboring hind. . . .
> Your means are ample, Heaven a heart bestow!
> So health and peace shall be your portion here;
> And yon bright sky, to which my soul aspires,
> Shall bless you with eternity of joy.[18]

Montano advises his son that bestowing God's compassion on enslaved peoples will make them loyal producers. For Montano, being a good steward also involves improving the lives of the enslaved laborers, so they remain subservient and loyal. Paternalistic stewardship here serves as an apology for the institution of slavery, reinforcing the status of the field hands as exploitable commodities and further stripping enslaved people of their right to humanity, equality, and liberty.

As is well-established, Caribbean planters justified their exploitation of enslaved laborers by arguing that slaves were subhuman individuals in need of paternalistic supervision. Enslaved laborers for the British, like the Spanish before them, as Yusoff and Saidiya Hartman argue, came to embody the replaceability endemic to a commodity. Tiffany King and Hartman maintain that "the fungibility of the commodity makes the captive body an abstract and empty vessel vulnerable to the projection of others' feelings, ideas, desires and values."[19] Grainger's instruction to Montano resembles the argument that Granville Sharp makes in his 1769 pamphlet *A Representation of the Injustice and Dangerous Tendency of Tolerating Slavery*, wherein Sharp asserts that "colonial masters in England cannot be justified in seizing and detaining such slaves unless they shall be able to prove that a Negro Slave is neither man, woman, nor child."[20] George Boulukos argues that while British planters like Montano believed that "slavery was inherently objectionable, and that blacks were commonly viewed as fully human, this did not dissuade them from eventually after some awkward groping" taking up Granville Sharp's rhetorical challenge (in his pamphlet) to divest "the Negro of his . . . humanity."[21] This denial of republican natural rights to enslaved laborers led to further unrest between slaves and their masters.

A wise planter does not abuse his enslaved laborers because he understands that content laborers will not revolt. Grainger thus attempts to situate a stark contrast among the multifaceted experiences of slaveholding and slavery. Grainger observes,

> In time a numerous gang of sturdy slaves,
> Well-fed, well-cloath'd, all emulous to gain
> Their master's smile who treated them like men;

> Blackn'd the Cane-lands: which with vast increase,
> Beyond the wish of avarice, paid his toil.[22]

Montano's crops flourish because his slave "gangs" are well fed and properly clothed. He treats his enslaved workers like laboring people, unlike the neighboring growers, who treat their slaves like fungible commodities. In this way, Grainger suggests that sufficient food and clothing is enough to uplift commodities into people and thus ensure enslaved compliance.

For Montano, the Caribbean environment works both for and against the plantation and its workers, "concurrently nurturing young cane shoots while steadily sapping the bodies of those enslaved."[23] Montano's slaves simultaneously fight against the harsh environment through clearing fields, digging holes, fighting dehydration and sunburn—while working within the environment of the sun and water for the growth of the sugarcane, "gravity and wind to work the mill, fire to process the cane, and microbes to ferment sugar into alcohol."[24] Considering these contradictory cultivation processes together under a broader conception of work, Grainger's georgic affords the reader a deeper understanding, as Neill Ostavall and Vaughn Scribner argue, of "how humans and agroecosystems came together to uphold a profitable cruel, and laborious Atlantic world."[25] The majority of work on early modern sugar plantations occurred at the intersection of human and environmental labor. Ostavall and Scribner assert that "early modern sugar agroecosystems were organized around the goal of creating products for blissfully unaware consumers in order to extract as much profit as possible from the work of humans and the environment, often with devastating outcomes for both."[26] For Grainger, the work of cultivating a plantation included controlling every facet of the Caribbean environment, both natural and human.

Grainger employed the georgic not only to instruct planters in sustainability, but also to reform the exploitation of the Caribbean agroecosystem and improve enslaved peoples working conditions. Establishing himself as a plantation manager, Grainger purchased slaves to rent out as extra field hands for harvesting and extracting cane juice. Caribbean planters would purchase gangs of slaves at minimal cost, which led to overworking. The extreme working conditions resulted in large amounts of crops being produced on small amounts of land. The high cost of provisions meant that many planters and overseers found it less expensive to work enslaved people to death, with new slave laborers imported from Africa, rather than providing adequate nourishment. John Gilmore observes that these enslaved "jobbing gangs" faced especially dire straits, because the hiring planter would work them "even harder than his own slaves because he had no capital stake in the survival of a rented laborer."[27]

The typical workday of a slave was around twelve hours and involved exhaustive hard hoeing as well as toting heavy baskets of manure. The common belief among planters was that time and work were intricately connected. The planter believed that the harder and longer slaves worked, the more productive their plantation would be, "as laborers would produce more and be too busy to rebel."[28] Ashton Warner, an enslaved eighteenth-century African, testified that the work of refining was arduous and that "many enslaved Africans sink under it and become ill, but if they complain, their complaints are not readily believed, or are considered only a pretense to escape from labour."[29] Grainger challenged the belief that overworking slaves would prevent revolt; rather, he believed that proper care and reasonable work schedules would alleviate insurrection, and in so doing, he provides a band-aid for a hemorrhaging wound.

Grainger's attempts to improve enslaved people's working conditions ultimately failed because the planters refused to see enslaved bodies as nothing more than working machines. Christopher F. Loar observes that "*The Sugar Cane* is notorious for its evisceration of the agency [of slaves], treating them not as autonomous human agents but as material inputs."[30] As Loar asserts, Grainger's advice to planters' blurs "distinctions between the human and the material so absolutely that enslaved labor becomes another naturalized material in the process of production; rendered unable to speak for themselves."[31] Grainger employs the concept of fungibility to convince planters to not overwork their slaves. However, fungibility fails as a strategy for managing laborers because, unlike a strict economic emphasis on the one-to-one exchangeability of a commodity, the Black body maintains figurative and metaphorical value, especially under the aegis of plantation economics. The labor of enslaved persons was crucial to the processes of crop rotation because it was Black bodies that transitioned between the cultivation of not only sugarcane, but also rice and indigo, which occurred on the same plot of land. King writes, "Black bodies thus became the intercessors of intermediary figures between multiple agricultural processes, spaces, and temporalities."[32] The mechanized expectations and deployment of enslaved bodies thus amounts to a degradation of both enslaved laborers and plantation landscape.

This commodification of the bodies of enslaved persons is made further apparent in the performance of enslavement that Grainger reports. When purchasing slaves, Grainger advises,

> Must thou from Africk reinforce thy gang?
> Let health and youth their every sinew firm;
> Clear roll their ample eye; their tongue be red;
> Broad swell their chest; their shoulders wide expand;
> Not prominent their belly; clean and strong.
> Their thighs and legs in just proportion rise.[33]

Grainger encourages planters to examine each slave they intend to purchase. Planters need to consider the physique of each potential laborer to determine to which task they will be best suited. However, Grainger's description of the purchasing and valuing of slaves reinforces how Black people are devalued on the auction block. Katherine McKittrick observes that the "documentation of the best qualities of the slave such as age, health, and labor specialty; the intimate bodily examination and demonstration; the pricing and bargaining—these activities spatialize and naturalize inequality."[34] The selling of enslaved persons on the auction block staged and violently forced acts of Black performativity such as hopping, jumping, and trotting "rendering the body of the enslaved as seemingly naturally suited to bondage and white pleasure-gazes."[35] The auction block for Grainger, as McKittrick argues, conceals the pain of Black bodies through reaffirming antiblack public expectations: a hopping, jumping, and trotting seemingly healthy Black body is a nonsuffering body. The auction block violently transforms human beings into fungible commodities through economic exchange.

In yet another series of warped logical arguments, Grainger attempts to ameliorate the environmental conditions of enslaved persons in Book Four by convincing planters to provide care for slaves while arguing that the lives of laborers would only be improved by trusting in their master's benevolence. This attempt to create a harmonious cooperative relationship is reflected when Grainger writes,

> Genius of Africk! . . .
> O attend my song.
> A muse that pities thy distressful state;
> Who sees, with grief, thy sons in fetters bound;
> Who wished freedom to the race of man.[36]

While I read Grainger's desire to emancipate the enslaved as sincere, yet he quickly moves away from directly addressing the plight of the enslaved and turns his attention to the owners providing instruction on how to properly care for their chattel. Grainger encourages the planter to act more like a caring physician and less like an abusive overseer toward those under his charge. Grainger asks,

> What care the jetty African requires?
> Yes, thou wilt deign to hear; a man thou art
> Who deem'st nought foreign that belongs to man.[37]

But, as in the previous case, the care of physicians is predicated on a plantation logic that demands the simultaneous optimization of enslaved bodies and the degradation of those bodies through labor.

For Grainger, humane treatment is the only means by which the planter will receive the desired output from the slaves. David S. Shields argues that Grainger advocated "for the material improvement of mankind brought about by imperial expansion yet he was compelled by economic circumstances to employ and justify slavery."[38] The physician believed that agricultural improvement would be extended to the lives of slaves. In Book Four as well as Book One, the poet encourages the planter to treat his slaves just as he would treat other valued tools and livestock. The system of amelioration he proposes is designed to make slaves content servants. Grainger proposes nearly a system of servile feudalism for plantations, wherein slaves participate in the agrarian economy. As part of his social reform agenda, Grainger proposes a freehold land system. Grainger instructs,

> Suffice not this; to every slave assign
> Some mountain-ground: or if waste broken land
> To thee belong, that broken land divide.
> This let them cultivate, one day each week;
> And there raise yams, and there cassada's root
> From a good daemon's staff cassada sprang.[39]

It is the humane thing to allow enslaved persons to benefit from the abundant staples that can be grown in the rich soil. For field hands to function productively, slaves must have enclosed land to grow yams and the "cassada" root. Beth Fowkes Tobin observes that, for Grainger, the enslaved, like the cane shoots, are both instruments of the planters will and free agents.[40] Grainger argued for a system of indentured slavery in which captives were treated as feudal serfs. However, the planters were interested in amelioration only as a means to eke out as much slave labor as possible. While Grainger employed the georgic to advocate for conservation of land and labor, planters employed methods of agricultural improvement to over-farm their plantations, which led to further abuse of enslaved laborers.

## FRENEAU'S GEORGIC SATIRES: THE ECOLOGICAL DEGRADATION OF AGRICULTURAL IMPROVEMENT AND AMELIORATION

In the late 1770s and 1780s, almost twenty years after Grainger published his West Indian georgic, the American abolitionist Philip Freneau began to write satire criticizing British attempts at conservation and amelioration. Fleeing the political turmoil of the American Revolution in 1776, Freneau sailed to the island of St. Croix where he resided on Captain Hanson's plantation until 1778. During his time in the West Indies, Freneau witnessed the ecological degradation of the sugar plantations as a result of attempts at stewardship, amelioration, and the further abuse of slaves. Responses to Grainger's system of amelioration, as Steven W. Thomas observes, were contradictory. Supporters of the British Amelioration Act claimed that they were making slavery more humane, providing benevolent stewardship that would ultimately lead to abolition.[41] However, in contrast to Montano's idyllic planation populated by industrious slaves, in his georgic satires

Freneau paints a picture of a polluted landscape populated by exhausted laborers. First published in 1778 and revised in 1786, Freneau's poem "The Beauties of Santa Cruz" responds to Grainger's portrayal of the plantation as an ecologically balanced Eden portraying British attempts at sustainable agriculture, as Sean X. Goudie observes, as a "postlapsarian tragedy deriving from the islands material realities of European slavery and recurring natural disasters that erode the unsustainable pastoral edifice that the poet speaker seeks to construct."[42] Freneau describes the Caribbean as an Edenic frontier that is constantly threatened by overfarming and the exploitation of slave labor.

In contrast to Grainger, Freneau believed that preserving the West Indian ecosystem was incompatible with the cultivation of sugarcane and plantation slavery. In a description of enslaved laborers on the sugar plantations published in a magazine circulated among the colonies to accompany "The Beauties of Santa Cruz," Freneau writes, "A description of the cruelties the poor slaves endure would be too irksome and unpleasant to me, and to those who have not beheld it would be incredible. Sufficient be it to say that no class of mankind in the known world undergo so complete a servitude as the common negroes in the West Indies."[43] Like Grainger, Freneau praises the pastoral charms of the tropical environment of Santa Cruz. However, for the abolitionist poet, the plantation does nothing to improve the welfare of the enslaved; the overworking of slaves to produce cash crops leads to ecological degradation. In "Santa Cruz's" progressive moments, Freneau urges his republican readers "to resist repeating European colonialism's mercantile excesses and abuses on arriving from the war-ridden continent" to the West Indies.[44] The portrayals of the socially dead West Indian slave and the alluring sugarcane are employed to demonstrate how British plantation agriculture and slavery have led to the ecological degradation of the island. Freneau, by Goudie's assessment, argues that "by avoiding these figures, the Sons of Columbia emerge as exemplary custodians of the island's natural resources and disseminators of republican values across the island, salvaging this one island at least from British tyranny in mutual and reciprocal measure."[45] Cultivation of the sugarcane has had deleterious effects not only on slaves, but also on white planters, who have degraded the environment through over-tilling.

Expressing his concern for the degradation of the cane fields and compassion for the suffering enslaved who work tirelessly in the sugarcane fields, Freneau observes,

> See yonder slave that slowly bends this way,
> With years, and pain, and ceaseless toil opprest,
> Though no complaining words his woes betray
> Perhaps in chains he left a helpless offspring there,
> Perhaps a wife, that he must see no more,
> Perhaps a father, who his love did share.

> Curs'd be the ship that brought him o'er the main,
> And curs'd the hands who from his country tore,
> May she be stranded, ne'er to float again,
> May they be shipwreck'd on some hostile shore.[46]

Challenging Grainger's portrayal of Montano's happy swains hoeing in the cane fields, Freneau argues that the "ceaseless toil" of cultivation exhausts the slaves' energy. While Grainger portrays enslaved bodies as efficient machines, Freneau views the captive laborers as an oppressed people abducted from their homeland. The georgic satirist argues that amelioration has failed to improve the working conditions of plantation slaves.

Freneau believed, similar to the militant abolitionist views of Grainger's critics Samuel Johnson and John Singleton, that no amount of benevolence could reform the abuse of Caribbean slaves. Unlike Montano's estate in Grainger, in Freneau's "To Sir Toby" (1791), the plantation is not an ecologically balanced utopia where the enslaved and their enslavers work together. Rather, the Caribbean plantation is a degraded pasture where African slaves oppressively toil. Freneau opens the poem with an abolitionist invective against plantation slavery by comparing Jamaica, as did Singleton, to Virgil's underworld from the *Aeneid* and a sulfurous hell:

> If there exists a hell—the case is clear—
> Sir Toby's slaves enjoy that portion here:
> Here are no blazing brimstone lakes—tis true;
> But kindled Rum too often burns as blue;
> In which some fiend, whom nature must detest,
> Steeps Toby's brand and marks poor cudjoe's breast.[47]

The overworked enslaved laborers are literally drowning in a sea of kindled rum. The slave owner Sir Toby, unlike Grainger's Montano, does not treat his slaves with purported dignity; he treats them like property. Sir Toby literally brands his slaves as property. In reading this moment, Philip Gould suggests that, "the initial image of the slave's branding, for example, not only suggests his new identity, or name, but connects such brutality to the debased habits of consumption engendered by the West Indian export economy."[48] The slave master will never see his laborers as anything more than mechanized bodies; Toby is concerned only with the amount of labor that he can get from his slaves rather than their welfare.

Grainger's georgic alleged that, by providing care and nourishment for slaves, masters would come to see their field hands as human. The benevolent planter would allow his workers to share in the productive bounty of the cane crop. However, Freneau argues that attempts at amelioration have not improved slaves' lives. He writes,

> Why were they brought from Eboe's sultry waste.
> To see that plenty which they must not taste—
> Food, which they cannot buy, and dare not steal;
> Yams and potatoes—many a scanty meal![49]

The enslaved have been brought to the Caribbean to work the land, but they cannot fully benefit from the productive harvest of the crop. Enslaved laborers refine sugar and produce rum, but the planter will not allow the slaves to benefit from the yield. The planter may provide a modest amount of land, as Grainger instructed, but the yams and potatoes that they can grow are hardly enough for sustenance or sustainability.

In conversation with Grainger, Freneau notes that slaveholders have abandoned a mindfulness and commitment to the health of enslaved laborers; the planter is so driven by a desire for profit that he simply overworks his slaves to death. Freneau details,

> O'er yond' rough hills a tribe of females go,
> Each with her gourd, her infant, and her hoe;
> Scorched by a sun that has no mercy here,
> Driven by a devil, whom men call overseer
> In chains, twelve wretches to their labours haste;
> Twice twelve I saw with iron collars graced![50]

Here Freneau describes the abuse that enslaved women must endure as they work. There is no one to take care of the enslaved mother's child, so she must hoe and carry her infant. The jobbing gangs that Grainger described are scorched by the heat of the midday sun. For Freneau, the conjoined violences enacted against the enslaved body and the overworked earth destroy the Caribbean landscape.

## TEACHING GRAINGER AND FRENEAU IN THE AMERICAN LITERATURE CLASSROOM

In my American Literature to 1865 survey course and upper division course, Eighteenth-Century Colonial Literature and Environment, I introduce the georgics of Grainger and Freneau as a product of Enlightenment debates about slavery, conservation, and Natural Rights. As a form of nature writing, the georgic exposes the destructive impact of British agrarian colonialism on the ecology of the Caribbean. Grainger and Freneau's ecopoetry raises a host of important questions about how sugar cultivation and slave labor affected the ecology of the Caribbean and how Virgil's methods of sustainable agriculture were employed and justified to promote conservation of land and improve methods of cultivation.

One of the challenges of teaching these texts is getting students to understand the critical connection between the exploitation of natural resources, farmland, and the overworking of slave labor. Students benefit from a *comparative empires approach*, which helps them to understand how the European desire to make a profit off the land and exploit natural resources led to the widespread abuse of enslaved labor. In *The Sugar-Cane*, Grainger employs the concept of stewardship of land and labor to promote Caribbean agriculture as an alternative to the Spanish attempt to degrade the landscape and enslaved peoples through mining.

Kathryn Yusoff's discussion of race and geologic extraction in *A Billion Black Anthropocenes or None* provides a theoretical framework for discussing the Black Legend model of Spanish colonialism, which resulted in the commodification and abuse of slave labor. For Grainger, the British model of plantation slavery was in theory a more humane model of colonialism, in which practicing agricultural stewardship would result in a more benevolent form of slavery. Studying Yusoff's theory of racial categorization helps students understand how slavery developed in the West Indies and how British colonists attempted to promote reform. Grainger's attempt to instruct planters in Virgil's methods of sustainable agriculture ultimately failed because British growers desire to produce cash crops led to the same type of commodification of slave labor that the Spanish had practiced. Saidiya Hartman's concept of Black fungibility in *Scenes of Subjection* is particularly helpful for developing an understanding of how enslaved bodies on plantations came to embody the interchangeability and replaceability endemic to the definition of a commodity.

As an illustration of how British colonial stewardship and sustainable agriculture commodified Black bodies, instructors might have students examine William De Brahm's *Map of South Carolina and Georgia* (1757) alongside Grainger's georgic to more fully understand how enslaved cultivation of cash crops led to the ecological degradation of West Indian farmland. De Brahm's map portrays Black bodies not as human laborers, but as material inputs facilitating cultivation processes. The map illustrates how the overuse of enslaved labor led to the type of ecological degradation of Caribbean farmland that Philip Freneau describes in his georgic satires at the end of the eighteenth century. Freneau's description of a dystopian degraded Caribbean landscape in his poetry is an applied example of how plantation agriculture has led to the exploitation of human and nature. Grainger attempted to improve the laboring conditions of enslaved persons on Caribbean plantations through amelioration; however, Freneau's poetry emphasizes the need for widespread abolition of the institution of slavery. Freneau's Caribbean georgics demonstrate how amelioration has failed to improve the laboring conditions of enslaved persons. While sustainable agriculture was an attempt to improve the working conditions of slaves, the policy had the opposite effect, reinforcing racial hierarchies and exposing abuse. These same questions and concerns are worth considering today, especially as calls for sustainable farming, agriculture, and labor remain just as present and can very well encode some of these identical politics and social violences.

## NOTES

1. Kathryn Yusoff, *A Billion Black Anthropocenes or None* (Minneapolis: University of Minnesota Press, 2018), 5.
2. David S. Shields, *Oracles of Empire: Poetry, Politics, and Commerce in British America 1690–1750* (Chicago: University of Chicago Press, 1990), 77.
3. Shields, *Oracles of Empire*, 77.
4. Courtney Weiss Smith, *Empiricist Devotions: Science, Religion, and Poetry in Early Eighteenth-Century England* (Charlottesville: University of Virginia Press, 2016), 177.
5. James Grainger, *The Sugar-Cane* [1764], in *Caribbeana: An Anthology of English Literature of the West Indies 657–177*, ed. Thomas W. Krise (Chicago: University of Chicago Press, 1999), 169 (1.7).
6. Anthony Low, *The Georgic Revolution* (Princeton, NJ: Princeton University Press, 1985), 12.
7. Kevis Goodman, *Georgic Modernity and British Romanticism* (London: Cambridge University Press, 2004), 21.
8. Grainger, *Sugar-Cane*, 169 (1.1–1.6).
9. Cristobol Silva, "Georgic Fantasies: James Grainger and the Poetry of Colonial Dislocation," *ELH* 83, no. 1 (2016): 139.
10. Joshua Bennett, *Being Property Once Myself: Blackness and the End of Man* (Cambridge, MA: Harvard University Press, 2020), 2.
11. Bennett, *Being Property Once Myself*, 2.
12. Bennett, *Being Property Once Myself*, 3.
13. Shaun Irlam, "Wish You Were: Exporting England in James Grainger's *The Sugar Cane*," *ELH* 68, no. 2 (2001): 386.
14. Silva, "Georgic Fantasies," 139.
15. Richard Frohock, *Heroes of Empire: The British Imperial Protagonist in America, 1596–1764* (Newark: University of Delaware Press, 2004), 170.
16. Frohock, *Heroes of Empire*, 170.
17. Frohock, *Heroes of Empire*, 170.
18. Grainger, *Sugar-Cane*, 196 (1.631–1.641).
19. Tiffany Lethabo King, "Racial Ecologies: Black Landscapes in Flux," in *Racial Ecologies*, ed. Leilani Nishime and Kim D. Hester Williams (Seattle: University of Washington Press, 2018), 65.
20. Sharp quoted in George Boulukos, *The Grateful Slave: The Emergence of Race in Eighteenth-Century British and American Culture* (London: Cambridge University Press, 2008), 100.
21. Boulukos, *Grateful Slave*, 101.
22. Grainger, *Sugar-Cane*, 195 (1.611–1.614).
23. Neill Ostavall and Vaugh Scribner, "The Devil Was in the Englishman That He Makes Everything Work: Implementing the Concept of Work to Reevaluate Sugar Production and Consumption in the Early Modern British Atlantic World," *Agricultural History Society* 92, no. 4 (2018): 463.
24. Ostavall and Scribner, "Devil Was in the Englishman," 463.
25. Ostavall and Scribner, "Devil Was in the Englishman," 463.
26. Ostavall and Scribner, "Devil Was in the Englishman," 473.
27. John Gilmore, *The Poetics of Empire: A Study of James Grainger's The Sugar Cane* (London: Athlone Press 2000), 17.
28. Ostavall and Scribner, "Devil Was in the Englishman," 474.

29. Ostavall and Scribner, "Devil Was in the Englishman," 474.
30. Christopher F. Loar, "Georgic Assemblies: James Grainger, John Dyer, and Bruno Latour," *Philological Quarterly* 97, no. 2 (2018): 252.
31. Loar, "Georgic Assemblies," 252.
32. King, "Racial Ecologies," 71.
33. Grainger, *Sugar-Cane*, 240 (4.72–4.77).
34. Katherine McKittrick, *Demonic Grounds: Black Women and the Cartographies of Struggle* (Minneapolis: University of Minnesota Press, 2006), 73.
35. McKittrick, *Demonic Grounds*, 73.
36. Grainger, *Sugar-Cane*, 238 (4.1, 4.12–4.16).
37. Grainger, *Sugar-Cane*, 238 (4.25–4.36).
38. Shields, *Oracles of Empire*, 73.
39. Grainger, *Sugar-Cane*, 251 (4.445–4.455).
40. Beth Fowkes Tobin, *Colonizing Nature: The Tropics in British Arts and Letters, 1760–1820* (Philadelphia: University of Pennsylvania Press 2004), 56.
41. The British Amelioration Act, passed in 1798, sought to further regulate plantation slavery in the Caribbean by mandating medical care and imposing penalties and fines for neglect and abuse of slaves. This legislation was designed to improve the social conditions of the enslaved; however, the imposing of regulations on planters only caused them to continue to overwork the enslaved and contributed to the degradation of the plantation landscape.
42. Sean X. Goudie, *Creole America: The West Indies and the Formation of Literature and Culture in the New Republic* (Philadelphia: University of Pennsylvania Press, 2006), 117.
43. Samuel B. Bandara, "Philip Freneau: First American Poet to Write About the Caribbean," *Inter-American Review of Bibliography* 36, no. 1 (1986): 156.
44. Goudie, *Creole America*, 123.
45. Goudie, *Creole America*, 123.
46. Philip Freneau, "The Beauties of Santa Cruz," in *The Poems of Philip Freneau Written Chiefly During the Late War* (1776) (Evans Early American Text Creation Partnership, 2011), 72–74, 145,https://quod.lib.umich.edu/cgi/t/text/text-idx?cc=evans;c=evans;idno=N15445.0001.001;node=N15445.0001.001:31;rgn=div1;view=text, (accessed September 12, 2021).
47. Philip Freneau, "To Sir Toby" (1784), *The Poems of Philip Freneau: Poet of the American Revolution*, ed. Fred Lewis Pattee (Princeton, NJ: Princeton University Library, 1902–1907), 258.
48. Philip Gould, *Barbaric Traffic: Commerce and Antislavery in the Eighteenth-Century Atlantic World* (Cambridge, MA: Harvard University Press, 2003), 54.
49. Freneau, "To Sir Toby," 259.
50. Freneau, "To Sir Toby," 259.

11

# JOHN THELWALL AND L. M. MONTGOMERY WRITE THE GREEN CITY

Kate Scarth

L. M. (LUCY MAUD) MONTGOMERY'S Anne Shirley welcomes a new season by declaring, "It's spring! . . . Even the most unlovely streets are transfigured by arms of bloom reaching over old board fences and a ribbon of dandelions in the grass that borders the sidewalks."[1] Anne draws attention here to the humblest urban nature. Dandelions, flowers usually dismissed and eradicated as weeds, can be transformative: unlovely urban spaces can be "transfigured," made bearable, beautiful, and perhaps inspiring through natural processes, if we pay close attention, as Anne does.

This essay explores such ordinary experiences of urban green spaces—parks, graveyards, gardens, fields, trees, flowers, weeds—and their impact on urban residents' health and well-being. Urban green spaces, in providing benefits, even necessities, like fresh air, room to exercise, and community-gathering places, can offer a "cultural fix": the use of social space to mitigate social crises, connected, for example, to public health, climate change, and affordable housing, as Kevin Loughran explains.[2] Eighteenth-century radical writer and orator John Thelwall (1764–1834) recognized the connection between what we would now call urban green spaces and environmental justice, specifically the importance of widespread access to fresh air and greenery in towns and cities. His polymathic influence on late-eighteenth- and nineteenth-century culture has been recently recuperated by Judith Thompson, Michael Scrivener, and Yasmin Solomonescu, among others.[3] I argue for Thelwall's enduring relevance to the environmental humanities, in particular how *The Peripatetic* (1793), his multi-genre walking journey through parts of Greater London and southern England, explores urban environmental justice, with applicability to literary cityscapes of other times and places. As a case study, I offer a Thelwallian reading of urban green spaces in the novel *Anne of the Island* (1915) by L. M. Montgomery (1874–1942), Canadian icon, international bestseller, and beloved creator of *Anne of Green Gables* (1908).

In an era of rapid global urbanization, of "the barren wilderness erased, / Or disappearing"[4]—dating at least from Thelwall's (and William Wordsworth's) late eighteenth century—people routinely encounter nature that is not wilderness, that is urban or suburban, industrial or polluted, small scale or infrequently experienced. Therefore, knowing, caring for, and safeguarding non-wilderness nature—including the green spaces of modern urban life—is a question of environmental justice, having serious implications for the democratic and widespread maintenance of human health, well-being, community, and belonging. This essay belongs to the urban threads of ecocriticism and the environmental humanities, as outlined, for example, by the scholars gathered in Michael Bennett and David Warfield Teague's *The Nature of Cities: Ecocriticism and Urban Environments*.[5] In eighteenth-century environmental humanities, scholars, notably Romanticists Ashton Nichols and Scott Hess, call for consideration of urban nature, including human-inhabited and even unpleasant or polluted spaces.[6] Thelwall turns attention to such spaces and to the routine practices that negatively impact the environment, such as sprawl, improvement (as an eighteenth-century form of gentrification), and air pollution. It is in these urban workaday spaces that people increasingly "learn their love of nature and their desire to care for nature."[7]

My comparative reading explores *The Peripatetic* as a theory of urban environmental justice, while also enriching Montgomery studies: conventionally seen as a Wordsworthian writer of natural and rural spaces, Montgomery emerges, with Thelwall, as a critic of urban environmental challenges. This juxtaposition of Thelwall and Montgomery leads to a more capacious understanding of Romanticism, its influence on Montgomery, and its enduring relevance for understanding the intersection of environmental and humanities inquiry. Reading Thelwall and Montgomery together reveals that the two writers' shared concerns with democratic access to urban green spaces, a deep knowledge of and care for place and people (including those that are impervious to praise according to aesthetic or social convention), and a resistance to, or rejection of, dominant but inequitable approaches to urban planning, building, and control. Both writers confront a range of urban issues, including affordable housing, cultural heritage preservation, gentrification, social isolation, and ill health, offering "cultural fixes" for these challenges via access to urban nature. For Thelwall and Montgomery, two processes—the securing of affordable housing and the act of walking—enable people to access and benefit from urban green spaces and thus inhabit cities and towns in ways that support individuals, communities, and the environments in which they live to flourish.

## PAIRING THELWALL AND MONTGOMERY

Since the 1980s, L. M. Montgomery's reputation has experienced a scholarly renaissance with the publication of her journals, chronicling the competing demands of a professional woman writer's life in the early twentieth century, including spousal and maternal duties, legal disputes, and mental health challenges; meanwhile interdisciplinary scholars continue to explore Montgomery's life, writing, and legacy.[8] Her writing, including *Anne of Green Gables* (translated into more than thirty languages), inspires literary pilgrimages, a healthy tourism industry, and countless adaptations in TV, film, theatre, dance, new media, fiction, and children's books.

A Thelwallian reading of Montgomery's novel *Anne of the Island* (1915)—the third book in the *Anne of Green Gables* series—sheds light on the socio-environmental justice currents present not only in her legacy, and in later adaptations of her work—like CBC/Netflix's *Anne with an E,* with its attention to Black, Indigenous, and LGBTQ+ rights—but also in her own fiction. Eighteenth-century writers, including Alexander Pope, Oliver Goldsmith, Walter Scott, and William Wordsworth, figure in Montgomery's reading and literary influences,[9] but Thelwall's political radicalism is an unlikely bedfellow for a writer largely perceived as creating domestic fiction and children's literature set in idyllic rural communities. Indeed, while scholarship on place and landscape has flourished in Montgomery studies, the places in question are decidedly rural, usually the Ontario or Prince Edward Island countryside.[10] My pairing of Thelwall and Montgomery positions Montgomery as *also* an urban writer and, more specifically, as an urban writer like Thelwall, championing democratic access to green space. Thelwall is troubled by the plight of Greater London's poor and working classes, as their homes and agrarian livelihoods are destroyed by affluent suburban sprawl, while Montgomery focuses on finding a healthy urban home that supports the first generation of Canadian female university students' education and holistic well-being.

For both Thelwall and Montgomery, urban issues—a healthy home, the possibility of sustaining oneself—*are* environmental issues. Hess specifically criticizes strains of twenty-first-century environmental thinking and activism that have a "Wordsworthian" tendency to define "nature as a space of leisure and contemplation rather than of work or inhabitance" and as a space of "flight from the social and the everyday . . . rather than an attempt to ground our lives ecologically in the here and now."[11] Such an approach to both nature and the city is problematic: "Instead of encouraging us to inhabit our everyday relationships, experience, and environments fully, this idealization of nature in opposition to the social world ironically promotes modern consumer fantasies of escape and transcendence, both from our own home place(s) and from other people. Defining nature as a place where humans are not, we divert our environmental attention away from the inhabited sites, everyday practices, and social and material structures that generate our biggest and most destructive environmental impact."[12] Urban environmental issues cannot be ignored in favor of spectacular and seemingly pristine wilderness; as Nichols states, drawing on a long tradition of ecological thinkers, "It is necessary now for human to save it all (the whole world) and make it all into something valuable—urban and natural, civilized and wild—to recognize the interconnectedness of every place and every aspect of human and nonhuman life."[13] Suburban environments, as a meeting point of urban and rural, home and garden, culture and nature, embody, for Thelwall and Montgomery, the greatest possibilities for creating environments rooted in well-being and democratic ideals.

Connecting Thelwall with Montgomery also introduces new intersections between her writing and the Romantic era. In scholarship about Montgomery's engagement with British Romanticism, Wordsworth and conventional Romantic ideology, with its emphasis on imagination, the self, and nature, dominate.[14] Montgomery is unlikely to have read or even heard of Wordsworth's contemporary, the radical romantic, Thelwall, who worked in experimental forms (such as *The Peripatetic*, published only once, in 1793, until Thompson's 2001 edition) and in ephemeral ones (the political speech, for example), and whose radical politics, leading to imprisonment, has obscured his work and legacy in contrast to his still well-known friends Wordsworth and Samuel Taylor Coleridge.[15] A pairing of Thelwall and Montgomery, rather than Wordsworth and Montgomery, places Montgomery in the context of a broader Romanticism, one that includes the individual but that is also sociable *and* community-minded, that is wild, rural, *and* urban, that is imaginative *and* political.[16]

## "SO MANY COMFORTABLE RETREATS": URBAN GREEN SPACES AND AFFORDABLE HOUSING

Thelwall repeatedly critiques urban sprawl, including its perpetuation and exacerbation of social inequality; *The Peripatetic*'s narrator Sylvanus describes, for example, "this overgrown Metropolis" with "its spreading streets and rising palaces; the trophies, if I may so express myself, of public misery and oppression."[17] Thelwall is particularly concerned with the equity implications of the issue known in the twenty-first century as access to affordable housing. The growing ranks of affluent suburbanites build their "rising palaces" and opulent villas, and they "improve" the surrounding landscape with extensive lawns made fashionable, and ubiquitous, by landscape architect "Capability" Brown. Such lawns allow for the sweeping views popularized by the picturesque aesthetic, but they also, as is the case here, necessitate the eradication of agricultural laborers' homes and communities—a process Sylvanus, no mincer of words, calls an "exterminating" act.[18] This suburban version of agricultural enclosure diminishes the physical environment, creating the homogeneous landscape of "Commercial Retirement," in which "the fields, the hedges, and the garden grounds," reflect "the triumphs of city elegance over the wild charms of nature."[19] Laborers, evicted from land swallowed by a rapidly expanding London, become disconnected from their social and natural environment, from the land that had supported them as agricultural workers. In the suburban landowner's capitalistic logic, "air and the cheerful verdure of the field are luxuries too great for Poverty."[20] Health and sustenance are obliterated; these dispossessed families will likely end up as urban slum dwellers in "a miserable habitation within the smoky confines of some increasing town," with their once healthy children "drooping and inactive."[21] Thelwall thereby paints urban issues—improvement/gentrification, eviction, overcrowding, socioeconomic segregation, air pollution, community destruction, ill health—as environmental health issues.

Suburban landowners have an authoritarian, transcendent view of their new property, transforming a functioning, long-entrenched, and multifaceted (economic, social, emotional, environmental) system into a space servicing and displaying the status of one person. In doing so, these landowners evince a failure of environmental imagination; they are at a remove from this community and this land, seeing it only from their own perspective, as an aestheticized landscape, not as a known, lived-in location (i.e., as a spot "dear" to the community and its collective memory).[22] The lone Wordsworthian wanderer is transformed into a suburban landowner,[23] securing nature to himself through property ownership while resisting, and destroying, other kinds of environments—sites of community, labor, and sustenance. Hess associates these latter locales with laboring-class poet John Clare, and Wordsworth's own sister, Dorothy Wordsworth, who, alongside descriptions of daffodils and mountains, her brother's favored topics, recorded in her journals pub lunches, payments to workers, and group walks.[24] Thelwall's censure of suburban improvement is not only a critique but is also part of the "cultural fix" he offers.

Thelwall offers more positive examples; he praises Greater London communities that both exemplify already extant healthy environments with access to green spaces and can be models for further development. Anticipating Leigh Hunt's suburban Hampstead poems of the Regency period,[25] narrator Sylvanus presents as ideal suburban living the harmonious mingling of nature and culture, of flourishing greenery and human habitation, of retreat and community: "the thick and frequent shade of trees, smiling in all their native exuberance of spray and foliage, and the modest little cottages peeping out occasionally from under their more towering branches."[26] Thelwall draws on an arsenal of aesthetic, sentimental, and spiritual language to counter examples of sprawl governed by the display of wealth, status, and fashion that eradicates dynamic communities and landscapes.

Sylvanus explicitly contradicts the dominant culture of fashion and conspicuous consumption, finding the potential for pride, health, and comfort in seemingly shabby villages that have yet to be discovered by improving, upwardly mobile suburban residents: "I can never help admiring the neglected neighbourhood [neglected by improvement], where decent Poverty may find so many comfortable retreats, be hid from gazing scorn, and enjoy a purer gale than the choaked city can afford."[27] British society has failed these people: neglected, they live in poverty and, when attention is paid to their neighborhood, destitution and ostracization are amplified, as in the example of the landscaping suburbanite. *The Peripatetic* challenges common assumptions that poverty equals discomfort and shame, that wealthy spaces are morally and atmospherically superior. As Judith Thompson writes, "Thelwall's satirical pastoral aims to assert the value and legitimacy of working-class interests by attacking the system that denies them; and in particular, the canons of taste that banish, sanitize, or demonize 'wretched' or 'common' objects and subjects in the name of aesthetic purity or moral ascendency."[28] In other words, these poor villagers have carved out their own "cultural fix" in their retreat from and resistance to an inequitable, often destructive, wider culture. In Sylvanus' democratic argument, the lower rather than the upper classes create the "fix," since it is the poor villagers, rather than the opulent landowners, who provide a model for urban housing. *The Peripatetic* offers readers alternative frameworks for reading and preserving suburban spaces, and perhaps even for building new ones that prize fresh air, comfort, community, and health over the usually ascendant, but environmentally devastating, model of opulent sprawl.

More than one hundred years later, Montgomery is equally concerned with access to safe, pleasant housing in a rapidly changing city. In *Anne of the Island*, Anne and her friends struggle to find a satisfactory home in the city because of their socioeconomic status as poor country girls (to paraphrase Anne) and the lack of urban housing with adequate green-space access. They ultimately find a suburban cottage, Patty's Place, the only unimproved home on a gentrified street of luxe mansions.

*Anne of the Island* is the chronicle of Anne Shirley's four university years in Kingsport (a fictional Halifax, Nova Scotia, Canada), and Patty's Place is the rented home she happily shares with friends during the last three years of her degree: it is rooted in friendship and comfort, as when Anne pictures "Patty's Place, its cosy open fire, Aunt Jamesina's mirthful eyes, the three cats, the merry chatter of the girls, the pleasantness of Friday evenings when college friends dropped in to talk of grave and gay."[29] Reading Patty's Place alongside places in *The Peripatetic* draws out Montgomery's engagement with affordable housing and environmental well-being.[30]

Kingsport is a historic port town growing in ways that, like Romantic-era London's development, have implications for social equality and access to affordable housing. Spofford Avenue (Young Avenue in "real-life" Halifax)[31] is a wealthy enclave, where Kingsport "sprouts out into modernity": "Nobody can build on [Spofford Avenue] unless he's a millionaire."[32] As with many of the newly developed properties that Thelwall observes, the display of status and fashion is a defining feature of Spofford Avenue: on a street of new millionaires' homes, old-fashioned, modest Patty's Place stands out, "ma[king] its nearest neighbour, the big lawn-encircled palace of a tobacco king, look exceedingly crude and showy and ill-bred by contrast."[33] While Patty's Place is the original home once located on a country road, it has been surrounded by affluent suburbanization. Like Thelwall had observed in Greater London, diverse, productive plant life—Patty's Place has an apple orchard and "dear, old-fashioned, unworldly flowers and shrubs"[34]—has been replaced by more homogeneous landscaping elsewhere on Spofford Avenue. Lawns, signifying wealth through the display of uncultivated open space not needed for economic or productive reasons, were first popularized in England by landscape architect "Capability" Brown and then imported to smaller-scale suburban lots around the world, particularly to North America, and continue to be a powerful status symbol.[35] The creation of lawns were a cause of the laborers' eviction in *The Peripatetic*, and, while lacking this extreme community extermination, social disintegration is also observable on Spofford Avenue. Aunt Jamesina, the young women's housekeeper, humorously points to the lack of neighborliness when she refuses to leave their cats alone for the holidays, with "no decent neighbours," only millionaires, to feed them.[36] As with Thelwall's wealthy suburban dwellers, the Spofford Avenue millionaires cut themselves off from the neighborhood not only by the lawns that encircle their houses, but also by their (perceived) unwillingness to sustain the community's human and nonhuman members.

In their first year of university, Anne and friends have lived in urban boardinghouses, lacking ideal Romantic access to urban nature: one friend's lodgings has a backyard, likely reserved for household chores like laundry, rather than a garden—it is "bleak and lonesome," "unholy," "the ugliest place in the world"—while Anne's room looks out on a graveyard.[37] These outdoor spaces' inadequacy is underlined by their Gothic elements: one is devoid of beauty and spiritual communion and one is a resting place for the dead (albeit a site that later becomes a refuge for Anne, as discussed below). Tired of boardinghouse life—the lack of nature and beauty, landladies' arbitrary rules, an absence of homeyness—the friends spend much of their first-year house-hunting to no avail: they cannot afford anything satisfactory.[38] And so, like in *The Peripatetic*, access to affordable housing, as an issue linked to environmental classism, becomes a concern of *Anne of the Island*.

Patty's Place represents a potential "cultural fix," an alternative to inadequate urban housing and unaffordable affluent suburbia. Patty's Place provides both proximity to the university where Anne and friends are trailblazing female postsecondary students *and* immersion in suburban nature. This nature includes the old-fashioned, beautiful flowers mentioned above, as well as sheltering trees, the apple orchard "in place of a back yard," access to the park (a fictional version of Halifax's Point Pleasant Park), and a generally rural feel, connecting the young women to their country homes ("The whole place might have been transplanted from some remote country village").[39] Here they can cultivate companionship, community, and comfort. Patty's Place is home, a retreat from the (urban) world, as is clear when one of the young women worries that "our jolly little nest here would be broken up—and we poor callow nestlings thrown out on the cruel world of boardinghouses again."[40] This metaphor, linking nature, the built environment, animal life, and a social space, emphasizes the importance of both green space and human companionship, of both nonhuman and human life, in creating a satisfactory urban home. From their Patty's Place base, these young women flourish socially, academically, and romantically. Phillipa Gordon, the beautiful, flighty, and only wealthy member of the Patty's Place cohort, undergoes the most development: from crippling indecisiveness, she learns how to make up her own mind when she falls in love with the future minister, Jonas Blake, marrying him even though he is poor.[41] Interestingly, given Montgomery's attention to the socio-spatial makeup of the city, the Blakes's married life will begin in the Patterson Street slums, offering spiritual and material sustenance to some of the city's most impoverished residents; the characters' shaping of the city and the possibilities of "cultural fixes" do not therefore end with the novel. The transformative potential of Patty's Place all begins when Anne and friends first discover it on a walk to the park, and indeed the act of walking is a fundamental way of accessing and experiencing urban spaces for both Thelwall and Montgomery.

## "I PURSUED MY TRANQUIL AND CHEERFUL WAY ALONG THE FIELDS": WALKING IN URBAN GREEN SPACES

For Thelwall and Montgomery, the home is a base for accessing other urban green spaces, so mobility, and especially walking, with its physical, psychological, and political benefits, is significant. Both writers identify urban developments in mobility and transportation that negatively impact health and well-being. Thelwall's Sylvanus is a critic of sedentary lifestyles: the late eighteenth century's growing numbers of white-collar workers—clerks working in City of London offices and "modern students"—are confined to desks, and "are in some degree compelled, by the institutions of society, to *bury* themselves in large cities."[42] As a result, their health is "debilitated," and the city becomes a metaphorical coffin: London is for Sylvanus explicitly the "grave of health," leading eventually, for those immobile worker-bees, to a literal grave.[43]

Sylvanus also creatively satirizes the elite, proudly parading in their carriages, an eighteenth-century status symbol analogous to a luxury car, or even a private plane, today: they are "shut up in glass cages, with varnished covers over their heads, like preparations of stuffed monsters in the cabinet of the natural historian," while the carriage journeys create "the clouds of dust that marked the winding courses of the roads," so that these "more-favoured sons of Fortune, were suffocating themselves for the benefit of the air."[44] Ironically, the journey here creates air pollution, which the commuters are ostensibly trying to escape, thereby negatively affecting the health of those seeking semirural retirement and anticipating the air pollution caused by twentieth- and twenty-first-century commutes by car. Again, a lack of physical mobility is linked to death—the carriage passengers are preserved specimens under a glass covering—and they have lost their humanity; indeed, they are monstrous in their detachment from the world around them. The wealthy can afford to create a glass shield to insulate themselves from the world and are not inhabiting the environment but "zooming" through it. Thelwall provides mounting evidence of the resulting ignorance about and alienation from these spaces, making it easier to engage in "monstrous" acts like evicting cottagers and poisoning the air.

Walking, meanwhile, is a solution to sedentary lifestyles and offers freedom of movement, thought, intellect, and spirit, as when Sylvanus describes "quitting the public road, in order to pursue a path, faintly tracked through the luxuriant herbage of the fields, and which left [him] at liberty to indulge the solitary reveries of a mind, to which the volume of nature is ever open at some page of instruction and delight."[45] Unlike the wealthy who travel by carriage, walking in Greater London's fields away from the dusty roads allows Sylvanus to escape urban noise and air pollution: "I felt a glow of health and vivacity, which the bustle and loaded atmosphere of the metropolis never yet afforded."[46] Given Thelwall's radical politics, it is not surprising that walking for him has political implications, as scholars have outlined and as is evidenced by Thelwall's use of "liberty" above.[47] Such democratic connotations are underlined when Sylvanus is "contemplating the distant prospect of green-swelling hills, that rushed immediately upon our view, and enjoying the *enfranchisement* of vision from the dull captivity of brick walls and square panes of glass."[48] Through the language of electoral reform (i.e., "enfranchisement"), Thelwall links the act of walking, which allows one to physically escape the "captivity" of the crowded city, the narrow and smoky street, the dusty commuter road, the cramped house, the office desk, and the enclosed carriage, to experience spatial, intellectual, and physical freedom. In an age before public health or affordable housing and rental policies, Londoners must assume responsibility for their health through activities, like walking, that are available to them. Walking supports people to create their own relationship with land without having to own property; therefore, it is an act that can resist landowners' power.[49] Indeed, walking, an activity available to the propertyless, has advantages over the conveyances of the affluent—traveling via carriage is distinctly not healthy, according to Thelwall. While the discourse of the sanctity of private property and the legal implications of trespassing put very real limits, both then and now, on walking, Thelwall's advocacy of the peripatetic is part of a long democratic history of walking. This history includes reclaiming access to now enclosed common land and expanding access to green space, including the establishment of public urban parks, beginning with the Derby Arboretum in 1840 (and the gradual opening up of older parks, like London's Kensington Gardens, to the masses), which led to the first "freedom-to-roam" bill of the 1880s and then to the 1949 Act of Parliament to establish national parks.[50]

Thelwall makes clear that walking's many benefits are available in Greater London. Sylvanus steps off the road, walking in the fields (St. George's Fields) within sight of the dusty commuter roads—"I pursued my tranquil and cheerful way along the fields, and smiled to behold, at irregular distances, to the right, and to the left, the clouds of dust that marked the winding courses of the roads"[51]—and so this is decidedly a suburban green space: these are fields accessible to the urbanite. One does not need to own a vast, manicured lawn or travel to wild, sublime, or picturesque landscapes, conventionally associated with (Romantic) nature, such as Percy Bysshe Shelley's Alps or Wordsworth's Lake District, to access tranquility and cheerfulness. Humble urban green spaces, not unlike Anne's dandelions, are celebrated here, as even the fields carved up by new commuter roads provide respite from the urban grind and pollution.

In line with Thelwall's democratic aims, Sylvanus and friends' excursions through fields and villages are sociable and community minded. In Socratic fashion, *The Peripatetic*'s characters converse, debate, and change their minds, each telling their own stories; there is no authority figure or expert.[52] Walking counteracts the distancing, alienating practices of vast, empty lawns and contained carriages, as it allows people to interact with others, learn each other's stories, see each other as fellow humans and, as a result, break down class barriers.[53] For Thelwall, "there is no human being with whom it is not worth while to spend a quarter of an hour."[54] Walking, moreover, shows solidarity with the common people, who lack access to more expensive modes of transportation; moreover, empathy is more possible than in other modes of transportation, as walking puts everyone, quite literally, on level ground.[55]

Between Thelwall's and Montgomery's time, trains and cars had decreased physical and increased mechanical mobility, and Montgomery too is concerned with urban mobilities, with the alienating processes of modern long-distance travel and the dangers of sedentary lifestyles. Anne's first encounter with Kingsport is in the overwhelming "blue-white glare of the crowded [train] station," which is "a howling wilderness of strangers!" where "Anne felt horribly bewildered."[56] Later, as a serious student, Anne sometimes experiences the physical effects of a desk-bound routine, becoming "pale," and so when on an afternoon of leisure, she is "contrasting the claims of indoors and out," she chooses not "tame domestic joys" but the park where she can be "alone and free and wild."[57] Over and over, walking emerges as a fix whereby Anne and her friends move freely through the city, particularly its urban green spaces: the garden and orchard of Patty's Place, the Old St. John's graveyard, the park.

Early in her time in Kingsport, while Anne is living in the urban boarding-house, the nearby graveyard, surrounded by "busy, bustling, modern thoroughfares," is "the only get-at-able" green space.[58] The graveyard coupled with the gothically rendered boardinghouse backyard evokes Thelwall's language of modern students being buried alive in London. Indeed, Anne's graveyard is a conflicted space: on one hand, it is a place of death and decay, and, on the other—reflecting nineteenth-century beautification initiatives that turned cemeteries into parks—it is a place of retreat in nature.[59] Indeed, walking to and through Kingsport's Old St. John's graveyard (the equivalent of Halifax's St. Paul's graveyard, also called the Old Burying Ground) provides mental and physical solace.[60] After a dispiriting first day of university "surrounded by crowds of strangers" where she felt "insignificant," the graveyard is the only, if unlikely, space "to go to get cheered up" for this nature-loving heroine.[61] Yet her friend clarifies that "Old St. John's is a darling place" with a Thelwall-worthy rationale: the graveyard is a landmark, "one of the sights of Kingsport"; it is a place to walk, explore, and enjoy "for a pleasure exertion"; a green oasis with "rows of trees all through it"; and a place of instruction (both as a potential place to study and to reflect on local and world history, due, for example, to the "beautiful monument to the memory of Nova Scotian soldiers who fell in the Crimean War").[62]

Just as Thelwall advised, walking allows for unexpected encounters with hitherto strangers and thus for sociability and community building. Although too intimidated to approach each other on the alienating university campus, where hierarchies among the students are rigid, Anne and her friend Priscilla, during their walk in the green graveyard, spontaneously encounter the lively Phillipa Gordon, soon-to-be friend and future housemate. By the end of the walk, Anne can declare, "I'm glad we met her, and I'm glad we went to Old St. John's. I believe I've put forth a tiny soul-root into Kingsport soil this afternoon. I hope so. I hate to feel transplanted."[63] Again an organic metaphor highlights the important interconnections between human society *and* green space, even a graveyard, in creating a sense of belonging and inhabitance in a new, seemingly unfriendly city.

Montgomery, like Thelwall before her, shifts attention to alternative landmarks and more democratic histories than those recorded in the more obvious or opulent statues, plaques, and tombstones like the Crimean War monument (a large memorial dominated by a lion that guards the entrance to both the real and imagined versions of this graveyard). Montgomery imports Thomas Gray's honoring of humble lives in "Elegy Written in a Country Churchyard" to Kingsport, as she describes the modesty of the people interred here, someone's "favorite child," a servant, and a "middy" (or midshipman, a junior-ranking naval officer), and of their grave markers—most are "roughly chiselled brown or gray native stone."[64] These humble lives connect to Anne's own modest background as a "poor" "country girl," moving to the city as an early female university student, despite financial constraints and social disapproval.[65] Her first day at university made her feel that she "would go down to [her] grave unwept, unhonored and unsung."[66] As she leaves the Old Burying Ground, Anne acknowledges the young "middy,"[67] dropping flowers by his tombstone; he, like Anne herself, with her character's continued popularity, and like those commemorated in Gray's poem, does not go completely unnoticed or unhonored.

Here we see a young university student becoming deeply connected to her new urban home and its green spaces. Since cities are increasingly where people experience and learn to love nature, how can the study of history, of literature, of the environmental humanities enhance that experience and connection for Anne's student counterparts in today's universities? Further, how can students' reading of and movement through cities foster knowledge and stewardship of place and, specifically, urban nature?

## READING AND WRITING THE GREEN CITY WITH STUDENTS

Thelwall's *The Peripatetic* offers teachers and students a guidebook not only to late-eighteenth-century London but also to other places, in other times. Both the act of walking and the act of reading in/about particular places provide specificity to engage meaningfully with social issues, to encounter and be challenged by the past and present, the personal and political, the social and natural, the real and imagined. *The Peripatetic* can show students how to know and even care about urban spaces, including green ones, no matter how marginal (socially, economically, ecologically, or aesthetically) they appear. *The Peripatetic* can be a pedagogical tool or framework at the center of experiential approaches to teaching and learning that supports students to move through both texts and urban spaces.

An example of how *The Peripatetic* can guide students' mobility, engaged in literature, culture, and politics, is an experiential study-abroad class, The New Peripatetic: Arts of Activism, Roots of Reform, that Judith Thompson and I conceptualized at Dalhousie University. This study-abroad course had been approved by the university for implementation in England in June 2017, but did not proceed. While I think our proposed course has potential as a study-abroad course, as we had intended it to run, a local or virtual version of this course mitigates the financial and other hurdles of a course delivered internationally. Here I draw on our aims for this course, to have been held in England, focusing on English writers, while demonstrating how Thelwall provides a malleable pedagogy that can be applied to different authors, times, and places, such as L. M. Montgomery's early-twentieth-century Canadian Maritimes. Such a class explores the origins and practice of pedestrian travel as an art of activism by deploying Thelwall's "politico-sentimental" peripatetic mode. Students follow in the footsteps of writers—as relevant to the local context and course aims, L. M. Montgomery, in my case, for example—understanding these writers within landscapes, such as Halifax, Nova Scotia, and social movements, like cemetery beautification, that shaped them and that they shaped in turn. Far from merely a passive form of "literary tourism," the class challenges students to think, talk, walk, and act outside the classroom. Along the road, like Thelwall and his contemporaries, students discuss the history and politics of tourism, as well as crossing boundaries between eras, cultures, and communities and considering the relationships between art and activism, physical mobility and social mobilization.

Thelwall asks his readers to look anew at the ordinary, everyday spaces in which we live and at the literature that shapes our understanding of those spaces, of our world. As a specific example, I'll consider memorialization, a topic that interested Thelwall, as Steve Poole indicated in his opening remarks at the 2021 John Thelwall Society Annual General Meeting, and that is particularly timely as statues on both sides of the Atlantic come down in 2020–2021. In *The Peripatetic*, Thelwall demands that readers question who gets memorialized and how and, significantly, who does not, repeatedly drawing readers' attention to modest abodes and endeavors. Montgomery, as we have seen, honors humble imperial agents, like the middy, but she also celebrates and romanticizes British imperialism. Students could read the *Anne of the Island* graveyard scene alongside other past and present debates about memorialization in Halifax green spaces, thus revealing the wider contexts of Montgomery's representation and exploring the politicized history of green spaces, such seemingly innocuous (i.e., beautiful, recreational) sites. One significant Halifax example is Africville, an African Nova Scotian community destroyed by the municipal government in the 1960s, which was transformed in the early twenty-first century into a controversial dog park and is the site of the reconstructed Seaview United Baptist Church, housing the Africville Museum.[68] The African Nova Scotian community dates back to the eighteenth century; another local example with roots in that century is that of Edward Cornwallis, the controversial founder of Halifax as a British settlement in 1749, whose statue was removed in 2018 from the square that bore his name until June 2021. The site is now Peace and Friendship Park, in recognition of the eighteenth-century treaties signed between British settlers and Indigenous peoples of what are now the Canadian Maritimes.[69] Reading local memorialization in urban green spaces through Thelwall's lens would elicit discussion of accessibility, audience, tourism, social justice, community, and the power of language, to name a few topics.

Students can consider central themes like memorialization and green spaces through the course's main assignment: a collaborative, creative update of *The Peripatetic*. This update directs students, like Thelwall before them, to reflect on differences and similarities in the ways human activity and culture are enacted on landscape, between the past—including Thelwall's time and, say, Montgomery's Halifax—and the present, namely the city through which the students move. Inspired by Thelwall's multi-genre miscellany, students, through their twenty-first-century peripatetic, intellectually and artistically engage with literature and space through a range of modes, including scholarly reflection—on historical research or via the close reading of texts, for example—and through the creation and inclusion of other media: found objects, poetry, photography, drawings, audio or video recordings, journal entries. Students build on their existing strengths, as well as try out new skills, thus creating scope for student-directed and mutual learning. In the Thelwallian spirit of learning from people and objects encountered in situ, students gain insight into the work of local scholarly, creative, educational, or heritage preservation groups through workshops, lectures, or site tours. The class as a whole, with advice from these local groups, considers how to curate the modern peripatetic for public exhibition online or at an in-person event. Students, in turn, gain personally and socially useful skills in, for example, public engagement and heritage and environmental preservation.

So, where do your students live? What texts are they reading? How does local history interact with the present? What local initiatives could they engage with? John Thelwall's *The Peripatetic* provides a flexible research methodology and experiential pedagogy to support culturally and environmentally engaged exploration of these questions. It offers generative possibilities for exploring humans' relationships to the landscapes in which they live, thus affirming the enduring relevance of the eighteenth-century environmental humanities in the classroom and beyond.[70]

## NOTES

1. L. M. Montgomery, *Anne of Windy Poplars* (Toronto: Seal Books, 1983), 78.
2. Kevin Loughran, "Urban Parks and Urban Problems: An Historical Perspective on Green Space Development as a Cultural Fix," *Urban Studies* 57, no. 11 (2008): 1.
3. See, for example, Michael Scrivener, *Seditious Allegories: John Thelwall and Jacobin Writing* (University Park: Pennsylvania State University Press, 2001); Judith Thompson, *John Thelwall in the Wordsworth Circle: The Silenced Partner* (New York: London: Palgrave, 2012); Yasmin Solomonescu, *John Thelwall and the Materialist Imagination* (London: Palgrave, 2014).
4. William Wordsworth, "The Excursion," in *Poems of William Wordsworth, Volume 2: Collected Reading Texts from The Cornell Wordsworth*, ed. Jared Curtis (Tirril, UK: Humanities-Ebooks, 2011), 8, 130–131.
5. Michael Bennett and David W. Teague, eds. *The Nature of Cities: Ecocriticism and Urban Environments* (Tucson: University of Arizona Press, 1999).
6. Scott Hess, *William Wordsworth and the Ecology of Authorship: The Roots of Environmentalism in Nineteenth-Century Culture* (Charlottesville: University of Virginia Press, 2012), and B. Ashton Nichols, *Beyond Romantic Ecocriticism: Toward Urbanatural Roosting* (London: Palgrave, 2011).
7. Nichols, *Beyond Romantic Ecocriticism,* 78.
8. See Emily Woster and Kate Scarth, "Welcome to the *Journal of L.M. Montgomery Studies*," *Journal of L.M. Montgomery Studies* 1, no. 1 (2019): 1–23.
9. See L. M. Montgomery, *The Complete Journals of L.M. Montgomery: The PEI Years, 1889–1900* (Oxford: Oxford University Press, 2017), 223 (Pope); 161 (Goldsmith); 144, 306, 448 (Scott); 241, 417 (Wordsworth).
10. See especially *L.M. Montgomery and the Matter of Nature(s)*, ed. Rita Bode and Jean Mitchell (Montreal: McGill-Queen's University Press, 2018). A small body of scholarship on

urban Montgomery is emerging; see, for example, Lesley D. Clement, "Toronto's Cultural Scene: Tonic or Toxin for a Sagged Soul?" in *L.M. Montgomery's Rainbow Valleys: The Ontario Years, 1911–1942,* ed. Rita Bode and Lesley D. Clement (Montreal: McGill-Queen's University Press, 2015), 238–260.

11. Hess, *William Wordsworth and the Ecology of Authorship*, 17.
12. Hess, *William Wordsworth and the Ecology of Authorship*, 17.
13. Nichols, *Beyond Romantic Ecocriticism*, 49.
14. For discussion of Montgomery and Wordsworth, see Elizabeth Rollins Epperly, *The Fragrance of Sweet-Grass: L.M. Montgomery's Heroines and the Pursuit of Romance* (Toronto: University of Toronto Press, 2014) and Margaret Steffler, "Brian O'Connal and Emily Byrd Starr: The Inheritors of Wordsworth's 'Gentle Breeze,'" in *Windows and Words: A Look at Canadian Children's Literature in English*, ed. Aïda Hudson and Susan-Ann Cooper (Ottawa: Ottawa University Press, 2003), 87–96. Epperly also discusses Scott and Montgomery.
15. Judith Thompson, "A Note on the Text," in *The Peripatetic*, by John Thelwall (Detroit: Wayne State University Press, 2001), 53–55.
16. Recent examples of this broader Romanticism include James Chandler and Kevin Gilmartin, eds., *Romantic Metropolis: The Urban Scene of British Culture, 1780–1840* (Cambridge, UK: Cambridge University Press, 2005), and Jon Mee, *Conversable Worlds: Literature, Contention, and Community 1762 to 1830* (Oxford: Oxford University Press, 2011).
17. John Thelwall, *The Peripatetic*, ed. Judith Thompson (Detroit: Wayne State University Press, 2001), 153.
18. Thelwall, *The Peripatetic*, 135.
19. Thelwall, *The Peripatetic*, 337.
20. Thelwall, *The Peripatetic*, 136.
21. Thelwall, *The Peripatetic*, 136.
22. Thelwall, *The Peripatetic*, 135.
23. For discussion of the wandering Wordsworthian soul made suburban dweller/gardener, see Elizabeth Jones, "Keats in the Suburbs," *Keats-Shelley Journal* 45 (1996): 23–43, especially p. 31.
24. Hess, *William Wordsworth and the Ecology of Authorship.*
25. See, for instance, Gillen D'Arcy Wood, "Leigh Hunt's New Suburbia: An Eco-Historical Study in Climate Poetics and Public Health," *Interdisciplinary Studies in Literature and Environment* 18, no. 3 (2011): 527–552.
26. Thelwall, *The Peripatetic*, 337–338.
27. Thelwall, *The Peripatetic*, 134.
28. Thompson, introduction to *The Peripatetic*, 36.
29. Montgomery, *Anne of the Island*, 131.
30. Analysis of Patty's Place as a suburban home is the subject of my article, "From Anne of Green Gables to Anne of the Suburbs: Lucy Maud Montgomery Reimagines Home in *Anne of the Island*," *Women's Writing* 27, no. 2 (2017): 150–164. I also discuss L. M. Montgomery's urban writing, with a focus on PEI's Charlottetown and Summerside, in "Born for City Life? L.M. Montgomery's Urban Prince Edward Island," *Island Magazine* (2020): 53–64.
31. Sue Lange, "L. M. Montgomery's Halifax: The Real Life Inspiration for *Anne of the Island*," *Shining Scroll* (2011): 3–9; L. M. Montgomery Institute, University of Prince Edward Island, https://www.lmmontgomery.ca/islandora/object/lmmi%3A4622 (accessed August 14, 2020).
32. Montgomery, *Anne of the Island*, 23, 48.
33. Montgomery, *Anne of the Island*, 48.
34. Montgomery, *Anne of the Island*, 48.
35. See, for example, Virginia Jenkins, *The Lawn: A History of an American Obsession* (Washington, DC: Smithsonian Books, 1994), 4–5. For discussion of the lawn in British Romanticism, see Timothy Morton, "Wordsworth Digs the Lawn," *European Romantic Review* 15,

no. 2 (2004): 317–327. Green lawns' considerable land, water, and chemical requirements have serious environmental and health implications (as does tobacco, making Montgomery's choice of Anne's neighbor prescient). For environmental discussion of the lawn, see F. Herbert Bormann, Diana Balmori, Gordon T. Geballe, and Lisa Verngaard, eds. *Redesigning the American Lawn: A Search for Environmental Harmony* (New Haven, CT: Yale University Press, 2001).
36. Montgomery, *Anne of the Island*, 131.
37. Montgomery, *Anne of the Island*, 31.
38. Montgomery, *Anne of the Island*, 68.
39. Montgomery, *Anne of the Island*, 48, 49, 49, 48, 48.
40. Montgomery, *Anne of the Island*, 172.
41. Montgomery, *Anne of the Island*, 44, 159.
42. Thelwall, *The Peripatetic*, 81, emphasis added.
43. Thelwall, *The Peripatetic*, 81, 337.
44. Thelwall, *The Peripatetic*, 132, 81.
45. Thelwall, *The Peripatetic*, 78.
46. Thelwall, *The Peripatetic*, 81.
47. See, for instance, Thompson, introduction to *The Peripatetic*; Scrivener, *Seditious Allegories*, 209; Solomonescu, *John Thelwall and the Materialist Imagination*, 34; Cristoph Houswitschka, "Peripatetics of Citizenship in the 1790s," in *Mediating Identities in Eighteenth-Century England: Public Negotiations, Literary Discourses, Topography*, ed. Isabel Karremann and Anja Muller (Farnham, UK: Ashgate, 2011), 178. For discussion of eighteenth-century urban walking's ties to individual experience, everyday life, and democratic resistance, see Catharina Löffler, *Walking in the City: Urban Experiences and Literary Psychogeography in Eighteenth Century London* (Wiesbaden: J.B. Metzler, 2017).
48. Thelwall, *The Peripatetic*, 132, emphasis added.
49. Houswitschka, "Peripatetics of Citizenship," 179.
50. "Derby Parks: History," *In Derby* (2017), https://www.inderby.org.uk/parks/derbys-parks-and-open-spaces/derby-arboretum/history/ (accessed September 27, 2020), and "History of Our National Park," Peak District National Park. https://www.peakdistrict.gov.uk/learning-about/about-the-national-park/our-history (accessed September 27, 2020).
51. Thelwall, *The Peripatetic*, 81.
52. Judith Thompson, introduction to *The Peripatetic*, 11, 26, 34; Scrivener, *Seditious Allegories*, 214; Houswitschka, "Peripatetics of Citizenship," 182.
53. Solomonescu, *John Thelwall and the Materialist Imagination*, 38.
54. Thelwall, *The Peripatetic*, 96.
55. Thelwall, *The Peripatetic*, 12.
56. Montgomery, *Anne of the Island*, 20.
57. Montgomery, *Anne of the Island*, 162.
58. Montgomery, *Anne of the Island*, 24, 26.
59. See Hannah M. Lane, "'The Garden of the Dead': The Old Burial Ground, Cemetery Reform, and Cultural Memory," in *The Creative City of Saint John*, ed. Gwendolyn Davies, Peter Larocque, and Christl Verduyn (Halifax: Formac, 2018), 84–90.
60. See Kate Scarth and Christopher Shalom, "Old Burying Ground," Halifax Literary Landmarks. https://halifaxliterarylandmarks.ca/tours/show/4 (accessed August 14, 2020).
61. Montgomery, *Anne of the Island*, 24, 25, 26.
62. Montgomery, *Anne of the Island*, 22.
63. Montgomery, *Anne of the Island*, 35.
64. Montgomery, *Anne of the Island*, 24, 27, 34.
65. Montgomery, *Anne of the Island*, 76, 12.
66. Montgomery, *Anne of the Island*, 25.
67. Montgomery, *Anne of the Island*, 34.

68. See this 2014 article about the Africville dog park controversy, "Africville Dog Park Upsets Black Community," *CBC-Nova Scotia*, July 13, 2014, https://www.cbc.ca/news/canada/nova-scotia/africville-dog-park-upsets-black-community-1.2705535; and this 2016 article about the closing of the dog park, Patrick Odell, "Fences around Former Dog Park at Africville Park Removed," *Global News*, February 26, 2016, https://globalnews.ca/news/2543797/fences-around-former-dog-park-at-africville-park-removed/. For a history of Africville, see Matthew McRae, "The Story of Africville," *Canadian Museum for Human Rights*, https://humanrights.ca/story/the-story-of-africville (accessed July 7, 2021). For information about the Africville Museum, see *Africville Heritage Trust*, https://africvillemuseum.org/ (accessed July 7, 2021).
69. See Gareth Hampshire, "Peace and Friendship Park Officially Replaces Cornwallis Park Name," *CBC-Nova Scotia*, June 9, 2021, https://www.cbc.ca/news/canada/nova-scotia/peace-friendship-cornwallis-park-daniel-paul-1.6058513.
70. Many thanks to Judith Thompson and Lesley Clement, Thelwall and Montgomery experts, respectively, not only for their excellent feedback on drafts of this essay but more generally for literary conversation and friendship. This chapter draws on research supported by the Social Sciences and Humanities Research Council.

# ACKNOWLEDGMENTS

*Eighteenth-Century Environmental Humanities* demonstrates what is possible when we think outside our distinct disciplines with an eye for interdisciplinary growth and a commitment to collegiality. Together we can break down the guarded silos of our fields, with the goal of speaking *with* rather than *at* one another. This collection would have been impossible without our mentors, students, colleagues, and loved ones who inspire us to do our very best work. This is just the beginning.

Gratitude is owed to the two anonymous readers, as well as the series editors, for their insights and commitment to our vision. Suzanne Guiod has spearheaded this project and Pam Dailey has shepherded its growth. Thank you for believing in this work.

# BIBLIOGRAPHY

Adams, Eliphalet. *A Discourse Occasioned by the Late Distressing Storm*. New London, 1717.
Adamson, Joni. "Introduction." In *Humanities for the Environment*, edited by Joni Adamson and Michael Davis, 3–19. New York: Routledge, 2016.
Adamson, Joni, and Scott Slovic. "The Shoulders We Stand On: An Introduction to Ethnicity and Ecocriticism." *MELUS Ethnicity and Ecocriticism* 34, no. 2 (Summer 2009): 5–24.
"Africville Dog Park Upsets Black Community." *CBC-Nova Scotia*. July 31, 2014. https://www.cbc.ca/news/canada/nova-scotia/africville-dog-park-upsets-black-community-1.2705535.
Ahmed, Sara. "Willful Stones." *Feminist Killjoys*, January 26, 2016. https://feministkilljoys.com/2016/01/29/willful-stones/.
Alaimo, Stacy. *Bodily Natures: Science, Environment, and the Material Self.* Bloomington: University of Indiana Press, 2010.
Albrecht, Glenn A. "Exiting the Anthropocene and Entering the Symbiocene." *Minding Nature* 9, no. 2 (Spring 2016): 12–16.
Alfano, Andrea. "Nor'easters May Become More Intense with Climate Change." *Scientific American*, January 26, 2015. http://www.scientificamerican.com/article/nor-easters-may-become-more-intense-with-climate-change/.
Allewaert, Monique. *Ariel's Ecology: Plantations, Personhood, and Colonialism in the American Tropics*. Minneapolis: University of Minnesota Press, 2013.
American Meteorological Society. *Bulletin of the American Meteorological Society* 99, no. 1 (January 2018).
———. Press release. December 13, 2017. https://www.ametsoc.org/ams/index.cfm/about-ams/news/news-releases/human-influence-on-climate-led-to-several-major-weather-extremes-in-2016/.
Anonymous. "West-Indian Eclogues." In *The Critical Review: or, Annals of Literature*, vol. 64, 434–435. London: A. Hamilton, 1787.
Anstey, Peter. "The Creation of the English Hippocrates." *Medical History* 55, no. 4 (2011): 457–478.
Ariosto, Ludovico. *Orlando Furioso: A New Verse Translation*. Translated by David R. Slavitt. Cambridge, MA: Belknap Press, 2009.
Armitage, David. "In Defense of Presentism." In *History and Human Flourishing*, edited by Darrin McMahon, 1–26. Oxford: Oxford University Press, forthcoming.
———. "The Red Atlantic." *Reviews in American History* 29, no. 4 (2001): 479–486.
Atkinson, Dwight. "The *Philosophical Transactions of the Royal Society of London*, 1675–1975: A Sociohistorical Discourse Analysis." *Language in Society* 25, no. 3 (1996): 333–371.
*Attribution of Extreme Weather Events in the Context of Climate Change*. Washington: National Academies Press, 2016.
Augustine, Patrick J. "The Significance of Place in Textual and Graphical Representation: The Mi'kmaq on Lennox Island, Prince Edward Island, and the Penobscot on Indian Island, Maine." Island studies master's thesis, University of Prince Edward Island, 2009.
Austin, Algernon. *Achieving Blackness: Race, Black Nationalism, and Afrocentrism in the Twentieth Century*. New York: New York University Press, 2006.

Baepler, Paul, ed. *White Slaves, African Masters: An Anthology of American Barbary Captivity Narratives*. Chicago: University of Chicago Press, 1999.
Baker, Nick. "The Recent History of Australia's Climate Change Wars." *SBS News*. January 1, 2020. https://www.sbs.com.au/news/the-recent-history-of-australia-s-climate-change-wars.
Bandara, Samuel B. "Philip Freneau: First American Poet to Write about the Caribbean." *InterAmerican Review of Bibliography* 36, no. 1 (1986): 151–179.
Baptist, Edward E. *The Half That Has Never Been Told: Slavery and the Making of American Capitalism*. New York: Basic Books, 2016.
Barad, Karen. *Meeting the Universe Halfway: Quantum Physics and the Entanglement of Matter and Meaning*. Durham, NC: Duke University Press, 2007.
Barnard, John Levi. *Empire of Ruin: Black Classicism and American Imperial Culture*. Oxford: Oxford University Press, 2018.
Baskin, Jeremy. *Geoengineering, the Anthropocene and the End of Nature*. London: Palgrave Macmillan, 2019.
Bellin, Jacques-Nicholas. "Remarks by M. Bellin in Relation to His Maps Drawn for P. Charlevoix's History of New France." *The Gentleman's Magazine*, February 1746.
Bennet, Jane. *Vibrant Matter: A Political Ecology of Things*. Durham, NC: Duke University Press, 2010.
Bennett, Joshua. *Being Property Once Myself*. Cambridge, MA: Harvard University Press, 2020.
Bennett, Michael, and David W. Teague, eds. *The Nature of Cities: Ecocriticism and Urban Environments*. Tucson: University of Arizona Press, 1999.
Bennett, Paula. "Phillis Wheatley's Vocation and the Paradox of the 'Afric Muse.'" *PMLA* 113, no. 1 (1998): 64–76.
Benton, Lauren. *A Search for Sovereignty: Law and Geography in European Empires, 1400–1900*. Cambridge, UK: Cambridge University Press, 2011.
Berger, John. "The Changing View of Man in Portrait." In *Selected Essays and Articles*, 35–40. London: Pelican, 1972.
Best, Stephen, and Saidiya Hartman. "Fugitive Justice." *Representations* 92, no. 1 (2005): 1–15.
Binhammer, Katherine. "Is the Eighteenth Century a Colonizing Temporality." *Eighteenth-Century Fiction* 33, no. 2 (2021): 199–204.
Birch, Tony. "'The Lifting of the Sky': Life Outside the Anthropocene." In *Humanities for the Environment*, edited by Joni Adamson and Michael Davis, 195–209. New York: Routledge, 2016.
Blaser, Mario. *Storytelling Globalization from the Chaco and Beyond*. Durham, NC: Duke University Press, 2010.
Blaser, Mario, and Marisol de la Cadena, eds. *A World of Many Worlds*. Durham, NC: Duke University Press, 2018.
Blum, Hester. "Introduction: Oceanic Studies." *Atlantic Studies* 10, no. 2 (2013): 151–155.
———. "The Prospect of Oceanic Studies." *Publications of the Modern Language Association of America* 125, no. 3 (2010): 670–677.
———. *The View from the Masthead: Maritime Imagination and Antebellum American Sea Narratives*. Chapel Hill: University of North Carolina Press, 2008.
Bly, Antonio T. "'On *Death's* Domain Intent I Fix My Eyes': Text, Context, and Subtext in the Elegies of Phillis Wheatley." *Early American Literature* 53, no. 2 (2018): 317–341.
Bode, Rita, and Jean Mitchell, eds. *L. M. Montgomery and the Matter of Nature(s)*. Montreal: McGill-Queen's University Press, 2018.
Boia, Lucian. *The Weather in the Imagination*. Translated by Roger Leverdier. London: Reaktion Books, 2005.
Bolster, Jeffrey W. "Opportunities in Marine Environmental History." *Environmental History* 11, no. 3 (2006): 567–597.
Boochkin, Murray. *Post-Scarcity and Anarchism*. Palo Alto, CA: Ramparts Press, 1971.

Bormann, F. Herbert, Diana Balmori, Gordon T. Geballe, and Lisa Verngaard, eds. *Redesigning the American Lawn: A Search for Environmental Harmony*. New Haven, CT: Yale University Press, 2001.

Bouchard, Jack. *Terra Nova: Food, Water and Work in an Early Atlantic World* (New Haven, CT: Yale University Press, forthcoming).

Boulukos, George. *The Grateful Slave: The Emergence of Race in Eighteenth-Century British and American Culture*. London: Cambridge University Press, 2008.

Bradley, Karin, and Johan Hedrén. *Green Utopianism: Perspectives, Politics and Micro-Practices*. New York: Routledge, 2014.

Braverman, Richard. *Plots and Counterplots, Sexual Politics and the Body Politic in English Literature*, 1660–1730. Cambridge, UK: Cambridge University Press, 2008.

Brayton, Daniel. *Shakespeare's Ocean: An Ecocritical Exploration*. Charlottesville: University of Virginia Press, 2012.

Brickhouse, Anna. *The Unsettlement of America: Translation, Interpretation, and the Story of Don Luis de Velasco, 1560–1945*. Oxford: Oxford University Press, 2015.

Bromberg, K. D., and M. D. Bertness. "Reconstructing New England Salt Marsh Losses Using Historical Maps." *Estuaries* 28, no. 6 (2005): 823–832.

Brown, Ras Michael. "'But the Mermaid Did Not Rise Up': The Death of a Simbi in the Carolina Lowcountry." *Southern Quarterly* 47, no. 1 (2010): 120–150.

Brückner, Martin. *The Social Life of Maps in America, 1750–1860*. Chapel Hill: University of North Carolina Press, 2017.

———, ed. *Early American Cartographies*. Chapel Hill: University of North Carolina Press, 2011.

Bucknell, Clare. "The Mid-Eighteenth-Century Georgic and Agricultural Improvement." *Journal for Eighteenth-Century Studies* 36, no. 3 (January 2013): 335–352.

Bumsted, J. M. *Land, Settlement, and Politics on Eighteenth-Century Prince Edward Island*. Montreal: McGill-Queen's University Press, 1987.

Callagher, Catherine. "The Rise of Fictionality." In *Novel*, vol. 1, edited by Franco Moretti, 336–363. Princeton, NJ: Princeton University Press, 2006.

Campbell, Claire. "'A Window Looking Seaward': Finding Environmental History in the Writing of L. M. Montgomery." In *Greater Gulf: Essays on the Environmental History of the Gulf of St. Lawrence*, edited by Claire Campbell, Edward MacDonald, and Brian Payne, 283–318. Montreal: McGill-Queen's University Press, 2020.

———. "National Interest and Continental Geography: Teaching about Canada through an Environmental Lens." In *Reflections from Abroad: The Construction of Canadian Identity*, edited by Christopher Kirkey and Richard Nimijean. New York: Palgrave Macmillan, 2022.

Campbell, Claire, and Robert Summerby-Murray, eds. *Land and Sea: Environmental History in Atlantic Canada*. Fredericton: Acadiensis Press, 2013.

Campbell, Claire, Edward MacDonald, and Brian Payne, eds. *The Greater Gulf: Essays on the Environmental History of the Gulf of St. Lawrence*. Montreal: McGill-Queen's University Press, 2020.

Cardena, Marisol de la. *Earth Beings: Ecologies of Practice across Andean Worlds*. Durham, NC: Duke University Press, 2015.

Caretta, Vincent. "Olaudah Equiano or Gustavus Vassa? New Light on an Eighteenth-century Question of Identity." *Slavery and Abolition* 20, no. 3 (December 1999): 96–105.

———. *Phillis Wheatley: Biography of a Genius in Bondage*. Athens: University of Georgia Press, 2011.

Carroll, Siobhan. *An Empire of Air and Sea: Uncolonizable Space in the British Imagination, 1750–1850*. Philadelphia: University of Pennsylvania Press, 2015.

———. "On Erasmus Darwin's *The Botanic Garden*, 1791–1792." *BRANCH: Britain, Representation and Nineteenth-Century History* (April 2016). https://branchcollective.org/?ps_articles=siobhan-carroll-on-erasmus-darwins-the-botanic-garden-1791-1792.

Carter, Sue. "Anne with an E Petition Hits One Million Signatures, Breaks Record." *Quill & Quire*, July 27, 2020. https://quillandquire.com/omni/anne-with-an-e-petition-hits-one-million-signatures-breaks-record/.

Casid, Jill. "Inhuming Empire: Islands as Colonial Nurseries and Graves." In *The Global Eighteenth Century*, edited by Felicity A. Nussbaum, 279–308. Baltimore, MD: Johns Hopkins University Press, 2003.

Catto, Norm, Kate MacQuarrie, and Marianne Hermann. "Geomorphic Response to Late Holocene Climate Variation and Anthropogenic Pressure, Northeastern Prince Edward Island, Canada." *Quaternary International* 87, no. 1 (2002): 101–117.

Cavitch, Max. *American Elegy: The Poetry of Mourning from the Puritans to Whitman*. Minneapolis: University of Minnesota Press, 2007.

Césaire, Aimé. *Discourse on Colonialism*. Translated by Joan Pinkham. New York: Monthly Review Press, 2000.

Chandler, James, and Kevin Gilmartin, eds. *Romantic Metropolis: The Urban Scene of British Culture, 1780–1840*. Cambridge, UK: Cambridge University Press, 2005.

Cheshire, Paul. "Classical Elements: Darwin, Gilbert, Blake, and Coleridge." *Wordsworth Circle* 50, no. 2 (2019): 147–165.

———. "The Hermetic Geography of William Gilbert." *Romanticism* 9, no. 1 (2003): 82–93.

———, ed. *William Gilbert and Esoteric Romanticism: A Contextual Study and Annotated Edition of* The Hurricane. Liverpool: Liverpool University Press, 2018.

Chiles, Katy. *Transformable Race: Surprizing Metamorphoses in the Literature of Early America*. Oxford: Oxford University Press, 2014.

Chow, Jeremy. "Taken by Storm: *Robinson Crusoe* and Aqueous Violence." In *Robinson Crusoe after 300 Years*, edited by Glynis Ridley and Andreas Mueller, 108–134. Lewisburg, PA: Bucknell University Press, 2021.

Clark, Andrew Hill. *Three Centuries and the Island: A Historical Geography of Settlement and Agriculture in Prince Edward Island, Canada*. Toronto: University of Toronto Press, 1959.

Clark, E., and H. M. Tsai. "Islands: Ecologically Unequal Exchange and Landesque Capital." In *Ecology and Power: Struggles over Land and Material Resources in the Past, Present and Future*, edited by A. Hornborg, Brett Clark, and Kenneth Hermele, 52–67. New York: Routledge, 2013.

Clarkson, Thomas. *The History of the Rise, Progress, and Accomplishment of the Abolition of the African Slave-Trade by the British Parliament*. 2 vols. Shoe-Lane: R. Taylor and Company, 1808. London: Frank Cass, reprint, 1968.

Clement, Lesley D. "Toronto's Cultural Scene: Tonic or Toxin for a Sagged Soul?" In *L.M. Montgomery's Rainbow Valleys: The Ontario Years, 1911–1942*, edited by Rita Bode and Lesley D. Clement, 238–260. Montreal: McGill-Queen's University Press, 2015.

Cohen, Jeffery Jerome. *Stone: An Inhuman History*. Minneapolis: University of Minnesota Press, 2015.

———, ed. *Prismatic Ecologies: Ecotheory beyond Green*. Minneapolis: University of Minnesota Press, 2013.

Cohen, Jeffrey Jerome, and Lowell Duckert, eds. "Introduction: Welcome to the Whirled." In *Veer Ecology: A Companion for Environmental Thinking*, 1–15. Minneapolis: University of Minnesota Press, 2017.

Cohen, Margaret. *The Novel and the Sea*. Princeton, NJ: Princeton University Press, 2010.

Coleman, Deidre. *Romantic Colonization and British Anti-Slavery*. Cambridge, UK: Cambridge University Press, 2005.

Collet, Dominik, and Maximilian Schuh, eds. *Famines during the "Little Ice Age" (1300–1800): Socionatural Entanglements in Premodern Societies*. Cham: Springer, 2018.

Cook, Benjamin I. *Drought: An Interdisciplinary Perspective*. New York: Columbia University Press, 2019.

Cook, William W., and James Tatum. *African American Writers and Classical Tradition*. Chicago: University of Chicago Press, 2011.

Corley, Ide. "The Subject of Abolitionist Rhetoric: Freedom and Trauma in 'The Life of Olaudah Equiano.'" *Modern Language Studies* 32, no. 2 (Autumn 2002): 139–156.
Cosgrove, Denis. "Geography within the Humanities." In *Envisioning Landscapes, Making Worlds: Geography and the Humanities*, edited by Stephen Daniels, Dydia DeLyser, J. Nicholas Entrikin, and Doug Richardson, xxii–xxv. New York: Routledge, 2011.
Costello, Leo. *J.M.W. Turner and the Subject of History*. Farnham, UK: Ashgate, 2012.
Cowper, William. "On the Ice-Islands Seen Floating in the Germanic Ocean." In *The Poetical Works of William Cowper. Complete Edition, with Memoir, Explanatory Notes, Etc.*, 508–509. New York: John Wurtele Lovell, n.d.
Crawford, Dorothy H. *Deadly Companions: How Microbes Shaped Our History*. New York: Oxford University Press, 2007.
Crowley, John E. *Imperial Landscapes: Britain's Global Visual Culture, 1745–1820*. New Haven: Yale University Press, 2011.
Culler, Jonathan. "Apostrophe." *Diacritics* 7, no. 4 (Winter 1977): 59–69.
Curley, Rosemary. "Wildlife Matters: A Historical Overview of Public Consciousness of Habitat Loss on Prince Edward Island." In *Time and a Place: An Environmental History of Prince Edward Island*, edited by Edward MacDonald, Joshua MacFadyen, and Irene Novaczek, 59–81. Charlottetown: McGill-Queen's University Press and Island Studies Press, 2016.
Curry, Patrick. *Prophecy and Power: Astrology in Early Modern England*. Cambridge, UK: Polity Press, 1989.
Dampier, William. *A New Voyage Round the World*. London: James Knapton, 1697.
———. *Voyages and Descriptions*. London: Printed for James Knapton, 1699.
Darwin, Erasmus. *The Botanic Garden. A Poem, in Two Parts, with Philosophical Notes*. 2nd edition. New York: T. & J. Swords, 1807.
Daston, Lorraine, and Peter Galison. "Truth-to Nature." In *Objectivity*, 55–114. Princeton, NJ: Zone Books, 2007.
Day, Thomas. *The Dying Negro: A Poem*. 3rd ed. London: W. Flexney, 1775.
Defoe, Daniel. *Robinson Crusoe*. Edited by John Richetti. London: Penguin Classics, 2001.
———. *Roxana*. Edited by John Mullan. Oxford: Oxford University Press, 1996.
Degroot, Dagomar. *The Frigid Golden Age: Climate Change, the Little Ice Age, and the Dutch Republic, 1560–1720*. Cambridge, UK: Cambridge University Press, 2018.
Dekker, George. *The Fictions of Romantic Tourism: Radcliffe, Scott, and Mary Shelley*. Stanford, CA: Stanford University Press, 2005.
Deleuze, Gilles, and Félix Guattari. *A Thousand Plateaus: Capitalism and Schizophrenia*. Minneapolis: University of Minnesota Press, 1987.
DeLoughrey, Elizabeth. *Allegories of the Anthropocene*. Durham, NC: Duke University Press, 2019.
Dennis, John. *Liberty Asserted: A Tragedy*. London: George Stahan and Bernard Lintott, 1704.
Den Ouden, Amy E. *Beyond Conquest: Native Peoples and the Struggle for History in New England*. Lincoln: University of Nebraska Press, 2005.
De Quincey, Thomas. "On Christianity as an Organ of Political Movement." In *Theological Essays and Other Papers*, 1–51. Boston: Ticknor, Reed, and Fields, 1854.
"Derby Parks: History." *In Derby*. 2017. https://www.inderby.org.uk/parks/derbys-parks-and-open-spaces/derby-arboretum/history/ (accessed September 13, 2021).
Des Barres, J. W. F. *The Atlantic Neptune*. London, 1777 and 1800.
Dolphijn, Rick, and Iris Van der Tuin. *New Materialisms: Interviews & Cartographies*. Ann Arbor, MI: Open Humanities Press, 2012. http://dx.doi.org/10.3998/ohp.11515701.0001.001.
Donegan, Kathleen. *Seasons of Misery: Catastrophe and Colonial Settlement in Early America*. Philadelphia: University of Pennsylvania Press, 2013.
Dovell, Karen Lerner. "The Interaction of the Classical Traditions of Literature and Politics in the Work of Phillis Wheatley." *New Essays on Phillis Wheatley*, edited by John C. Shields, 35–55. Knoxville: University of Tennessee Press, 2011.

Doyle, Laura. *Freedom's Empire: Race and the Rise of the Novel in Atlantic Modernity, 1640–1940.* Durham, NC: Duke University Press, 2008.

Duckert, Lowell. *For All Waters: Finding Ourselves in Early Modern Wetscapes.* Minneapolis: University of Minnesota Press, 2017.

Dwyer, Allan. "Liminality and Change in an Atlantic Borderland: Notre Dame Bay, Newfoundland in the Eighteenth Century." In *Land and Sea: Environmental History in Atlantic Canada*, edited by Claire Campbell and Robert Summerby-Murray, 27–44. Fredericton: Acadiensis Press, 2013.

Edelson, S. Max. "Colonizing St. John Island: A History in Maps." *Early Canadian History*, November 14, 2018. https://earlycanadianhistory.ca/2018/11/14/colonizing-st-john-island-a-history-in-maps/.

———. "The New Map of Empire." *Map Scholar*. University of Virginia. http://mapscholar.org/empire/ (accessed September 13, 2021).

———. *The New Map of Empire: How Britain Imagined America before Independence.* Cambridge, MA: Harvard University Press, 2017.

Edney, Matthew. *Cartography: The Ideal and Its History.* Chicago: University of Chicago Press, 2019.

Egberts, Linde, and Meindert Schroor, eds. *Waddenland Outstanding: History, Landscape and Cultural Heritage of the Wadden Sea Region.* Amsterdam: Amsterdam University Press, 2018.

Eisenstadt, Peter R. "The Weather and Weather Forecasting in Colonial America." PhD diss., New York University, 1990.

Ellison, Julie. *Cato's Tears and the Making of Anglo American Emotion.* Chicago: University of Chicago Press, 2001.

Ellison, Katherine. "Why Climate Change Skeptics Are Backing Geoengineering." *Wired.* March 28, 2018. https://www.wired.com/story/why-climate-change-skeptics-are-backing-geoengineering.

Epperly, Elizabeth Rollins. *The Fragrance of Sweet-Grass: L.M. Montgomery's Heroines and the Pursuit of Romance.* Toronto: University of Toronto Press, 2014.

Equiano, Olaudah. *The Interesting Narrative of Olaudah Equiano.* Edited by Vincent Caretta. New York: Penguin Classics, 1995.

Erling, Monica. "Wilding." In *Encyclopedia of Race and Crime*, edited by Helen Taylor Greene and Shaun L. Gabbidon. Los Angeles: SAGE Publications, 2009.

Ewing, Eve L. "Eve Ewing on Phillis Wheatley." *New York Times: The 1619 Project*, August 18, 2019.

"Explaining Extreme Events from a Climate Perspective." *Bulletin of the American Meteorological Society* (Special Supplements 2012).

Exquemelin, Alexander. *Bucaniers of America.* London: Printed for William Crooke, 1684.

———. *The Buccaneers of America.* Mineola, NY: Dover, 1969.

Farhat, Abbas, Aitazaz A. Farooque, and Hassan Afzaal. "Homogeneity in Patterns of Climate Extremes between Two Cities—A Potential for Flood Planning in Relation to Climate Change." *Water* 12, no. 3 (2020): 782.

Fauci, Anthony. "Dr. Fauci: The Virus Decides When the Country Will Open Back Up." CNN. April 10, 2020. Video, 2:26. https://www.cnn.com/videos/health/2020/04/10/dr–anthony-fauci-coronavirus-response-newday-vpx.cnn.

Fernández, Néstor, Laetitia M. Navarro, and Henrique M. Pereira. "Rewilding: A Call for Boosting Ecological Complexity in Conservation." *Conservation Letters* 10, no. 3 (2017): 276–278.

Finley, Gerald E. "HERIOT, GEORGE." In *Dictionary of Canadian Biography*, vol. 7. University of Toronto/Université Laval, 2003. http://www.biographi.ca/en/bio/heriot_george_7E.html (accessed September 13, 2021).

Fleming, James Rodger. *Fixing the Sky: The Checkered History of Weather and Climate Control.* New York: Columbia University Press, 2010.

Forbes, Donald L., George S. Parkes, Gavin K. Manson, and Lorne A. Ketch. "Storms and Shoreline Retreat in the Southern Gulf of St. Lawrence." *Marine Geology* 210, no. 1 (2004): 169–204.

Ford, James Edward. "An African Diasporic Critique of Violence." *Systems of Life*, edited by Richard A. Barney and Walter Montag, 56–80. New York: Fordham University Press, 2018.

Foucault, Michel. *The Order of Things: An Archaeology of the Human Sciences*. New York: Vintage Books, 1994.

Foulke, Robert. *The Sea Voyage Narrative*. New York: Routledge, 2002.

Fowkes Tobin, Beth. *Colonizing Nature; The Tropics in British Arts and Letters, 1760–1820*. Philadelphia: University of Pennsylvania Press, 2004.

Freneau, Philip. "The Beauties of Santa Cruz." In *The Poems of Philip Freneau Written Chiefly During the Late War* (1776). Ann Arbor: Evans Early American Text Creation Partnership, reprint 2011.

———. "To Sir Toby." *The Poems of Philip Freneau: Poet of the American Revolution*. Edited by Fred Lewis Pattee. Princeton, NJ: Princeton University Library, 1902–1907.

Frohock, Richard. *Buccaneers and Privateers: The Story of the English Sea Rover, 1675–1725*. Newark: University of Delaware Press, 2012.

———. "Exquemelin's Buccaneers: Violence, Authority, and the Word in Early Caribbean History." *Eighteenth-Century Life* 34, no. 1 (2010): 56–72.

Frohock, Richard R. *Heroes of Empire*. Newark: University of Delaware Press, 2004.

Fuentes, Agustín, and Natalie Porter. "Kinship." In *Critical Terms for Animal Studies*, edited by Lori Gruen, 182–196. Chicago: University of Chicago Press, 2019.

G. W. *Plan of Lot 31* (No. 2). Public Archives and Records Office, 1784.

"Garangula, Speech to New France's Governor La Barre." In *The Literatures of Colonial America*, edited by Ivy Schweitzer and Susan Castillo. New York: Wiley-Blackwell, 2001.

Garrard, Greg. "Environmental Humanities: Notes Towards a Summary for Policymakers." In *The Routledge Companion to the Environmental Humanities*, edited by Ursula Heise, Jon Christensen, and Michelle Niemann, 462–472. New York: Routledge, 2017.

Gates, Henry Louis, Jr. *The Signifying Monkey: A Theory of African-American Literary Criticism*. Oxford: Oxford University Press, 1988.

———. *The Trials of Phillis Wheatley*. New York: Civitas Books, 2010.

Gay, Ross. "A Small Needful Fact." https://poets.org/poem/small-needful-fact (accessed February 27, 2021).

Gilbert, William. *The Hurricane: A Theosophical and Western Eclogue to Which Is Subjoined a Solitary Effusion in a Summer's Evening*. In *William Gilbert and Esoteric Romanticism: A Contextual Study and Annotated Edition of* The Hurricane. Edited by Paul Cheshire. Liverpool: Liverpool University Press, 2018.

Gillis, John R. *The Human Shore: Seacoasts in History*. Chicago: University of Chicago Press, 2013.

Gilmore, John. *The Poetics of Empire: A Study of James Grainger's The Sugar Cane*. London: Athlone Press, 2000.

Glissant, Édouard. *Caribbean Discourse: Selected Essays*. Translated by J. Michael Dash. Charlottesville: University Press of Virginia, 1989.

Godwin, Joscelyn. *The Theosophical Enlightenment*. Albany: State University of New York Press, 1994.

Golinski, Jan. *British Weather and the Climate of Enlightenment*. Chicago: University of Chicago Press, 2007.

———. "The Weather in Eighteenth-Century Britain." In *Weather, Culture, Climate*, edited by Sarah Strauss and Benjamin S. Orlove, 17–38. Oxford: Berg, 2003.

Goodman, Kevis. *Georgic Modernity and British Romanticism*. London: Cambridge University Press, 2004.

Gould, Philip. *Barbaric Traffic: Commerce and Antislavery in the Eighteenth-Century Atlantic World*. Cambridge, MA: Harvard University Press, 2003.

Goudie, Sean X. *Creole America: The West Indies and the Formation of Literature and Culture in the New Republic*. Philadelphia: University of Pennsylvania Press, 2006.

Graff, E. J. "Boston's Winter from Hell." *New York Times*, February 20, 2015, A-27.

Grainger, James. *The Sugar-Cane*. In *Caribbeana: An Anthology of English Literature of the West Indies 1657–1777*, edited by Thomas W. Krise, 166–260. Chicago: University of Illinois Press, 1999.

Greelee, John. "Eight Islands on Four Maps: The Cartographic Renegotiation of Hawaiʻi, 1876–1959." *Cartographica* 50, no. 3 (2015): 119–140.

Greenwood, Emily. "The Politics of Classicism in the Poetry of Phillis Wheatley." *Ancient Slavery and Abolition*, edited by Edith Hall, Richard Alston, and Justine McConnell, 153–180. Oxford: Oxford University Press, 2011.

Griffiths, Devin. *The Age of Analogy: Science and Literature between the Darwins*. Baltimore, MD: Johns Hopkins University Press, 2016.

Groch, Sherryn. "Space Mirrors, Fake Volcanoes: The Radical Plans to Fix the Climate." *Sydney Morning Herald*, February 14, 2020. https://www.smh.com.au/national/space-mirrors-fake-volcanoes-the-radical-plans-to-fix-the-climate-20200122-p53tq3.html.

Grove, Jean M. "The Onset of the Little Ice Age." In *History and Climate: Memories of the Future?*, edited by P. D. Jones, A.E.J. Ogilvie, T. D. Davies, and K. R. Briffa, 153–186. New York: Springer Science+Business Media, 2001.

Gruen, Lori. *Entangled Empathy: An Alternative Ethic for Our Relationships with Animals*. New York: Lantern Books, 2016.

Grusin, Richard, ed. *Anthropocene Feminism*. Minneapolis: University of Minnesota Press, 2017.

Grydehøj, Adam. "Islands as Legible Geographies: Perceiving the Islandness of Kalaallit Nunaat (Greenland)." *Journal of Marine and Island Cultures* 7, no. 1 (2018): 1–11.

Grydehøj, Adam, Yaso Nadarajah, and Ulunnguaq Markussen. "Islands of Indigeneity: Cultural Distinction, Indigenous Territory and Island Spatiality." *Area* 52, no. 1 (2020): 14–22.

Gunderson, Ryan, Diana Stuart, and Brian Petersen. "The Political Economy of Geoengineering as Plan B: Technological Rationality, Moral Hazard, and New Technology." *New Political Economy* 24, no. 5 (2019): 696–715.

Hadot, Pierre. *The Veil of Isis: An Essay on the History of the Idea of Nature*. Translated by Michael Chase. Cambridge, MA: Belknap Press, 2006.

Hairston, Eric Ashley. "The Trojan Horse: Memory, Transformation, and Afric Ambition in *Poems on Various Subjects, Religious and Moral*." In *New Essays on Phillis Wheatley*, edited by John C. Shields, 57–94. Knoxville: University of Tennessee Press, 2011.

Halberstam, Jack. *Wild Things: The Disorder of Desire*. Durham, NC: Duke University Press, 2020.

Hall, David. *Days of Wonder, Days of Judgment: Popular Religious Belief in Early New England*. Cambridge, MA: Harvard University Press, 1989.

Hall, Michael G. *The Last American Puritan: The Life of Increase Mather, 1639–1723*. Middletown, CT: Wesleyan University Press, 1988.

Hamblyn, Richard, ed. "Introduction." *Daniel Defoe's* The Storm. London: Penguin Books, 2003.

Hampshire, Gareth. "Peace and Friendship Park Officially Replaces Cornwallis Park Name." *CBC-Nova Scotia*. June 9, 2021. https://www.cbc.ca/news/canada/nova-scotia/peace-friendship-cornwallis-park-daniel-paul-1.6058513.

Hanna, Mark G. *Pirate Nests and the Rise of the British Empire, 1570–1740*. Chapel Hill: University of North Carolina Press, 2015.

Hansen, James, Makiko Sato, and Reto Ruedy. "Perception of Climate Change." *PNAS* 109, no. 37 (September 2012): E2415–E2423.

Haraway, Donna. "Situated Knowledges: The Science Question in Feminism and the Privilege of Partial Perspective." *Feminist Studies* 14, no. 3 (1988): 575–599.

———. *Staying with the Trouble: Making Kin in the Chthulucene*. Durham, NC: Duke University Press, 2016.

Hare, E. H. "The Origin and Spread of Dementia Paralytica." *Journal of Mental Science* 105, no. 440 (1959): 594–626.

Harman, Graham. *The Quadruple Object*. Alresford, UK: Zero Books, 2011.

Harney, Stefano, and Fred Moten. *The Undercommons: Fugitive Planning and Black Study*. New York: Minor Compositions, 2013.

Harper, Kristin N., Molly K. Zuckerman, Megan L. Harper, John D. Kingston, and George J. Armelagos. "The Origin and Antiquity of Syphilis Revisited: An Appraisal of Old World Pre-Columbian Evidence for Treponemal Infection." *American Journal of Physical Anthropology* 146, no. S53 (2011): 99–133.

Harris, Richard, and Martin Luymes. "The Growth of Toronto, 1861–1941: A Cartographic Essay." *Urban History Review / Revue D'histoire Urbaine* 18, no. 3 (1990): 244–255.

Harris, Will. "Phillis Wheatley: A Muslim Connection." *African American Review* 48, no. 1–2 (2015): 1–15.

Hartley, L. P. *The Go-Between*. New York: Stein and Day, 1953.

Hartman, Saidiya V. *Lose Your Mother: A Journey along the Transatlantic Slave Route*. New York: Farrar, Straus, and Giroux, 2007.

Hassol, Susan Joy, Simon Torok, Sophie Lewis, and Patrick Luganda. "(Un)Natural Disasters: Communicating Linkages between Extreme Events and Climate Change." *Bulletin of the World Meteorological Organization* 65, no. 2 (2016): 2–9.

Hatvany, Matthew. "'Wedded to the Marshes': Salt Marshes and Socio-Economic Differentiation in Early Prince Edward Island." *Acadiensis* 30, no. 2 (2001): 40–55.

Hayden, Lucy. "Classical Tidings from the Afric Muse: Phillis Wheatley's Use of Greek and Roman Mythology." *College Language Association Journal* 35, no. 4 (1992): 432–447.

Heidegger, Martin. "The Origin of the Work of Art." In *Poetry, Language, Thought*, 15–86. Translated by Alfred Hofstadter. New York: Harper, 1975.

Heise, Ursula. "Introduction: Planet, Species, Justice—and the Stories We Tell about Them." In *The Routledge Companion to the Environmental Humanities*, edited by Ursula Heise, Jon Christensen, and Michelle Niemann, 1–10. New York: Routledge, 2017.

Heller-Roazen, Daniel. *The Enemy of All: Piracy and the Law of Nations*. Brooklyn: Zone Books, 2009.

Hess, Scott. *William Wordsworth and the Ecology of Authorship: The Roots of Environmentalism in Nineteenth-Century Culture*. Charlottesville: University of Virginia Press, 2012.

Hickman, Jared. *The Black Prometheus: Race and Radicalism in the Age of Atlantic Slavery*. Oxford: Oxford University Press, 2016.

"History of Our National Park," Peak District National Park. https://www.peakdistrict.gov.uk/learning-about/about-the-national-park/our-history (accessed September 13, 2021).

Hornsby, Stephen. *British Atlantic, American Frontier: Spaces of Power in the Early Modern British America*. Hanover, NH: University Press of New England, 2005.

———. *Surveyors of Empire: Samuel Holland, J.F.W. Des Barres, and the Making of the Atlantic Neptune*. Montreal: McGill-Queen's University Press, 2011.

Horowitz, James. "Ovid in Restoration and Eighteenth-Century England." *A Handbook to the Reception of Ovid*, edited by John F. Miller and Carole E. Newlands, 355–370. New York: Wiley Blackwell, 2014.

Houswitschka, Christoph. "Peripatetics of Citizenship in the 1790s." In *Mediating Identities in Eighteenth-Century England: Public Negotiations, Literary Discourses, Topography*, edited by Isabel Karremann and Anja Muller, 177–192. Farnham, UK: Ashgate, 2011.

Howe, Ellic. *Astrology and the Third Reich*. Wellingborough, UK: Aquarian Press, 1984.

Huggan, Graham, and Helen Tiffin. *Postcolonial Ecocriticism: Literature, Animals, Environment*. 2nd ed. London: Routledge, 2015.

Hume, Angela, and Gillian Osborne. *Ecopoetics: Essays in the Field*. Iowa City: University of Iowa Press, 2018.

Hunt, Lynn. "Against Presentism." *Perspectives on History*. May 1, 2002. https://www.historians.org/publications-and-directories/perspectives-on-history/may-2002/against-presentism.

Hunter, Kathryn Montgomery. *Doctor's Stories: The Narrative Structure of Medical Knowledge*. Princeton, NJ: Princeton University Press, 1991.

Hutchinson, Jonathan. "An Address on Syphilis as an Imitator." *British Medical Journal* 1, no. 953 (1879): 541–542.

Iannini, Christopher P. *Fatal Revolutions: Natural History, West Indian Slavery, and the Routes of American Literature*. Chapel Hill: University of North Carolina Press, 2012.

Ing, M. B. *Carte de l'Accadie et Pais Voisins pour servir a l'Histoire Generale des Voyages: Par M.B. Ingr. de la Marine*. Prince Edward Island: Prince Edward Island Museum and Heritage Foundation, 1757. https://www.islandimagined.ca/islandora/object/imagined:208348/-/Carte%20de%20l'Accadie%20et%20Pais%20Voisins%20pour%20servir%20a%20l'Histoire%20Generale%20des%20Voyages:%20Par%20M.B.Ing%20de%20la%20Marine (accessed 13 September 2021).

Intergovernmental Panel on Climate Change. *Second Assessment, Climate Change*. 1995. https://archive.ipcc.ch/pdf/climate-changes-1995/ipcc-2nd-assessment/2nd-assessment-en.pdf.

Irlam, Shaun. "Wish You Were: Exporting England in James Grainger's *The Sugar Cane*." *ELH* 68, no. 2 (2001): 377–396.

Jacobus, Mary. "'Distressful Gift': Talking to the Dead." *South Atlantic Quarterly* 106, no. 2 (2007): 393–418.

James, C.L.R. "Preface to the First Edition." In *The Black Jacobins*, viiii–xi. 2nd ed. London: Vintage, 1963.

Janković, Vladimir. *Reading the Skies: A Cultural History of English Weather, 1650–1820*. Chicago: University of Chicago Press, 2000.

Jeffers, Honorée Fanonne. *The Age of Phillis*. Middletown, CT: Wesleyan Press, 2020.

Jefferson, Thomas. *Notes on the State of Virginia*. Edited by David Waldstreicher. Boston: Beacon, 2002.

Jenkins, Virgina. *The Lawn: A History of an American Obsession*. Washington, DC: Smithsonian Books, 1994.

Jennings, Regina. "African Sun Imagery in the Poetry of Phillis Wheatley." *Pennsylvania English* 22, no. 1–2 (2000): 68–76.

Jeremiah, Edward T. *The Emergence of Reflexivity in Greek Language and Thought: From Homer to Plato and Beyond*. Leiden: Brill, 2012.

Johns, Alessa, ed. *Dreadful Visitations: Confronting Natural Catastrophe in the Age of Enlightenment*. New York: Routledge, 1999.

Johnson, Alexander. *The First Mapping of America: The General Survey of British North America*. Bristol, UK: I.B. Tauris, 2017.

Jones, Elizabeth. "Keats in the Suburbs." *Keats-Shelley Journal* 45 (1996): 23–43.

Jones, Richard F. "Eclogue Types in English Poetry of the Eighteenth Century." *Journal of English and Germanic Philology* 24, no. 1 (1925): 33–60.

Jordan, June. "The Difficult Miracle of Black Poetry in America: Something Like a Sonnet for Phillis Wheatley." In *Some of Us Did Not Die: New and Selected Essays by June Jordan*, 174–186. London: Penguin Books, 2002.

Judd, Richard. *Second Nature: An Environmental History of New England*. Amherst: University of Massachusetts Press, 2014.

Kaufman, Paul. "*The Hurricane* and the Romantic Poets." *English Miscellany* 21 (1970): 99–116.

Keegan, Bridget. *British Labouring-Class Nature Poetry, 1730–1837*. New York: Palgrave, 2008.

Keenleyside, Heather. "The Rise of the Novel and the Fall of Personification." In *Eighteenth-Century Poetry and the Rise of the Novel Reconsidered*, edited by Kate Parker and Courtney Weiss Smith, 105–133. Lewisburg, PA: Bucknell University Press, 2014.

Keenlyside, David, and Helen Kristmanson. "The Palaeo-Environment and the Peopling of Prince Edward Island: An Archaeological Perspective." In *Time and a Place: An Environmental History of Prince Edward Island*, edited by Edward MacDonald, Joshua MacFadyen, and Irene Novaczek, 59–81. Charlottetown: McGill-Queen's University Press and Island Studies Press, 2016.

Keller, Catherine, and Mary-Jane Rubenstein, eds. *Entangled Worlds: Religion, Science, and New Materialisms*. New York: Fordham University Press, 2017.

Kerényi, Karl. *Goddesses of Sun and Moon*. Translated by Murray Stein. Dallas: Spring Publications, 1979.

Kincaid, Jamaica. *A Small Place*. New York: Farrar, Straus and Giroux, 2000.

King-Hele, Desmond. *Erasmus Darwin and the Romantic Poets*. Basingstoke, UK: Macmillan, 1986.

King, Tiffany Lethabo. *The Black Shoals: Offshore Formations of Black and Native Studies*. Durham, NC: Duke University Press, 2019.

———. "Racial Ecologies: Black Landscapes in Flux." In *Racial Ecologies*, edited by Leilani Nishime and Kim D. Hester Williams, 65–75. Seattle: University of Washington Press, 2018.

Klein, Bernhard, and Gesa Mackenthun, eds. *Sea Changes: Historicizing the Ocean*. New York: Routledge, 2004.

Knapp, Steven. *Personification and the Sublime: Milton to Coleridge*. Cambridge, MA: Harvard University Press, 1985.

Koch, Alexander, Chris Brierly, Mark M. Maslin, and Simon Lewis. "Earth System Impacts of the European Arrival and Great Dying in the Americas after 1492." *Quaternary Science Reviews* 207, no. 1 (2019): 13–36.

Kohn, Eduardo. *How Forests Think: Toward an Anthropology beyond the Human*. Berkeley: University of California Press, 2013.

Kupperman, Karen Ordahl. "The Puzzle of the American Climate in the Early Colonial Period." *American Historical Review* 87, no. 5 (1982): 1262–1289.

LaFleur, Greta. *The Natural History of Sexuality in Early America*. Baltimore, MD: Johns Hopkins University Press, 2018.

Lamb, Jonathan. *Preserving the Self in the South Seas, 1680–1840*. Chicago: University of Chicago Press, 2001.

———. *The Things Things Say*. Princeton, NJ: Princeton University Press, 2011.

Lane, Hannah M. "'The Garden of the Dead': The Old Burial Ground, Cemetery Reform, and Cultural Memory." In *The Creative City of Saint John*, edited by Gwendolyn Davies, Peter Larocque, and Christl Verduyn, 84–90. Halifax: Formac, 2018.

Lange, Sue. "L. M. Montgomery's Halifax: The Real Life Inspiration for *Anne of the Island*." *Shining Scroll* (2011): 3–9.

Larrère, Catherine. "Un catastrophisme absolu?" In *L'invention de la catastrophe au XIIIe Siècle*, edited by Anne-Marie Mercier-Faivre and Chantal Thomas, 143–152. Geneva: Droz, 2008.

Lefebvre, Henri. *The Production of Space*. Malden, MA: Blackwell, 1991.

LeMenager, Stephanie. "The Humanities after the Anthropocene." In *The Routledge Companion to the Environmental Humanities*, edited by Ursula Heise, Jon Christensen, and Michelle Niemann, 473–481. New York: Routledge, 2017.

Lennox, Jeffers. *Homelands and Empires: Indigenous Spaces, Imperial Fictions, and Competition for Territory in Northeastern North America, 1690–1763*. Toronto: University of Toronto Press, 2017.

Levander, Caroline F., and Robert S. Levine. "Introduction." In *Hemispheric American Studies*, edited by Caroline F. Levander and Robert S. Levine, 1–17. New Brunswick, NJ: Rutgers University Press, 2007.

Lewis, Jayne Elizabeth. *Air's Appearance: Literary Atmosphere in British Fiction, 1660–1794*. Chicago: University of Chicago Press, 2012.

Lewis, Simon L, and Mark A Maslin. "A Transparent Framework for Defining the Anthropocene Epoch." *Anthropocene Review* 2, no. 2 (2015): 128–146.

Ligon, Richard. *A True and Exact History of the Island of Barbados*. London: Humphrey Moseley, 1657.

Linebaugh, Peter. *The Magna Carta Manifesto: Liberties and Commons for All*. Berkeley: University of California Press, 2008.

Linebaugh, Peter, and Marcus Rediker. *The Many-Headed Hydra: Sailors, Slaves, Commoners, and the Hidden History of the Revolutionary Atlantic*. Boston: Beacon Press, 2000.

Lipman, Andrew. *The Saltwater Frontier: Indians and the Contest for the American Coast*. New Haven, CT: Yale University Press, 2015.

Loar, Christopher F. "Georgic Assemblies: James Grainger, John Dyer, and Bruno Latour." *Philological Quarterly* 97, no. 2 (2018): 241–261.

Lockerby, Earle, and Douglas Sobey. *Samuel Holland: His Work and Legacy on Prince Edward Island*. Prince Edward Island: Island Studies Press, 2015.

Löffler, Catharina. *Walking in the City: Urban Experiences and Literary Psychogeography in Eighteenth Century London*. Weisbaden: J.B. Metzler, 2017.

Lopenzina, Drew. *Red Ink, Native Americans Picking Up the Pen in the Colonial Period*. Albany: SUNY Press, 2009.

Loscocco, Paula. *Phillis Wheatley's Miltonic Poetics*. New York: Palgrave Macmillan, 2014.

Loughran, Kevin. "Urban Parks and Urban Problems: An Historical Perspective on Green Space Development as Cultural Fix." *Urban Studies* 57, no. 11 (2008): 2321–2338.

Low, Anthony. *The Georgic Revolution*. Princeton, NJ: Princeton University Press, 1985.

Lowenthal, David. *The Past Is a Foreign Country*. Cambridge, UK: Cambridge University Press, 1985.

Lowes, John Livingston. *The Road to Xanadu: A Study in the Ways of the Imagination*. Boston: Houghton Mifflin, 1955.

Luby, Brittany, and Stephen Hay. "Coming Home to Indigenous Place Names in Canada." *Canadian Historical Review* 100, no. 1 (2019): 101–103.

Luciano, Dana. "The Inhuman Anthropocene." *Avidly*, March 22, 2015.

MacDonald, Edward, and Boyde Beck. "Lines in the Water: Time and Place in a Fishery." In *Time and a Place: An Environmental History of Prince Edward Island*, edited by Edward MacDonald, Joshua MacFadyen, and Irene Novaczek, 218–221. Charlottetown: McGill-Queen's University Press and Island Studies Press, 2016.

MacDonald, Edward, Joshua MacFadyen, and Irene Novaczek, eds. "Introduction: Promise and Premise: An Environmental History for Prince Edward Island." In *Time and a Place: An Environmental History of Prince Edward Island*, edited by Edward MacDonald, Joshua MacFadyen, and Irene Novaczek, 3–15. Charlottetown: McGill-Queen's University Press and Island Studies Press, 2016.

Mague, Stephen T. "In Search of the Shawmut Peninsula: Using Modern Cartographic Analysis to Discover the 'Original' Boston Shoreline." In *Remaking Boston: An Environmental History of the City and Its Surroundings*, edited by Anthony Penna and Conrad Wright, 75–102. Pittsburgh: University of Pittsburgh Press, 2009.

Malamud, Margaret. *African Americans and the Classics: Antiquity, Abolition, and Activism*. London: I.B. Tauris, 2016.

Mallory, Devona. "I Remember Mama: Honoring the Goddess-Mother while Denouncing the Slaveowner-God in Phillis Wheatley's Poetry." *New Essays on Phillis Wheatley*, edited by John C. Shields, 19–34. Knoxville: University of Tennessee Press, 2011.

Malm, Andreas. *Fossil Capital: The Rise of Steam Power and the Roots of Global Warming*. London: Verso, 2016.

Mann, Charles. *1493: Uncovering the New World That Columbus Created*. New York: Vintage, 2011.

Mann, Michael, E. Lloyd, and Elisabeth Oreskes. "Assessing Climate Change Impacts on Extreme Weather Events: The Case for an Alternative (Bayesian) Approach." *Climatic Change* 144, no. 2 (2017): 131–142.

"Map." Mi'kmaq Confederacy of Prince Edward Island. http://mcpei.ca/map/ (accessed September 13, 2021).

Markley, Robert. "'Casualties and Disasters': Defoe and the Interpretation of Climatic Instability." *Journal for Early Modern Cultural Studies* 8, no. 2 (2008): 102–124.

Martin, Craig. *Renaissance Meteorology: From Pomponazzi to Descartes*. Baltimore, MD: Johns Hopkins University Press, 2011.

Mather, Increase. *The Doctrine of Divine Providences Opened and Applied*. Boston, 1684.

———. *An Essay on Illustrious Providences*. Boston, 1684.

———. *The Voice of God in Stormy Winds. Considered, in Two Sermons, Occasioned by the Dreadful and Unparallel'd Storm, in the European Nations. Novemb. 27th. 1703*. Boston, 1704.

McCallam, David. *Volcanoes in Eighteenth-Century Europe: An Essay in Environmental Humanities*. Oxford: Oxford University Press, 2019.

McCoubrey, John. "Turner's *Slave Ship*: Abolition, Ruskin, and Reception." *Word and Image* 14, no. 4 (1998): 319–353.

McCulloch, M. M., D. Forbes, and R. D. Shaw. *Coastal Impacts of Climate Change and Sea-Level Rise on Prince Edward Island*. Dartmouth, NS: Geological Survey of Canada/Bedford Institute of Oceanography, 2002.

McKay, Marilyn. *Picturing the Land: Narrating Territories in Canadian Landscape Art, 1500–1950*. Montreal: McGill-Queen's University Press, 2011.

McKittrick, Katherine. *Demonic Grounds: Black Women and the Cartographies of Struggle*. Minneapolis: University of Minnesota Press, 2006.

McKittrick, Katherine, Frances H. O'Shaughnessy, and Kendall Witaszek. "Rhythm, or On Sylvia Wynter's Science of the Word." *American Quarterly* 70, no. 4 (2018): 867–874.

McNabb, Debra. "The Role of Land in the Development Horton Township, 1760–1775." In *They Planted Well: New England Planters in Maritime Canada*, edited by Margaret Conrad, 151–160. Fredericton: Acadiensis Press, 1988.

McNeill, J. R. "Introductory Remarks: The Anthropocene and the Eighteenth Century." *Eighteenth-Century Studies* 49, no. 2 (2016): 117–128.

McRae, Matthew. "The Story of Africville." *Canadian Museum for Human Rights*. https://humanrights.ca/story/the-story-of-africville (accessed September 13, 2021).

Mee, Jon. *Conversable Worlds: Literature, Contention, and Community 1762 to 1830*. Oxford: Oxford University Press, 2011.

Melville, Herman. *Benito Cereno*. Edited by Brian Yothers. Peterborough, ON: Broadview, 2019.

Mentz, Steve. *At the Bottom of Shakespeare's Ocean*. New York: Continuum, 2009.

———. *Break Up the Anthropocene*. Minneapolis: University of Minnesota Press, 2019.

———. *Shipwreck Modernity: Ecologies of Globalization, 1550–1719*. Minneapolis: University of Minnesota Press, 2015.

———. "Toward a Blue Cultural Studies: The Sea, Maritime Culture, and Early Modern English Literature." *Literature Compass* 6, no. 5 (2009): 997–1013.

Mignolo, Walter D., and Catherine E. Walsh, eds. *On Decoloniality: Concepts, Analytics, Praxis*. Durham, NC: Duke University Press, 2018.

Milton, John. *The Complete Poetry and Essential Prose of John Milton*. Edited by William Kerrigan, John Rumrich, and Stephen M. Fallon. New York: Modern Library, 2007.

Monmonier, Mark. *Coast Lines: How Mapmakers Frame the World and Chart Environmental Change*. Chicago: University of Chicago Press, 2008.

Montgomery, L. M. *Anne of Ingleside*. London: Virago, 2017.

———. *Anne of Windy Poplars*. Toronto: Seal Books, 1983.

———. *The Complete Journals of L.M. Montgomery: The PEI Years, 1889–1900*. Oxford: Oxford University Press, 2017.

Mooney, Chris. "What the Massive Snowfall in Boston Tells Us about Global Warming." *Washington Post*, February 10, 2015.

Morens, David M., and Anthony S. Fauci. "Emerging Pandemic Diseases: How We Got to COVID-19." *Cell* 182, no. 5 (2020): 1077–1092.

Morgan, Jennifer. *Laboring Women: Reproduction and Gender in New World Slavery*. Philadelphia: University of Pennsylvania Press, 2011.

Morton, Timothy. *The Ecological Thought*. Cambridge, MA: Harvard University Press, 2010.

———. "Wordsworth Digs the Lawn." *European Romantic Review* 15, no. 2 (2004): 317–327.

Moten, Fred, and Stefano Harney. *The Undercommons*. New York: Autonomedia, 2013.

Mulcahy, Matthew. *Hurricanes and Society in the Greater British Caribbean, 1624–1783*. Baltimore, MD: Johns Hopkins University Press, 2006.

———. "Weathering the Storms: Hurricanes and Risk in the British Greater Caribbean." *Business History Review* 78, no. 4 (2004): 635–663.

Muñoz, José Esteban. *Cruising Utopia: The Then and There of Queer Futurity*. New York: New York University Press, 2009.

Nardizzi, Vin, and Tiffany Jo Werth, eds. *Premodern Ecologies in the Modern Literary Imagination*. Toronto: University of Toronto Press, 2019.

National Academy of Science. *Attribution of Extreme Weather Events in the Context of Climate Change*. Washington, DC: National Academies Press, 2016.

Natural Resources Canada. *Canada's Marine Coasts in a Changing Climate*. Ottawa: Government of Canada, 2016.

Neill, Anna. "Buccaneer Ethnography: Nature, Culture, and Nation in the Journals of William Dampier." *Eighteenth-Century Studies* 33, no. 2 (2000): 165–180.

Neimanis, Astrida, Cecilia Åsberg, and Johan Hedrén. "Four Problems, Four Directions for Environmental Humanities: Toward Critical Posthumanities for the Anthropocene." *Ethics & the Environment* 20, no. 1 (2015): 67–97.

Nichols, B. Ashton. *Beyond Romantic Ecocriticism: Toward Urbanatural Roosting*. London: Palgrave, 2011.

Nishime, Leilani, and Kim D. Hester Williams, eds. *Racial Ecologies*. Seattle: University of Washington Press, 2018.

Nixon, Rob. *Slow Violence and the Environmentalism of the Poor*. Cambridge, MA: Harvard University Press, 2011.

Nussbaum, Felicity. "Introduction." In *The Global Eighteenth Century*, edited by Felicity Nussbaum, 1–18. Baltimore, MD: Johns Hopkins University Press, 2003.

Odell, Patrick. "Fences around Former Dog Park at Africville Park Removed." *Global News*, February 26, 2016. https://globalnews.ca/news/2543797/fences-around-former-dog-park-at-africville-park-removed/.

O'Neal, Sondra. "A Slave's Subtle War: Phillis Wheatley's Use of Biblical Myth and Symbol." *Early American Literature* 21 (1986): 144–165.

Oppermann, Serpil, and Serenella Iovino, eds. *Environmental Humanities: Voices from the Anthropocene*. New York: Rowman & Littlefield, 2017.

Orr, Bridget. *British Enlightenment Theatre: Dramatizing Difference*. Cambridge, UK: Cambridge University Press, 2019.

———. *Empire on the Stage, 1660–1714*. Cambridge, UK: Cambridge University Press, 2001.

Ostavall, Neill, and Vaugh Scribner. "The Devil Was in the Englishman That He Makes Everything Work: Implementing the Concept of Work to Reevaluate Sugar Production and Consumption in the Early Modern British Atlantic World." *Agricultural History Society* 92, no. 4 (2018): 461–490.

Outka, Paul. *Race and Nature from Transcendentalism to the Harlem Renaissance*. London: Palgrave Macmillan, 2008.

Ovid, *Metamorphoses*. Translated by A. D. Melville. Oxford: Oxford World Classics, 2009.

Owens, Deirdre Cooper. *Medical Bondage: Race, Gender and the Origins of American Gynecology*. Athens: University of Georgia Press, 2018.

Page, Michael R. *The Literary Imagination from Erasmus Darwin to H. G. Wells: Science, Evolution, and Ecology*. Farnham, UK: Ashgate, 2012.

Pak, Chris. *Terraforming: Ecopolitical Transformations and Environmentalism in Science Fiction*. Liverpool: Liverpool University Press, 2016.

Parikka, Jussi. *The Anthrobscene*. Minneapolis: University of Minnesota Press, 2014.

Parmenter, Jon. "After the Mourning Wars: The Iroquois as Allies in Colonial North American Campaigns 1676–1760." *William and Mary Quarterly* 64, no. 1 (2007): 39–76.

Parrish, Susan Scott. *American Curiosity: Cultures of Natural History in the Colonial British Atlantic World*. Chapel Hill: University of North Carolina Press, 2006.

Pastore, Christopher. *Between Land and Sea: The Atlantic Coast and the Transformation of New England*. Cambridge, MA: Harvard University Press, 2014.

Pedley, M. S. *The Commerce of Cartography: Making and Marketing Maps in Eighteenth-Century France and England*. Chicago: University of Chicago Press, 2005.

Peters, John Durham. *The Marvelous Clouds: Toward a Philosophy of Elemental Media*. Chicago: University of Chicago Press, 2015.

Pettorelli, Nathalie, Sarah M. Durant, and Johan T. du Toit, eds. *Rewilding*. Cambridge, UK: Cambridge University Press, 2019.

Philip, M. NourbeSe. *Zong!*. Middletown, CT: Wesleyan University Press, 2008.

Piana, Pietro, and Charles Watkins. "Questioning the View: Historical Geography and Topographical Art." *Geography Compass* 14, no. 4 (2020): 111–134.

Piper, Kelsey. "The Climate Renegade: What Happens When Someone Wants to Go It Alone on Fixing the Climate?" *Vox*, June 4, 2019. https://www.vox.com/the-highlight/2019/5/24/18273198/climate-change-russ-george-unilateral-geoengineering.

*Plan of Lot 30 (No.1)*. Public Archives and Records Office, 1784. https://www.islandimagined.ca/islandora/object/imagined:208320 (accessed September 13, 2021).

Pope, Alexander. *Essay to Man, Epistle II*. London: J. Wilford, 1733.

———. *The Rape of the Lock and A Key to the Lock*. London: Hesperus Press, 2004.

———, trans. *The Iliad*. 1715. London: J. Bumpus, 1820.

Pope, Peter E. "Historical Archaeology and the Maritime Cultural Landscape of the Atlantic Fishery." In *Method and Meaning in Canadian Environmental History*, edited by Alan MacEachern and William Turkel, 36–54. Toronto: Nelson, 2009.

Povinelli, Elizabeth. *Geontologies: A Requiem to Late Liberalism*. Durham, NC: Duke University Press, 2016.

Pratt, Mary Louise. *Imperial Eyes: Travel Writing and Transculturation*. New York: Routledge, 1992.

Preston, Diana, and Michael Preston. *A Pirate of Exquisite Mind: Explorer, Naturalist, and Buccaneer: The Life of William Dampier*. London: Corgi Books, 2005.

Quigley, Killian. "The Pastoral Sublime: William Diaper and Eclogue's Marine Frontier." *Eighteenth-Century Studies* 53, no. 1 (2019): 109–128.

Rankin, K. J., and Poul Holm. "Cartographical Perspectives on the Evolution of Fisheries in Newfoundland's Grand Banks Area and Adjacent North Atlantic Waters in the Sixteenth and Seventeenth Centuries." *Terrae Incognitae* 51, no. 3 (2019): 190–218.

Ravel, Jeffrey. "President's Column." *American Society for Eighteenth-Century Studies News Circular* 175 (June 30, 2020).

Rediker, Marcus. *Villains of All Nations: Atlantic Pirates in the Golden Age*. Boston: Beacon Press, 2004.

Reimen, Peter. "Oeconomy and Ecology in Early Modern England." *PMLA* 132, no. 5 (2017): 1117–1133

Reno, Seth. "The New Volcanoes of Industry." *18th-Century Common*, February 22, 2019. https://www.18thcenturycommon.org/volcano/.

Richardson, Robbie. *The Savage and Modern Self: North American Indians in Eighteenth-Century British Literature and Culture*. Toronto: University of Toronto Press, 2018.

Ritchie, Robert C. *Captain Kidd and the War against the Pirates*. Cambridge, MA: Harvard University Press, 1986.

Rix, Robert. *William Blake and the Cultures of Radical Christianity*. London: Routledge, 2007.

Rozwadowski, Helen M. *Fathoming the Ocean: The Discovery and Exploration of the Deep Sea*. Cambridge, MA: Belknap Press, 2005.

Rusert, Britt. *Fugitive Science: Empiricism and Freedom in Early African American Culture*. New York: New York University Press, 2017.

———. "Plantation Ecologies: The Experimental Plantation in and against James Grainger's *The Sugar Cane*." *Early American Studies* 13, no. 2 (2015): 341–373.

Sasser, Daryl. "The Weather and Theology: The Influence of the Natural World on Religious Thought in Puritan New England." PhD diss., Union Theological Seminary, 2010.

Saunders, Timothy. *Bucolic Ecology: Virgil's Eclogues and the Environmental Literary Tradition*. London: Bloomsbury, 2008.

Scarth, Kate. "Born for City Life? L.M. Montgomery's Urban Prince Edward Island." *Island Magazine* (2020): 53–64.

———. "From Anne of Green Gables to Anne of the Suburbs: Lucy Maud Montgomery Reimagines Home in *Anne of the Island*." *Women's Writing* 27, no. 2 (2017): 150–164.

Scarth, Kate, and Christopher Shalom. "Old Burying Ground." *Halifax Literary Landmarks*. https://halifaxliterarylandmarks.ca/tours/show/4 (accessed September 13, 2021).

Schiebinger, Londa. *Nature's Body: Gender in the Making of Modern Science*. Boston: Beacon Press, 1993.

———. *Plants and Empire: Colonial Bioprospecting in the Atlantic World*. Cambridge, MA: Harvard University Press, 2004.

Schmidt, Benjamin. "The Humanities Are in Crisis." *The Atlantic*, August 23, 2018. https://www.theatlantic.com/ideas/archive/2018/08/the-humanities-face-a-crisisof-confidence/567565/.

Schofield, Robert E. *The Lunar Society of Birmingham: A Social History of Provincial Science and Industry in Eighteenth-Century England*. Oxford: Clarendon Press, 1963.

Schuchard, Marsha Keith. "Rediscovering William 'Hurricane' Gilbert: A Lost Voice of Revolution and Madness in the Worlds of Blake and the Romantics." William Gilbert Home Page, April 25, 2019. http://www.williamgilbert.com/GILBERT_Schuchard.htm.

Schulz, Max F. *Paradise Preserved: Recreations of Eden in Eighteenth- and Nineteenth- Century England*. Cambridge, UK: Cambridge University Press, 1985.

Schweiger, Tristan J. "Grainger's West Indian Planter: Property and Authority in *The Sugar Cane*." *Eighteenth-Century Studies* 50, no. 4 (Summer 2017): 401–416.

Scrivener, Michael. *Seditious Allegories: John Thelwall and Jacobin Writing*. University Park: Pennsylvania State University Press, 2001.

Sewall, Samuel. *Diary of Samuel Sewall, 1674–1729*. Vol. 2. New York: Farrar, Straus, and Giroux, 1973.

Shapin, Steven. *A Social History of Truth: Civility and Science in Seventeenth-Century England*. Chicago: University of Chicago Press, 1994.

Shapin, Steven, and Simon Schaffer. "Seeing and Believing: The Experimental Production of Pneumatic Facts." In *Leviathan and the Air-Pump: Hobbes, Boyle, and the Experimental Life*, 22–79. Princeton, NJ: Princeton University Press, 1989.

Sharma, Nandita. "Strategic Anti-Essentialism: Decolonizing Decolonization." In *Sylvia Wynter: On Being Human as Praxis*, edited by Katherine McKittrick, 164–182. Durham, NC: Duke University Press, 2015.

Sharpe, Christina. Twitter, June 6, 2020. https://twitter.com/hystericalblkns/status/1269331358704259072.

———. *In the Wake: On Blackness and Being*. Durham, NC: Duke University Press, 2016.

Shields, David S. *Oracles of Empire: Poetry, Politics, and Commerce in British America 1690–1750*. Chicago: The University of Chicago Press, 1990.

Shields, John C. *The American Aeneas: Classical Origins of the American Self*. Knoxville: University of Tennessee Press, 2004.

———. *Phillis Wheatley's Poetics of Liberation*. Knoxville: University of Tennessee Press, 2008.

———. "Phillis Wheatley's Subversive Pastoral." *Eighteenth-Century Studies* 27, no. 4 (1994): 631–647.

———. "Phillis Wheatley's Use of Classicism." *American Literature* 52, no. 1 (1980): 97–111.

———., ed. *The Collected Works of Phillis Wheatley*. New York: Oxford University Press, 1988.

Shteir, Ann. "She Comes!—the GODDESS!": Narrating Nature in Erasmus Darwin's *The Botanic Garden*." In *Fact and Fiction: Literary and Scientific Cultures in Germany and Britain*, edited by Christine Lehleite, 73–96. Toronto: University of Toronto Press, 2016.

Silva, Cristobal. "Georgic Fantasies: James Grainger and the Poetry of Colonial Dislocation." *ELH* 83, no. 1 (2016): 127–156.

Sinanan, Kerry. "BLM 2020: Breathing, Resistance, and the War against Enslavement." *Age of Revolutions*, June 10, 2020. https://ageofrevolutions.com/2020/06/10/blm-2020-breathing-resistance-and-the-war-against-enslavement/.

———. "Lost Mothers in the Caribbean Plantation and Contemporary Black Maternal and Infant Mortality." In *Caribbean Literature in Transition*, Vol. 1: 1800–1920, edited by Evelyn O'Callaghan and Tim Watson, 390–408. Cambridge, UK: Cambridge University Press, 2021.

Sloane, Hans. *A Voyage to the Islands Madera, Barbados, Nieves, S. Christopher and Jamaica*. London, 1707.

Slovic, Scott. "Epistemologies and Politics in American Nature Writing: Embedded Rhetoric and Discrete Rhetoric." In *Green Culture: Environmental Rhetoric in Contemporary America*, edited by Carl George Herndl and Stuart C. Brown, 82–110. Madison: University of Wisconsin Press, 1996.

Smith, Anthony Paul. *A Non-Philosophical Theory of Nature: Ecologies of Thought*. New York: Palgrave Macmillan, 2013.

Smith, Courtney Weiss. *Empiricist Devotions: Science, Religion, and Poetry in Early Eighteenth-Century England*. Charlottesville: University of Virginia Press, 2016.

Smith, Mick. *Against Ecological Sovereignty: Ethics, Biopolitics, and Saving the Natural World*. Minneapolis: University of Minnesota Press, 2011.

Smith, Nicholas D. "Jacopo Sannazaro's *Eclogae Piscatoriae* (1526) and the 'Pastoral Debate' in Eighteenth-Century England." *Studies in Philology* 99, no. 4 (2002): 432–450.

Smith, Noah. "America Is Pumping Out Too Many Ph.Ds." *Bloomberg Opinion*, January 4, 2021. https://www.bloomberg.com/opinion/articles/2021-01-04/america-is-pumping-out-too-many-ph-d-s?srnd=opinion.

Solomonescu, Yasmin. *John Thelwall and the Materialist Imagination*. London: Palgrave, 2014.

Spigner, Nicole A. "Phillis Wheatley's Niobean Poetics." *Brill's Companion to Classics in the Early Americas*, edited by Matthew E. Duquès, Adam Goldwyn, and Maya Feile Tomes. Leiden: Brill, 2021.

Spillers, Hortense J. *Black, White, and in Color: Essays on American Literature and Culture*. Chicago: University of Chicago Press, 2003.

Spillers, Hortense. "Mama's Baby, Papa's Maybe: An American Grammar." In *Black, White, and in Color*, 203–229. Chicago: University of Chicago Press, 2003.

Sprat, Thomas. *The History of the Royal-Society of London, for the Improving of Natural Knowledge*. 2nd ed. London, 1702.

Stamm, Lola V. "Global Challenge of Antibiotic-Resistant *Treponema pallidum*." *Antimicrobial Agents and Chemotherapy* 54, no. 2 (2010): 583–589.

Stanfield, James Field. *The Guinea Voyage: A Poem, in Three Books*. London: James Philips, 1789.

"Statutes of the Royal Society Enacted in 1663." In *A History of the Royal Society with Memoirs of the Presidents, Compiled from Authentic Documents*, edited by Charles Richard Weld, 524–540. London, 1848.

Stedman, J[ohn] G[abriel]. *Narrative, of a Five Years' Expedition, against the Revolted Negroes of Surinam, in Guiana, on the wild coast of South America; from the year 1772, to 1777*. London: J. Johnson, 1796.

Steffler, Margaret. "Brian O'Connal and Emily Byrd Starr: The Inheritors of Wordsworth's 'Gentle Breeze.'" In *Windows and Words: A Look at Canadian Children's Literature in English*, edited by Aïda Hudson and Susan-Ann Cooper, 87–96. Ottawa: Ottawa University Press, 2003.

Steinberg, Philip E. *The Social Construction of the Ocean*. Cambridge, UK: Cambridge University Press, 2001.

Stewart, John. *An account of Prince Edward Island in the gulph of St. Lawrence, North America*. 1806.

Stewart, Robert. Letter "To George Washington from Robert Steward, 14 January 1764." *Founders Online*. National Historical Publications and Records Commission and the University of Virginia Press. https://founders.archives.gov/documents/Washington/02-07-02-0169 (accessed September 13, 2021).

Stigler, Stephen. "Richard Price, the First Bayesian." *Statistical Science* 33, no. 1 (2018): 117–125.

Stott, Peter A., D. A. Stone, and M. R. Allen. "Human Contribution to the European Heat Wave in 2003." *Nature* 432 (December 2004): 610–614.

Stratford, Elaine, Godfrey Baldacchino, Elizabeth McMahon, Carol Farbotko, and Andrew Harwood. "Envisioning the Archipelago." *Island Studies Journal* 6, no. 2 (2011): 113–130.

Summerby-Murray, Robert. "Four Hundred Years of Mapping in the Upper Bay of Fundy: Changing Coastal Environments and Economies, 1550–1950." *Journal of the Royal Nova Scotia Historical Society* 15 (2012): 1–23.

Sutter, Paul, and Paul M. Pressley, eds. *Coastal Nature, Coastal Culture: Environmental Histories of the Georgia Coast.* Athens: University of Georgia Press, 2018.

Takahata, Kimberly. "'Follow Me Your Guide': Poetic Empire in John Singleton's *A General Description of the West-Indian Islands.*" *Studies in Eighteenth-Century Culture* 49 (2020): 45–64.

Tampa, M., I. Sarbu, C. Matei, V. Benea, and S. R. Georgescu. "Brief History of Syphilis." *Journal of Medicine and Life* 7, no. 1 (2014): 4–10.

Tanasescu, Mihnea. "Field Notes on the Meaning of Rewilding." *Ethics, Policy and Environment* 20, no. 3 (2017): 333–349.

Tarlo, Harriet, and Judith Tucker. "'Drawing Closer': An Ecocritical Consideration of Collaborative, Cross-Disciplinary Practices of Walking, Writing, Drawing and Exhibiting." In *Extending Ecocriticism: New Readings in Literature, Visual Arts, Performance and On-Site Interpretation*, edited by Peter Barry and William Welstead, 47–69. Manchester, UK: Manchester University Press, 2017.

Thelwall, John. *The Peripatetic.* Edited by Judith Thompson. Detroit: Wayne State University Press, 2001.

Thomas, Steven W. "Doctoring Ideology: James Grainger's *The Sugar Cane* and the Bodies of Empire." *Early American Studies* 4, no. 1 (2006): 78–111.

Thomas, Sue. "Catastrophic History, Cyclonic Wreckage and Repair in William Gilbert's *The Hurricane* and Diana McCaulay's *Huracan.*" In *Tracking the Literature of Tropical Weather: Typhoons, Hurricanes, and Cyclones*, edited by Anne Collett, Russell McDougall, and Sue Thomas, 227–250. New York: Palgrave, 2017.

Thompson, Gordon E. "Methodism and the Consolation of Heavenly Bliss in Phillis Wheatley's Funeral Elegies." *CLA Journal* 48, no. 1 (2004): 34–50.

Thompson, Judith. "Introduction." *The Peripatetic.* Detroit: Wayne State University Press, 2001.

———. *John Thelwall in the Wordsworth Circle: The Silenced Partner.* New York: London: Palgrave, 2012.

———. "A Note on the Text." In *The Peripatetic*, 53–55. Detroit: Wayne State University Press, 2001.

Thomson, James. *The Seasons.* 1726–1746. Reprint edited by James Sambrook. Oxford: Clarendon, 1972.

Thorn, Jennifer. "'All Beautiful in Woe': Gender, Nation, and Phillis Wheatley's 'Niobe.'" *Studies in Eighteenth-Century Culture* 37, no. 1 (2008): 233–258.

———. "Phillis Wheatley's Ghosts: The Racial Melancholy of New England Protestants." *Eighteenth Century* 50, no. 1 (2009): 73–99.

Thresher, G. G. *Charlottetown Harbour.* Prince Edward Island Archives, ca. 1830.

Tobin, Beth Fowkes. *Colonizing Nature: The Tropics in British Arts and Letters, 1760–1820.* Philadelphia: University of Pennsylvania Press, 2004.

Tompkins, Kyla Wazana. "On the Limits and Promise of New Materialist Philosophy." *Lateral: Journal of the Cultural Studies Association* 5, no. 1 (2016): 1–6.

"Tour of Inspection Made by the Sieur de La Roque, Census 1752." Translation, in *Report of the Public Archives of Canada, 1905, Part 2.* Ottawa: S. E. Dawson, 1906.

Trenberth, Kevin. "Framing the Way to Relate Climate Extremes to Climate Change." *Climatic Change* 115, no. 2 (2012): 283–290.

Trenberth, Kevin E., John T. Fasullo, and Theodore G. Shepherd. "Attribution of Climate Extreme Events." *Nature Climate Change* 5, no. 8 (2015): 725–730.

Tsing, Anna. *The Mushroom at the End of the World.* Princeton, NJ: Princeton University Press, 2015.

Tsing, Anna, Jacob Copeman, and Alice Street. "Strathern beyond the Human: Testimony of a Spore." *Theory, Culture & Society* 31, no. 2–3 (2014): 221–241.

Turner, J.M.W. *The Slave Ship.* 1840. Oil on canvas. Museum of Fine Arts, Boston.

Van Duzer, Chet, and Lauren Beck. *Canada before Confederation: Maps at the Exhibition* Wilmington, DE: Vernon Press, 2017.

Vasseur, Liette, Mary Thornbush, and Steve Plante. "Climatic and Environmental Changes Affecting Communities in Atlantic Canada." *Sustainability* 9, no. 8 (2017): 1293.

Waldron, Ingrid R. G. *There's Something in the Water: Environmental Racism in Indigenous and Black Communities.* Black Point, NS: Fernwood Publishers, 2018.

Waldstreicher, David. "Ancients, Moderns, and Africans: Phillis Wheatley and the Politics of Empire and Slavery in the American Revolution." *Journal of the Early Republic* 37, no. 4 (2017): 701–733.

Wallace, Elizabeth. "The Things Things Don't Say: The Rape of the Lock, Vitalism, and New Materialism." *Eighteenth Century* 59, no. 1 (2018): 105–122.

Walsh, Catherine E. "The Decolonial *For*: Resurgences, Shifts, and Movements." In *On Decoloniality: Concepts, Analytics, Praxis,* edited by Walter D. Mignolo and Catherine E. Walsh, 15–32. Durham, NC: Duke University Press, 2018.

Walshe, R. A., and C. E. Stancioff. "Small Island Perspectives on Climate Change." *Island Studies Journal* 13, no. 1 (2018): 13–24.

Walters, Tracey. *African American Literature and the Classicist Tradition.* London: Palgrave Macmillan, 2007.

*The Watercolour World.* Watercolour World. https://www.watercolourworld.org (accessed October 1, 2020).

Weheliye, Alexander. *Habeas Viscus: Racializing Assemblages, Biopolitics, and Black Feminist Theories of the Human.* Durham, NC: Duke University Press, 2014.

Weisman, Richard. *Witchcraft, Magic, and Religion in Seventeenth-Century Massachusetts.* Amherst: University of Massachusetts Press, 1984.

Wertheimer, Eric. *Underwriting: The Poetics of Insurance in America, 1722–1872.* Stanford, CA: Stanford University Press, 2006.

Wheatley Peters, Phillis. "Messrs. Hussey and Coffin." In *Phillis Wheatley: Biography of a Genius in Bondage.* Edited by Vincent Caretta. Athens: University of Georgia Press, 2014.

———. "Niobe in Distress for Her Children, Slain by Apollo." *Phillis Wheatley: Complete Writings,* edited by Vincent Carretta, 52–59. London: Penguin Books, 2001.

———. "TO A LADY ON HER REMARKABLE PRESERVATION IN AN HURRICANE IN *NORTH-CAROLINA.*" In *The Collected Works of Phillis Wheatley,* edited by John C. Shields, 1–6. New York: Oxford University Press, 1988.

———. "To Maecenas." In *The Collected Works of Phillis Wheatley,* edited by John C. Shields. New York: Oxford University Press, 1988.

———. "To S.M., a young *African* Painter, on seeing his Works." *Phillis Wheatley: Complete Writings,* edited by Vincent Carretta, 59–60. London: Penguin Books, 2001.

White, Kyle Powys. "Indigenous Science Fiction for the Anthropocene: Ancestral Dystopias and Fantasies of Climate Change Crises." *Environment and Planning E: Nature and Space* 1–2 (2018): 224–242.

White, Kyle Powys, Chris Caldwell, and Marie Schaefer. "Indigenous Lessons about Sustainability Are Not Just for 'All Humanity.'" In *Sustainability: Approaches to Environmental Justice and Social Power,* edited by Julie Sze, 149–179. New York: New York University Press, 2018.

White, Richard. *The Middle Ground: Indians, Empires, and Republics in the Great Lakes Region, 1650–1815.* Cambridge, UK: Cambridge University Press, 1991.

Wickman, Thomas. "The Great Snow of 1717: Settler Landscapes, Deep Snow Cover, and Winter's Environmental History." *Northeastern Naturalist* 24, no. 7 (2017): H81–114.

———. *Snowshoe Country: An Environmental and Cultural History of Winter in the Early American Northeast*. Cambridge, UK: Cambridge University Press, 2018.

Wigginton, Caroline. *In the Neighborhood: Women's Publication in Early America*. Amherst: University of Massachusetts Press, 2016.

Wigglesworth, Michael. "God's Controversy with New England, Written at the Time of the Great Drought Anno, 1662, By a Lover of New England's Prosperity." *Electronic Texts in American Studies*, edited by Reiner Smolinksi. https://digitalcommons.unl.edu/cgi/viewcontent.cgi?article=1036&context=etas (accessed September 13, 2021).

Willett, Cynthia. *Interspecies Ethics*. New York: Columbia University Press, 2014.

Williams, Jeffrey J. "The New Humanities." *Chronicle Review*, November 14, 2019. http://connect.chronicle.com/rs/931-EKA-218/images/ChronicleReview_Endgame.pdf.

Wilson, Brett. *A Race of Female Patriots: Women and Public Spirit on the British Stage, 1688–1745*. Lewisburg, PA: Bucknell University Press, 2011.

Winship, Michael P. "Prodigies, Puritanism, and the Perils of Natural Philosophy: The Example of Cotton Mather." *William and Mary Quarterly* 51, no. 1 (1994): 92–105.

———. *Seers of God: Puritan Providentialism in the Restoration and Early Enlightenment*. Baltimore, MD: Johns Hopkins University Press, 1996.

Wisecup, Kelly. *Medical Encounters: Knowledge and Identity in Early American Literature*. Amherst: University of Massachusetts, 2013.

Woertendyke, Gretchen J. *Hemispheric Regionalism: Romance and the Geography of Genre*. New York: Oxford University Press, 2016.

Wood, Denis, and John Fels. *The Natures of Maps: Cartographic Constructions of the Natural World*. Chicago: University of Chicago Press, 2008.

Wood, Gillen D'Arcy. *Tambora: The Eruption That Changed the World*. Princeton, NJ: Princeton University Press, 2014.

———. "Leigh Hunt's New Suburbia: An Eco-Historical Study in Climate Poetics and Public Health." *Interdisciplinary Studies in Literature and Environment* 18, no. 3 (2011): 527–552.

Wordsworth, William. "The Excursion." In *Poems of William Wordsworth, Volume 2: Collected Reading Texts from The Cornell Wordsworth*, edited by Jared Curtis. Tirril, UK: Humanities-Ebooks, 2011.

Worthington, David, ed. *The New Coastal History: Cultural and Environmental Perspectives from Scotland and Beyond*. Cham: Springer International, 2017.

Woster, Emily, and Kate Scarth. "Welcome to the *Journal of L.M. Montgomery Studies*." *Journal of L.M. Montgomery Studies* 1, no. 1 (2019): 1–23.

Wynn, Graeme. "Museums, Laboratories, Showcases: Islands in Environmental History." In *Time and a Place: An Environmental History of Prince Edward Island*, edited by Edward MacDonald, Joshua MacFadyen, and Irene Novaczek, 36–56. Charlottetown: McGill-Queen's University Press and Island Studies Press, 2016.

———. "Reflections on the Environmental History of Atlantic Canada." In *Land and Sea: Environmental History in Atlantic Canada*, edited by Claire Campbell and Robert Summerby-Murray, 235–256. Fredericton: Acadiensis Press, 2013.

Wynter, Sylvia. "1492: A New World View." In *Race, Discourse, and the Origin of the Americas: A New World View*, edited by Vera Lawrence Hyatt and Rex Nettleford, 5–57. Washington, DC: Smithsonian Institution Press, 1995.

———. "Novel and History, Plot and Plantation." *Savacou* 5 (1971): 95–102.

———. "Unsettling the Coloniality of Being/Power/Truth/Freedom: Towards the Human, after Man, Its Overrepresentation—an Argument." *CR: The New Centennial Review* 3, no. 3 (2003): 257–337.

Wynter, Sylvia, and Katherine McKittrick. "Unparalleled Catastrophe for Our Species?: Or, to Give Humanness a Different Future: Conversations." In *On Being Human as Praxis*, edited by Katherine McKittrick, 9–89. Durham, NC: Duke University Press, 2014.

Yates, Julian. "Represent." In *Veer Ecology: A Companion for Ecological Thinking*, edited by Jeffrey Jerome Cohen and Lowell Duckert, 122–135. Minneapolis: University of Minnesota Press, 2017.

Yusoff, Kathryn. *A Billion Black Anthropocenes or None*. Minneapolis: University of Minnesota Press, 2018.

Yuzwa, Zachary. "The Fall of Troy in Old Huronia." In *Classical Reception in the Early Americas*, edited by Adam Goldwyn, Maya Feile Tomes, and Matthew Duquès. Leiden: Brill, 2021.

Zalasiewicz, Jan, et al., "The Working Group on the Anthropocene: Summary of Evidence and Interim Recommendations." *Anthropocene* 19 (2017): 55–60.

Zielinski, Gregory A., and Barry D. Keim. *New England Weather/New England Climate*. Hanover, NH: University Press of New England, 2003.

Zilberstein, Anya. *A Temperate Empire: Making Climate Change in Early America*. Oxford: Oxford University Press. 2016.

# NOTES ON CONTRIBUTORS

CHRISTOPHER ALLAN BLACK is a member of the American literature faculty and an instructor in the Core Literature and First Year Composition programs at the University of Memphis in Tennessee, where he teaches courses in eighteenth- and nineteenth-century American literature, African American literature, and world literature. His scholarship focuses on environment, slavery, social justice, and reform in colonial America.

CLAIRE CAMPBELL is an environmental historian of Canada and the North Atlantic world. She has written extensively on historic landscapes as sites of public as well as environmental history, while her current research considers Canada as a coastal nation and the presence of water and waterbodies—fresh and salt—in Atlantic cities. She is a professor of history at Bucknell University in Lewisburg, Pennsylvania, and teaches widely in environmental history, including classes on cities, rivers, and early North America.

JEREMY CHOW is an assistant professor of English at Bucknell University in Lewisburg, Pennsylvania. His research interweaves eighteenth-century studies, the environmental humanities, and queer and sexuality studies. His first book, *The Queerness of Water: Violent Entanglements in Troubled Ecologies*, traces the ways literary bodies of water participate in reshaping and unsettling masculinity and sexuality in the long eighteenth century.

MARIAH CRILLEY is an assistant professor of interdisciplinary studies at Virginia Commonwealth University in Richmond. Her research interests include early American literature, disability studies, medical humanities, and environmental studies. In her pedagogy, she is dedicated to empowering student agency through accessibility, transparency, and transfer.

MATTHEW DUQUÈS received his PhD in English from Vanderbilt University in 2013. He has taught at Vanderbilt, North Dakota State University, Princeton, and the University of North Alabama, where he received tenure in 2019. He currently teaches at St. Joseph's University in Philadelphia, Pennsylvania, and is coeditor of *Brill's Companion to Classics in the Early Americas* (2021).

ANNETTE HULBERT is a visiting assistant professor at Willamette University in Salem, Oregon. Her research puts the new perspectives made available by climate historiography to work for more traditional literary-historical concerns about social representation. Her dissertation, "Writing in the Storm: Britain's Literary Weather, 1667–1790," considers the question of why storms are so pervasive in literature of the period, engaging the environmental humanities, cultural studies of science, religious studies, and twenty-first-century sociological research about Hurricane Katrina.

SHELBY JOHNSON is an assistant professor of English at Oklahoma State University in Stillwater, where she researches and teaches on sexuality, race, and ecocriticism in the long eighteenth century and in Indigenous studies. Her first book, *The Rich Earth Between Us: The Intimate Grounds of Race and Sexuality in the Atlantic World*, takes up the ways Black and Native decolonial worldmaking practices contest the extractive economies and racialized intimacies of settler colonialism.

ELLIOT PATSOURA received his PhD from the University of Melbourne in 2021. His dissertation, "Beyond the Biogenetic Law: Organic Analogies in Philosophy and Psychoanalysis," traces the passage of the concept of "recapitulation" from the eighteenth-century philosophies of Johann Gottfried Herder, Immanuel Kant, and Friedrich Schelling, through the nineteenth-century Darwinism of Ernst Haeckel, and into the hands of Sigmund Freud in the early twentieth century. His research has been published in *Angelaki* and *Parrhesia*.

JASON M. PAYTON is the recruitment and enrollment coordinator for the Morehead Honors College at the University of Georgia in Athens, where he also served as an assistant professor of English. His research on early American and Atlantic literatures has appeared in *Early American Literature* and *American Literature and the New Puritan Studies*. He is currently completing a book on Caribbean buccaneer narratives from the seventeenth century to the present.

KATE SCARTH is the chair of L. M. Montgomery studies and an assistant professor in the Applied Communication, Leadership, and Culture (ACLC) program at the University of Prince Edward Island in Canada. Her scholarly interests include British Romanticism and Atlantic Canadian literature, and she has published on Jane Austen, Charlotte Smith, and L. M. Montgomery in publications including *Women's Writing*, *European Romantic Review*, and *The Island Magazine*. Scarth is the founding coeditor of the *Journal of L.M. Montgomery Studies*, and, with Trinna S. Frever, is collecting readers' and fans' stories of discovering and engaging with the world of L. M. Montgomery, through the "Your L.M. Montgomery Story" project.

ADAM W. SWEETING is an associate professor of humanities at Boston University. He is the author of *Reading Houses and Building Books: Andrew Jackson Downing and the Architecture of Popular Antebellum Literature* and *Beneath the Second Sun: A Cultural History of Indian Summer*. His teaching includes interdisciplinary courses in the humanities and a team-taught course on climate change.

AMI YOON is a PhD candidate in the Department of English and Comparative Literature at Columbia University. Her research focuses on the intersections among nineteenth-century poetry, natural history, and violence in America. Her writing is forthcoming in *Leviathan: A Journal of Melville Studies*.

# INDEX


abolition, 12, 39, 175, 181–182, 198, 202
activism, 3–4, 13, 81, 207, 216–217
Adams, Eliphalet, 55–57
aesthetics, African, 79; landscape, 72–73; for realizing an anticolonial imaginary, 171; of resistance, 77
agency: disease's, 13, 88, 90, 92, 99 (*see also* disease; illness); environmental, 140; human, 30, 91–92, 137, 139, 196; of human action combined with natural weather, 56; intra-active, 12; more-than-human, 74; in Native and African cosmologies, 72; nonhuman, 89, 99; objects and, 85n29; (sea)water's, 137–138, 147n5; textual, 46
Ahmed, Sara, 77
amelioration, 189, 191, 198, 200, 202, 204n41. *See also* British Amelioration Act
American Renaissance, 147
American Revolution, 174, 182, 198
Anthropocene, 4–6, 61–62, 108–109, 132n8, 172
anti-blackness, 40, 77
anticolonialism, 14, 170–175, 178, 180. *See also* decoloniality
Antigua, 144, 170, 172–173, 176–179, 185n40
apostrophe, 39–40, 42–43, 46, 50–51, 52n34

Barad, Karen, 72, 89, 91–92
Bayesian Statistics, 63–64
Bennett, Jane, 42, 89, 91–92, 137
Black authorship, 71, 73
Black Lives Matter, 13, 81–83
blue humanities, 13–14
British Amelioration Act, 198, 204n41. *See also* amelioration

Canadian Maritimes, 217
capitalism: Atlantic, 174; colonial, 41–42; early, imperial, 98; merchant, 139–140, 145; rise of, 85n31

Césaire, Aimé, 170, 177
Clarkson, Thomas, 39–41
climate anxiety, 23, 25, 28, 33
climate change: anthropogenic, 12, 53, 56–57, 61–62, 65, 88; and cartography, 111; discourses and knowledges of, 11; effects of, 172; and extreme events, 62–64; and the "golden spike," 5; and indigeneity, 15; and racial capitalism, 84n18; sea level rise and, 107; vulnerability in era of, 105, 130. *See also* Anthropocene
Cohen, Jeffrey Jerome, 2–3, 82
Coleridge, Samuel Taylor, 171, 208
colonialism: Black Legend model of Spanish, 202; British agrarian, 201; colonial expansion, 76, 85n28, 144; destructions engendered by, 170–171; inhuman forces of, 83; (neo)colonialism, 5; New World, 181; occludes, 87n72; settler, 15, 136, 175, 182, 190. *See also* colonization
colonization: as cause of environmental change, 13; laboring-class work of, 41; legacies of, 132n11; and the Little Ice Age, 154 (*see also* Little Ice Age); natural history's absolution of, 98; power of sea as threat, 146; responsibility for disease transmission, 88 (*see also* etiology); settler, 165–167. *See also* colonialism
Columbus, Christopher, 90, 96–97, 101n5
Corley, Ide, 44, 46, 52n34
coronavirus, 82, 88–89, 98, 99–100. *See also* pandemic
Costello, Leo, 40–41
crisis: climate, 25, 35n9, 74, 85n28, 106, 130; ecological, 146; environmental, 58; in the humanities, 3–6
critical race studies, 2, 7, 73

Dampier, William, 137, 140–146, 148–149; "Discourses of Winds," 137, 141–145; *New Voyage Round the World*, 140–141, 143, 148–149; *Voyages and Descriptions*, 137, 140–143, 145, 149

Darwin, Erasmus, 6, 11–12, 24–34, 35n10, 37n45, 37n50
decoloniality, 14–15, 16, 137, 170, 175–177, 181–182. *See also* anticolonialism; indigeneity
Defoe, Daniel, 50, 59
De Quincy, Thomas, 24
Diaper, William, 180
disability, 89, 98
disease: agency of, 89–90, 99; ecology and, 153, 160, 163, 168; enslavement and, 48; natural history and, 95–96, 98, 101n7; pandemic and, 88–91. *See also* etiology; illness; syphilis
displacement, 117, 157–158, 162
dispossession, 154, 161

eclogue, 15, 172, 179–180
ecocriticism, 3; and environmental justice, 39, 42, 72–73, 146, 206
ecologies: of anti-Black violence, 77; colonial, 82, 87n72; damaged, 154; entanglements between human flesh and, 84n12; new, 48, 77, 174 (*see also* Sharpe, Christina); of nonbeing, 82; produced, 41; transatlantic, 181
economies: colonial extractive, 72; libidinal, 76; plantation, 174
education: focalizing indigeneity in, 15; higher, 3; humanities-driven, 4; insularity of American, 129; in the L'nuey project, 133n12; Thelwallian interactive, 218; of young Indigenous generations, 161
eighteenth-century studies: and the Anthropocene, 5; and the blue humanities, 13; and contribution of new materialist frameworks to, 12; critiquing emergent modernity, 136; and the environmental humanities, 1–2, 10–11; and Indigenous representation, 14; and presentism, 9
elegy, 74–75, 81
emancipation, 41, 174
empathy, 7, 214
Enlightenment (historical period), 1, 55, 74, 154–155, 161–163, 181, 201
enslavement, 39, 72, 76, 98, 154, 163, 190, 196. *See also* slavery
environmental health, 209, 212–213, 220n35
environmental history, 13, 106, 108–109, 129–130, 131, 132n9, 132n10, 137, 167
environmental humanities: activism and, 165; American literature and, 64; climate and, 25, 34, 53, 106; as discipline, 1–4, 216; environmental history and, 129; new materialisms and, 72, 89; teaching, 218; urban, 206
environmental justice, 15–16, 39, 42, 114, 205–206, 207
equality, 171, 173, 194, 210
Equiano, Olaudah, 39, 44, 46, 48–50
erasure, 5, 14–15, 45, 50, 72
etiology, 89, 91, 99. *See also* disease; illness
exploitation, 15, 72, 74, 170, 190–191, 193–195, 199, 202
Exquemelin, Alexander Oliver, 137–141

First Peoples, 156, 162, 167. *See also* indigeneity
freedom, 153, 173, 175, 177, 180, 182, 213
French Revolution, 174, 182
Freneau, Phillip, 15, 189, 191, 198–202
Fugitive Slave Act, 40
fungibility: Black, 174, 202 (*see also* Hartman, Saidiya); of Black labor, 190; of enslaved persons, 189; failure as strategy to manage laborers, 196; libidinal economies of, 76 (*see also* economies)

genocide, 76, 85n28, 98, 174
geoengineering, 25–33
georgic, 15, 176, 189–202
Gilbert, William, 15, 170–182, 183n11, 185n40, 185n47
Global North, 154, 161–162, 167
Golden Age, 175
Grainger, James, 15, 181, 185, 189–202
Gray, Thomas, 215–216
Great Snow of 1717, 55–57
Great Storm of 1703, 23, 53–54, 59–60, 66n29
grief, 71–72, 74–75, 79–81

Haitian Revolution, 82–83, 175, 184n22
Halberstam, Jack, 184n28
Halifax, Nova Scotia (fictional, Kingsport), 210–212, 214–218
Haraway, Donna, 8
Harney, Stefano, 181, 185
Hartman, Saidiya, 194, 202
Haudenosaunee, 154. *See also* Iroquois
Hermeticism, 172, 183n11
homeland, 154; Iroquois, 155, 164; Mohawk, 156, 165, 157, 158–160, 165, 161, 168, 200
humoralism, 173
Huron, 153, 155–156, 160, 162, 165, 169n10

illness, 89–90, 91, 92, 95, 98, 99
inclusivity, 182

indigeneity: archive and, 133n12; erasure of, 14–15; imperialism and, 109, 164, 170, 183n11, 190, 217; Mi'kmaq, 111, 113, 124, 128; power of, 156, 174, 177 (*see also* First Peoples); study of, 2–3, 165, 167–168, 181–182; women and, 153–154, 157, 161–162 (*see also* homeland). *See also* Haudenosaunee; Iroquois; Mohawk (indigenous people)
Industrial Revolution, 5, 62
Iroquois, 154–155, 163–165, 169n10. *See also* Haudenosaunee

justice, 39, 42, 50, 170, 175, 182, 205–207

kin, nonhuman, 7, 71–72, 77, 80, 155, 157
King, Tiffany Lethabo, 73–74, 84n18, 136

landscape: aestheticized, 209; aesthetics of the eighteenth century, 72–73; Black bodies and natural, 13; Caribbean, 201–202; coastal, 114; colonizers changing of, 156 (*see also* colonization); colonized subjectivity and, 84f18; Edenic, 15; homogenous, 208; humans' relationships to the, 217–218; imperial, 184f29; as a palimpsest, 107; plantation, 176, 196, 204n41 (*see also* British Amelioration Act); polluted, 199; relationships to the, 218; Richard Wilson's, 78; scholarship on place and, 207; snowstorms' effect on, 57; Spanish degradation of, 202; Adam and Eve's enjoyment of, 176; students understanding writers within, 216
Ligon, Richard, 179
Little Ice Age, 11–12, 23–26, 34n1, 35n10, 54, 85n28, 66n29 (*see also* Great Storm of 1703), 154
London: pandemic, 167; print culture, 108; Royal Society and, 24, 27, 94; theater culture, 154, 163, 165–166, 168; Thelwall, 205, 207, 208–210, 212–214; weather, 60
loss, 40, 42, 71, 74, 75, 77, 81, 130, 144, 155, 158

marginalization, 15, 175, 216
maternity, 7, 72, 82. *See also* motherhood
Mather, Increase, 53, 58–62, 64–65
McKittrick, Katherine, 81, 197
Mentz, Steve, 5, 139, 146, 147n3
Middle Passage, 42, 44, 73
Mignolo, Walter, 181
Mi'kmaq (first nations people), 107, 111, 124, 128. *See also* indigeneity
Milton, John, 43, 162, 176
Mohawk (indigenous people), 14, 153, 155–157, 159, 162, 164–168. *See also* indigeneity
monoculture, 7–8, 176
monstrosity, 28–29, 34
Montgomery, L.M., 16, 107, 205–218; *Anne of Green Gables*, 205, 207; *Anne of the Island*, 205, 207, 210–212, 214–217; *Anne with an E*, 207
motherhood, 71–73, 76, 83. *See also* maternity
mourning, 72, 74, 75, 81; wars, 154–155

natural disasters, 56, 162, 164, 172, 181, 199
natural history: of Caribbean, 140, 171, 178–179; as discipline, 8, 93–95, 98, 101n20, 102n23; and ministry, 59; and ontology, 92; pandemic and, 88; Sloane's, 90, 95–97, 99–100
natural phenomena, 93, 161–162
new materialisms, 3, 8, 11–14, 51, 72, 74, 84n12, 89, 166
North Pole, 24, 26–27
Nova Scotia, 113, 114, 125, 129, 133n16, 205–218

Ovid, 71–72, 75–77, 80–81

pandemic, 3, 10, 82, 88–90, 96, 98–100, 110, 130, 154. *See also* coronavirus
pedagogy: collaborative, 16; inclusive and interdisciplinary, 3; and research, 17; scholarship and, 1–2, 80, 216, 218, 247
piracy, 14, 136–138, 140–149
plantation: economies, 174–177, 204n41; Isle Saint-Jean, 119; novel and, 87n72; stewardship, 189, 190–191, 193, 195–200, 202; as utopia, 15; West Indies, 172. *See also* enslavement; slavery
Pope, Alexander, 29, 43–44, 49, 166, 180, 207; *Essay on Man,* 44
presentism, 9–10
Prince Edward Island, 13, 15, 105–130, 205, 219n30
Providence, 39–40, 50, 53–54, 59, 61–62, 64, 65n3; divine providence and weather, 54, 55, 56, 57, 61, 64

Queen Anne's War, 155–156

rebellion, 72, 79, 82
reproduction, 72, 76, 81–82
republican natural rights, 194, 199, 201

research: archival, 14; ecocritical, 73; in the environmental humanities, 1, 3, 73; historical, 9, 218; methodologies of, 14; STEM research, 4; and teaching, 1, 16–17, 86n70
resilience, 75, 77, 154
resistance: Black, 74, 77, 81–82; to captivity, 162, 182; democratic, 220n47; ecological, 137–138, 140, 144; metaphors of, 41, 174, 206; of syphilis, 101n12; trauma and, 45, 210
Restoration (historical period), 1, 165–166
revolution, 5, 82, 170, 174
Romanticism (historical period), 1, 33, 35n12, 206, 208, 211–218, 219n16, 219n35
Royal Society, 23–24, 27, 88, 93–95; literary standards, 93–99; *Philosophical Transactions,* 60, 93, 95; virtual witnessing, 90, 93, 94, 95, 96, 97

Sharpe, Christina, 40–41, 48, 50, 73, 77, 82, 174
Shelley, Percy Bysshe, 214
Shirley, Anne, 16, 205–218
slavery: abolition of, 182; African women's sexuality, 72–75, 78, 85n27; labor of, 176; literary atmospheres, 38–41, 44, 48–50, 80–81, 180, 189–202; racial violence of, 142–143, 174; transatlantic, 42, 82, 98, 102n33, 172. *See also* enslavement
Sloane, Hans, 13, 88, 89, 89–90, 92, 93, 94–99, 100, 102f33; *A Voyage to the Islands of Madera, Barbados, S. Christophers and Jamaica,* 90, 94–99
South Pole, 26–28
sovereignty, 74, 80, 145; nature's, 26, 32–33; environmental, 137, 145, 148n13
sublime, 42–44, 50, 166, 214
syphilis: as agent, 89–90, 92, 100; in Americas, 101n5; etiology of, 101n12, 101n14, 102n37; in Sloane's *Voyage,* 94–99, 101n7; spread in Europe, 88, 91

teaching: across periods, 64; with archival maps, 128–130; art, 79; comparative empires approach to, 201–202; on early America, 129; eighteenth-century environmental humanities, 16–17; experiential, 216–217; inscription as augmenting, 128; piracy narratives, 146; research, 1; slavery and M. NourbeSe Philip's *Zong!,* 50
Thelwell, John, 16, 205–218; *The Peripatetic,* 205–206, 208–210, 212–214, 216–218
Thompson, Judith, 205, 208, 210, 216
Thomson, James, 41–42, 50
trauma, 44–45, 117, 153, 157, 162–163

urban environments, 205–218
utopia, 7–8, 11, 15, 154, 184n28, 193, 200

violence: anti-Black, 76–77, 142–143, 174, 201; colonial, 5, 80, 138, 140, 174–175 (*see also* colonialism); environmental, 42, 56, 61, 201; historical, 117; indigenous, 14, 162, 174 (*see also* indigeneity); physical, 141–142, 146, 193; social, 42, 82, 84n11, 203
Virgil, 43, 171–172, 179–180, 189, 191–192, 200–202

West Indies, 48, 88, 138, 142–143, 179, 189–190, 198–199, 202
Wheatley Peters, Phillis, 6, 12–13, 39, 42–50, 83; environmental scholarship and new materialism, 72–74; lyrics and elegies, 74–77; "On Messrs. Hussey and Coffin," 45–47; "Niobe in Distress for her Children" and Ovid's *Metamorphoses,* 71–72, 75–77, 80–81; Niobe's "blaspheming breath," 81–84; *Poems on Various Subjects, Religious and Moral,* 46, 71–72; Richard Wilson's *The Destruction of the Children of Niobe,* 77–80; "To a Lady on Her Remarkable Preservation," 46–47; "To Maecenas," 42–44, 46–47; "To S.M., a *Young* African painter, on seeing his Works," 79
Wigglesworth, Michael, 57–58, 60
wilding, 171, 175, 177, 184n28
Wilson, Richard, 71, 77–79
Wordsworth, Dorothy, 209
Wordsworth, William, 33, 171, 206–209, 214, 219n23
Wynter, Sylvia, 73, 85n31, 87n72, 137, 143, 177

Yaws, 90–91, 101n7, 102n37
Yusoff, Kathryn, 5, 72–73, 83, 84n18, 190, 194, 202

*Zong, The* (ship), 38–41, 50

Printed and bound by CPI Group (UK) Ltd, Croydon, CR0 4YY
16/09/2026
14963067-0004